Opening the Cage

NEW DIRECTIONS IN MATHEMATICS AND SCIENCE EDUCATION

Volume 23

Scope

Mathematics and science education are in a state of change. Received models of teaching, curriculum, and researching in the two fields are adopting and developing new ways of thinking about how people of all ages know, learn, and develop. The recent literature in both fields includes contributions focusing on issues and using theoretical frames that were unthinkable a decade ago. For example, we see an increase in the use of conceptual and methodological tools from anthropology and semiotics to understand how different forms of knowledge are interconnected, how students learn, how textbooks are written, etcetera. Science and mathematics educators also have turned to issues such as identity and emotion as salient to the way in which people of all ages display and develop knowledge and skills. And they use dialectical or phenomenological approaches to answer ever arising questions about learning and development in science and mathematics.

The purpose of this series is to encourage the publication of books that are close to the cutting edge of both fields. The series aims at becoming a leader in providing refreshing and bold new work—rather than out-of-date reproductions of past states of the art—shaping both fields more than reproducing them, thereby closing the traditional gap that exists between journal articles and books in terms of their salience about what is new. The series is intended not only to foster books concerned with knowing, learning, and teaching in school but also with doing and learning mathematics and science across the whole lifespan (e.g., science in kindergarten; mathematics at work); and it is to be a vehicle for publishing books that fall between the two domains—such as when scientists learn about graphs and graphing as part of their work.

Opening the Cage

Critique and Politics of Mathematics Education

Edited by

Ole Skovsmose
Aalborg University
State University of São Paulo, Denmark and Brazil

Brian Greer
Portland State University, USA

SENSE PUBLISHERS
ROTTERDAM/BOSTON/TAIPEI

A C.I.P. record for this book is available from the Library of Congress.

ISBN: 978-94-6091-806-3 (paperback)
ISBN: 978-94-6091-807-0 (hardback)
ISBN: 978-94-6091-808-7 (e-book)

Published by: Sense Publishers,
P.O. Box 21858,
3001 AW Rotterdam,
The Netherlands
https://www.sensepublishers.com

Cover illustration by Ole Skovsmose

Printed on acid-free paper

TABLE OF CONTENTS

PREFACE

We express our gratitude to all the people who have helped us in opening the cage. Co-operating with the authors has been very inspiring, and once more we have experienced that editing a book is a fascinating learning process.

We have had assistance from many other people as well. Thus, Denival Biotto Filho has helped us formatting several of the chapters. Sense Publishers have provided us with much support. Our wives, Miriam Godoy Penteado and Swapna Mukhopadhyay have brought much light to the whole project.

Ole Skovsmose
Brian Greer

SEEING THE CAGE? THE EMERGENCE OF CRITICAL MATHEMATICS EDUCATION

BRIAN GREER AND OLE SKOVSMOSE

The grounds for this book may be stated simply. Critical mathematics educators: have concerns about mathematics education as they observe it taking place in multiple contexts; believe that those concerns cannot be met by changes internal to mathematics education but must be addressed in sociopolitical contexts; accept the responsibility of making value judgments, while subjecting these judgments to perpetual critique; desire to promote change in accordance with those value judgments, envisaging alternatives to what is generally the case. These concerns, political consciousness, acceptance of responsibility, and desire and work for change generate the central theme of the book, namely critical agency as a synthesis of reflection and action directed towards the inherently political nature of mathematics education.

CRITIQUE OF MATHEMATICS EDUCATION

A cluster of related words, including "crisis", "criticism", and "critique", all derive from the Greek verb *krinein*, meaning "to decide" or "to judge". "Crisis" can mean, in general, "a time of intense difficulty or danger" or, more specifically in a medical context, "the turning point of a disease when an important change takes place indicating either recovery or death". Likewise, "criticism" has the general meaning of "the expression of disapproval of someone or something based on perceived faults or mistakes" and the more specific meaning of "the critical assessment of a literary or artistic work". "Critique" means "a detailed analysis and assessment". The medical meaning of "crisis" also carries the implication of a turning-point where alternative actions are possible. Thus, the diverse connotations of "crisis" refer to dangerous situations and analysis, diagnostic analysis, turning-points, and the opportunity for action.[1] Accordingly, we think of "critique" in terms of "critical agency", signifying that critique refers to both reflection and action, integrating the various aforementioned connotations.

Within the period during which we have been working in mathematics education, there have been major developments in the directions advocated in this book – at least within certain active, albeit marginalized, pockets of mathematics educators. The emergence of this social, cultural, political approach has been well summarized by Vithal (2003, Chapter 1). Among the important pioneering

O. Skovsmose and B. Greer (eds.), Opening the Cage: Critique and Politics of Mathematics Education, 1–20.

publications during the period 1973–1988 (several of which were written in languages other than English) are the following:

- Claudia Zaslavsky, particularly in her 1973 book, *Africa Counts*, was one of the first to valorise the diversity of non-academic mathematical practices of non-Western cultures.
- Peter Damerow, Ulla Elwitz, Christine Keitel, and Jürgen Zimmer (1974), provided a sociopolitical perspective on elementary mathematics education.
- Dieter Volk (1975) formulated a mathematics education from an emancipatory perspective and later (Volk, 1979), provided a panoramic view of the wide range of critical approaches to mathematics education that had been developed at that point.
- Stieg Mellin-Olsen (1977) contributed to a pioneering analysis of the socio-political analyses of mathematics education, which he developed further in *The Politics of Mathematics Education* (1987).
- Ole Skovsmose (1980, 1981a, 1981b) took steps towards formulating a critical mathematics education.
- Munir Fasheh (1982) raised many penetrating questions about the politics of mathematics education, such as the following:

  > Is it possible to teach mathematics effectively – that is, to enhance a critical attitude of one's self, society, and culture; to be an instrument in changing attitudes, convictions, and perspectives; to improve the ability of students to interpret the events of their immediate community, and to serve its needs better – without being attacked by existing authorities whether they are educational, scientific, political, religious, or any other form?

- Marilyn Frankenstein (1983) recognized the relevance of Paulo Freire's praxis to mathematics education and showed how Freirean principles could be adapted in teaching mathematics for social justice.
- Ubiratan D'Ambrosio delivered a plenary talk on *Socio-Cultural Bases for Mathematical Education* (D'Ambrosio, 1984) at the International Congress on Mathematical Education in Adelaide, to which an extra day had been added on the theme *Mathematics, Education, and Society*. A year later, he introduced the concept of ethnomathematics (D'Ambrosio, 1985).
- Alan Bishop (1988) authored a key early work casting mathematics as an activity practiced in sophisticated ways in all cultures.

Not all the development of critical mathematics education was within the protected world of academe. In South Africa, a movement was created called "People's mathematics" (Vithal, 2003, pp. 27–35) that was part of the People's Education movement during the anti-Apartheid struggle. Vithal (p. 29) takes as a starting point the presentation of politically charged curricular materials by Chris Breen (1986) at the white-dominated Mathematics Association of South Africa's annual conference. Other important stands were taken by, amongst many others, Jill Adler (1988) and Cyril Julie. The latter presented a paper at the first of three conferences on *Political Dimensions of Mathematics Education*, succeeded by the ongoing biennial conferences on *Mathematics Education and Society*. The theme of that first conference (Noss, Brown, Dowling, Drake, Harris, Hoyles, and Mellin-Olsen, 1990) was *Action and Critique*.

In the course of these developments, we may say that the critique of mathematics education has emerged from the restraint of being directed *internally*, that is to say concerned primarily with how mathematics is learned and taught, to being also *external*, concerned with the embeddedness of mathematics education and mathematics within historical, cultural, social, and political contexts, and the implications and ramifications thereof.

The inspiring influence of Freire as a central figure in the development of critical education continues to be recognized (e.g., Apple, Au, & Gandin, 2011b, Part IV) and is pervasive throughout this book. Nowhere is there a clearer statement than the following:

> This is a great discovery, education is politics! After that, when a teacher discovers that he or she is a politician, too, the teacher has to ask, What kind of politics am I doing in the classroom? That is, in favor of whom am I being a teacher? By asking in favor of whom am I educating, the teacher must also ask against whom am I educating. Of course, the teacher who asks in favor of whom I am educating and against whom, must also be teaching in favor of something and against something. This "something" is just the political project, the political profile of society, the political "dream". (Freire, 1987, p. 46)

Freire did not write explicitly about mathematics education. Indeed, in a conversation with D'Ambrosio, he commented that:

> In my generation of Brazilians from the North-East, when we referred to mathematicians, we were referring to something suited for gods or for geniuses. There was a concession for the genius individual who might do Mathematics without being a god. As a consequence, how many critical intelligences, how much curiosity, how many enquirers, how many capacities that were abstract to become concrete, have we lost? (Freire, D'Ambrosio, & Mendonça, 1997, p. 8)

Nevertheless, and in the spirit of Freire's insistence that his pedagogy be not copied, but *reinvented in context*, the relevance of his work and philosophy to mathematics education has been demonstrated and put into practice, notably by Frankenstein (1983, 1989), and more recently Gutstein (2006; Chapter 1, this volume).

The field of mathematics education, in general, has considerably matured, as reflected in the diversification of influential disciplines and related methodologies – broadly speaking, the balancing of technical disciplines by human disciplines such as sociology, sociolinguistics, anthropology, psychoanalysis, and of formal statistical methods by interpretative methods of research and analysis. Within the field, there is heightened cultural and historical awareness, both within and beyond academic mathematics, and an increased acknowledgment of the ubiquity and importance of "mathematics in action" and the implications for mathematics education, including more curricular prominence for probability, data handling, modelling, and applications. In relation to the political nature of the enterprise, there is greater attention to the relationships between knowledge, education, and power. Important concomitant developments that have both been influenced by, and have impacted, developments in mathematics education include the emphasis on cognition as both situated and cultural. The philosophical realm has been burst open by postmodern questionings of the assumptions of progress, notions of truth and objectivity, and transparency of language. A new philosophy of mathematics (e.g. Ernest, 2009) reflects historical changes in views of the ontology of mathematics and its relations to forms of life.

Much of the above historical sketch may be seen as manifesting the *humanization* (a word frequently used by Freire) of mathematics and mathematics education, encapsulated in the phrase "mathematics as a human activity". This phrase rejects the Platonic view of mathematics as existing independently of its creators and users, as does the conception of ethnomathematics as "the mathematics which is practised among identifiable cultural groups, such as national-tribal societies, labor groups, children of a certain age bracket, professional classes, and so on" (D'Ambrosio, 1985, p. 45). In relation to mathematics education, it is reflected in the understanding that teaching/learning is not a purely intellectual activity – students are not "cognitive angels".

Furthermore, (re)humanizing mathematics and mathematics education are inextricably political activities. The political nature of mathematics education follows from the recognition that improving (mathematics) education is a human problem, not a technical problem (Kilpatrick, 1981). Of profound political importance is the challenge to mathematicians and mathematics educators to accept ethical responsibilities, in particular as posed by D'Ambrosio (2010, p. 51): "It is clear that mathematics provides the foundation of the technological, industrial, military, economic and political systems and that in turn mathematics relies on these systems for the material bases of its continuing progress. It is important to question the role of mathematics and mathematics education in arriving at the present global predicaments of humankind."

The globalisation of mathematics education is clearly illustrated in this book, albeit not in a comprehensive manner (which would be impossible). The authors collectively represent considerable diversity (including amongst those based in universities in the USA). The first languages of the authors also are diverse, which prompts acknowledgment of the fact that the book is written in English. Thus, we agree with the sentiment expressed in similar circumstances by Apple et al., (2011a, p. 15): "Inevitably a book such as this, no matter how large, partly centers dominant voices even in its attempt to be conscious of that centering … Given the geopolitics of publishing and academic writing … and the role of English as an imperial project, right now we can but note this as part of a constitutive dilemma." One aspect of this hegemony is that pioneering works in critical mathematics education written in languages other than English (some of which were cited earlier) receive limited attention in Anglophone circles.

In an extremely deep and powerful analysis, Edward Said (1994) showed how culture, particularly literature, both reflected and reinforced unexamined assumptions of the peoples of colonizing and colonized countries, and also how resistance to imperialism developed in literature. To our knowledge, no analysis of comparable depth exists for mathematics; however there have been important acknowledgements of the role of mathematics in the imperial/colonial enterprise (e.g. Bishop, 1990; Urton, 1997), the old forms of which continue to affect ideology of former colonies, and new forms of which are rampant under the name of globalisation. These forms are primarily economic, in particular disaster capitalism (Klein, 2007). The roles of mathematics in these geopolitical movements deserve deeper scrutiny, for example, the claims by Milton Friedman and others of the Chicago school that their theory of economics has a sound mathematical foundation (Klein, 2007, pp. 61–62).

POLITICS OF MATHEMATICS EDUCATION

Critical education is situated within local, national, and global politics. What Michael Apple terms "conservative modernization" is currently the dominant ideology within education in the USA and controlled or aligned countries, a situation exacerbated under President Obama. As described by Apple et al., (2011a, p. 10), "conservative modernization" is an alliance (by no means free of internal contradictions) among neoliberals, neoconservatives, authoritarian populist religious conservatives, and the professional and managerial middle class. As with politics in general, what goes for the USA also applies in other industrialized countries, and to a greater or lesser extent, to many parts of the world. (Although it might be considered that we are disproportionately referring in our discussion to the situation in the USA as opposed to the rest of the world, the current political reality is that much of what applies in the USA applies widely beyond, albeit with variations.)

Historically and contemporarily, there are strong links between mathematics and war (which might be characterized as the ultimate example of attempting to solve human problems by technical means). A great deal of the world's intellectual talent in mathematics (and science) is used in the creation of better ways of killing, subjugating, or surveilling and controlling people, of which current deployment of flightless aircraft, "drones", provides a chilling example (Skovsmose, 2010). By contrast with the vision of D'Ambrosio cited above, that mathematicians and mathematics educators should be concerned with the critical problems facing humankind, mathematics education is typically laced with nationalism. (By way of example, think of the reactions in media and political circles in participating countries when results of international comparative studies of school performance in mathematics are announced.)

A major common thread discernible in the strategic alliance described above is what might be called the ideology of certainty. The constituents of that alliance have apparently unshakable convictions in such disparate belief systems as the power of the market, the inherent superiority of certain civilizations, various religions, social values, the power of technical and managerial techniques to solve problems and regulate people's lives. The desire for certainty seems to be a natural human yearning, often expressed through religion, or through positivist logic. Bertrand Russell, for example, sought certainty in logic only to reach, ultimately, the conclusion that "all human knowledge is uncertain, inexact, and partial" (Russell, 1992, p. 527). As Russell found, mathematics can offer a tantalizing illusion of certainty. As far as mathematicians are concerned, this mindset becomes seriously dangerous when they seek to project the certainty that they find internally within mathematics onto its applications (Skovsmose, 2005, p. 48), an attitude that is typically inculcated within mathematics classrooms. Such an attitude is also manifest in a search for procedures, rules, algorithms, that allow people to abdicate the responsibility of making judgments in complex social situations. People and institutions within mainstream mathematics education too often collude with the political establishment by wilfully remaining oblivious of the social and political contexts outside their self-constructed cage:

> It is unfortunate but true that there is not a long tradition within the mainstream of mathematics education of both critically and rigorously examining the connections between mathematics as an area of study and the larger relations of unequal economic, political, and cultural power. (Apple, 2000, p. 243)

At the core of our work in exposing mathematics education as an inherently political enterprise is the dialectic between reflection and action, that again is a Freirean theme, since "reflection – true reflection – leads to action. On the other hand, when the situation calls for action, that action will constitute an authentic praxis only if its consequences become the object of critical reflection" (Freire, 1997, p. 48). We use the term "critical agency" to express the dialectic between reflection and action. Through this term we also acknowledge that politics and critique of mathematics education are integrated. The notion of critical agency can be seen as a generalization of the notion of mathematical agency (e.g., Stemhagen, 2009). While mathematical agency is discussed largely in terms of students in school mathematics, it also is applicable to teachers, researchers, and, in the most general way, people dealing in their lives with mathematics in action. Our adoption of the term "critical agency" signals that we regard critique as not only reflective and contemplative, but also implying the goal of trying to make changes. Critical agency for students implies that students come to speak for themselves; that they become able to read and write the world, in Freire's metaphor, that has been taken up and applied to mathematics education by Gutstein (2006; Chapter 1, this volume). Critical agency for educators means searching out spaces, such as alternative schools and classroom environments, in which students become capable of, and have a disposition towards, critical analysis, in particular with mathematical tools, to address issues of importance to them personally, to their communities, and to humankind in general (Mukhopadhyay & Greer, 2001). It also includes the exploration of the possibilities of making changes, large or small, in daily classroom routines despite all the regulations and demands that restrict diversity of practice and content. A first step in any exploration of what is possible is to imagine that something in what is the case could be different, or that a state of affairs perceived as necessary is actually contingent. Thus, a pedagogical imagination forms part of a critical agency.

Critical agency refers to the whole socio-political context of education, including attempts to influence policy. De Corte, Greer, and Verschaffel (1996, p. 534) commented that "perhaps the enlightenment of political decision-makers, and other groups such as parents, administrators, and the public in general … is the biggest educational challenge." This statement is a direct expression of the political dimension of this agency.

Critical agency also implies openness to self-critique.

OUTLINE OF THE BOOK

The organization of the book into four parts was one of many possibilities; many chapters could sit equally aptly in different sections; themes interweave across the linear organization.

Part I: Mathematics education is politics

The title given to this section, of course, echoes Freire, as quoted earlier. The situatedness of mathematics education within historical, cultural, social, and political contexts is reflected in this section, and, indeed, throughout the book. Some manifestations of the global and national apparatus relating to education are obvious, such as the formulation and control by states of curricula, testing, teacher education requirements, statements of policy, and all the paraphernalia of educational systems. However, digging into the ideological substrates of such tangible external manifestations is complex and contentious. Even more complex and contentious is trying to lay bare the relationships, in relation to mathematics and mathematics education, among knowledge, power, hegemony, racism, inequity, social injustice.

Freire's conceptualisation of education as "a weapon in the struggle" forms part of the title of Chapter 1, by Eric Gutstein. Critical agency can be nurtured on a local level, in a particular situation (as Freire did). The situation in which Gutstein works includes the political activity of a community demanding a school promised to them, to the point of carrying out a hunger strike. Gutstein's praxis transcends the aim of working for equity *within* mathematics education to enacting a vision of equity *through* mathematics education, by teaching content and raising political consciousness together, as Freire did when teaching literacy. In his chapter in this volume, Gutstein further elaborates his reinvention of Freire's praxis within the context of teaching mathematics that is already clear in the title of his 2006 book *Reading and writing the world with mathematics*. In this chapter, he describes how the content for the class he taught was co-developed with the students in the form of generative themes: elections, displacement (gentrification, foreclosures, and immigration/ deportation), HIV/AIDS, criminalisation, and sexism. This work is taking place in the national US context of neoliberal efforts to privatise education, which have, if anything, accelerated under President Obama. As Gutstein says, "there are always spaces in which to act".

In Chapter 2, Alexandre Pais surveys the very extensive body of recent literature addressing equity in mathematics education (and see Atweh, Graven, Secada, & Valero, 2011) to form the basis for a very profound critique of mathematics education as a field. A central point of this critique is that, for the bulk of the literature on equity, the authors acknowledge that the problems being discussed are inextricably political, yet then generally proceed as if the solution can be found through changes internal to mathematics education. Drawing on Žižek's work, Pais argues trenchantly that this search for a solution within the system rather than changing the whole system is chimerical because inequity is an inherent feature of capitalist education. The fine-sounding slogan "mathematics for all" falls apart because of the same structural contradictions (and see Gutstein, 2009). Much research on equity focuses on what are called "achievement gaps" and how to "close" them; few think to interrogate the nature of the gaps (materialized as differences in scores on tests which are themselves in need of interrogation) or to ask the more important question "Why are the gaps there?" Thus, the verdict reached by Pais on a very considerable body of research is that it is almost all looking for the lost object where it is easiest to look ("pseudo-activity", to use Žižek's term), not where it is most likely to be found. Pais takes as his task the raising of awareness necessary to acknowledge and understand this fundamental mischaracterization of the "problem".

Munir Fasheh considers the macro-effects of mathematically influenced worldviews and conceptions of humankind on his native Palestine, part of the global effect of neoliberal economic theory (Klein, 2007). In an earlier paper, Fasheh (1997) commented on the fact that he had lived through four very different political regimes in Palestine, yet throughout those changes, mathematics did not change. Having learnt and taught academic mathematics, he was brought to awareness by coming to see the complexity of the mathematics of his mother as a dressmaker, bringing to mind the statement by Freire that "the intellectual activity of those without power is always characterized as non-intellectual" (Freire & Macedo, 1987, p. 122). Through the insight offered by this experience, he came to see himself, a scholar versed in Western mathematics, as a collaborator in cultural imperialism, the process whereby long-standing viable, place-based modes of living, and of education aligned with that living, are being destroyed. As he points out, this process was already clear to the Palestinian educator Khalil Sakakini at the end of the 19th century. In particular, Fasheh points to the dehumanisation that comes with grading and credentialing people (echoing a major theme in the chapter by Pais) that he characterizes as a form of cultural violence. He now works to develop education that is tied to the culture, the land, and sustainable local economy.

In Chapter 4, Brian Greer examines an overtly political exercise called the National Mathematics Advisory Panel set up by President George Bush. This project was remarkable in that it all but ignored the entire academic field of mathematics education. This disappearance was achieved, in particular, through the composition of the panel, and through the adoption of rigid criteria for what constitutes research. As a result, there is very little of what we would recognize as mainstream mathematics education research in the document, let alone anything remotely resembling critical mathematics education. The simplest way to characterize the whole exercise is as the endorsement by the Bush administration of one side in the ideological dispute commonly known in the USA as the "Math Wars". There is also a very strong nationalistic tone and warning of crisis in the framing of the panel's brief and in the report that resulted (one member in the final meeting commended his fellow-members on their patriotism); the workings of the military-industrial-academic complex can also be seen at various points. Of particular concern is that, with some honourable exceptions, and in particular an issue of *Educational Research* (December, 2008) devoted to the topic, in which strong criticisms were voiced, there was limited resistance on behalf of the mathematics education community, including the small number of mathematics educators that were on the panel. Thanks to the excellent conventions and laws in the USA about making information available, there is a vast amount of public-domain information relating to NMAP, and it offers a rich topic for a sociological analysis of the intersection of politics and education.

Part II: Borderland positions

Arising from geopolitical developments, military conflicts, popular uprisings, economic hegemony, population movements, the rising voices of subalterns, post-colonial resistance, and so on, the contemporary world reflects hybridity and diversity. These inherent characteristics of humankind relate not only to ethnicity, but also to culture, educational, economic, and political systems, mathematical practices, worldviews. As Skovsmose (Chapter 16, this volume) argues, mathematics education as a field has largely failed to reflect the diversity of situations within which mathematics is learnt and taught in schools and elsewhere, as illustrated by the chapters in this section and elsewhere in this book. In this respect, the simplism of the slogan "mathematics for all" becomes evident (see Pais, Chapter 2, this volume). To unpack it, we need to ask many questions, including: Does inclusion imply assimilation without accommodation? Whose mathematics and for what purposes? Is it *really* necessary for all? If access is granted, at what cost in cultural violence?

In Chapter 5, Marta Civil provides a survey of research on the mathematical education of immigrant children. Although, as she points out, immigrants constitute only 3% of the world's population, there is considerable political tension relating to them, particularly in the USA and Europe. As she points out, the circumstances of immigrant students raise in sharp focus issues of assimilation versus multiculturalism (which apply also to non-immigrant populations such as indigenous groups). Specifically with respect to mathematics, this tension relates to school mathematics versus the mathematics of the students and their communities, which naturally impacts interpersonal relationships between teachers and students. Again, the perception of mathematics as being "the same for everyone" makes it easier to avoid responsibility for mathematic teachers to educate themselves about the mathematical practices of other cultures. Further, pressure for linguistic assimilation interacts in complex ways with the teaching of mathematics – again an issue by no means only for immigrant children (Setati & Planas, Chapter 7, this volume).

Concretising Civil's survey, Sikunder Ali Baber offers a detailed case study of a student in a specific context, exemplifying Skovsmose's (2005) exhortation to consider not just the "background" of an individual, but also the "foreground", i.e. subjective perceptions of the future possibilities and opportunities open to her/him. In this chapter, Ali Baber first describes in detail the circumstances of a particular immigrant community, namely Pakistanis in Barcelona. As discussed in the chapter by Setati and Planas, the situation is further complicated by the politics of languages – in this case, Catalan and Spanish (and, increasingly, English). The complexity of the case study that follows illustrates not just diversity but also hybridity, and puts into sharp relief the nature of much research in mathematics education wherein the complexity of an individual is reduced to half a dozen values of often simplistically defined variables (e.g. socio-economic status). Baber concludes by making recommendations on the basis of his analysis for how teachers, schools, and policy-makers could better serve the needs of immigrant children in the kind of multicultural context that is becoming common in many parts of the world.

In both chapters just outlined, the importance of language in learning mathematics is clear. In Chapter 7, Mamokgethi Setati and Núria Planas consider the question of use of languages in mathematics classrooms in which the children speak two or more languages. They do so in two contrasting contexts, South Africa and Catalonia, the latter showing that English is not the only dominant language in such circumstances (though English is also making inroads in Catalonia). As the examples make strikingly clear, the choices of policies in such circumstances regarding language of teaching and learning are highly political. What happens in mathematics classrooms accordingly needs to be analysed, not simply in terms of the implications of the language of learning and instruction for development of mathematical cognition, but also in relation to sociopolitical aspects and power relationships. Extensive research on teachers' and students' perspectives illuminates these issues. It is clear from the research in South Africa that the dominance of English as the source of cultural capital is well established, and this takes precedence over questions of language as a cognitive tool and over questions of language and cultural identity; similar comments apply to the dominance of Catalan. These cases vividly exemplify the tension that occurs between acknowledging cultural diversity and trying to afford educational and economic opportunities through mastery of the dominant discourse.

In Chapter 8, Gelsa Knijnik and Fernanda Wanderer consider two examples of sets of mathematical practices in Brazilian communities deriving from what Wittgenstein termed "forms of life", integrating culture, world-view, and language. As with the contributions by Gutstein and Fasheh, they thus illustrate the teaching/learning/doing of mathematics in specific situations. The first example relates to a community of German settlers in Brazil in the mid-20th century compelled to adopt the Portuguese language. The second relates to the more recent Landless Movement that incorporates out-of-school mathematics embedded in the political situation of the landless peasants. In both cases, Knijnik and Wanderer describe how the state, in the pursuit of political unity and cultural homogeneity, forced these communities to conform to the norms of school mathematics by absorbing them into the state system "so that they will have access to the knowledge imparted to all people". Knijnik and Wanderer examine the cultural violence done thereby through what Hardt and Negri (2003) term the mechanism of "differential inclusion".

In Chapter 9, Danny Martin and Maisie Gholson present their reflections on what it means to be critical Black scholars in mathematics education in the USA. Most, if not all, of what they describe applies to Black scholars in general, but there are several ways in which the fact that their field is mathematics education brings certain issues into particular focus. The "ideology of no ideology" that supports the fiction of academics as ethically and politically neutral is particularly easy to project in relation to mathematics. Given the prominence of mathematics in testing, mathematics affords fertile ground for those who, indirectly or indirectly, continue to advance claims of intellectual inferiority of Black children – in particular, claims that black (and poor) children lack the capacity to engage in abstraction and formal mathematical thinking. And mathematics is an intellectual area with a particularly strongly entrenched epistemology, which needs to be challenged by alternative epistemologies. Above all, through both the form and the content of their chapter, Martin and Gholson remind us that the children who are studied and the scholars that do the research are individuals that deserve to be considered as more than a handful of statistically manipulable values of variables.

Intermezzo: Totakahini (The Tale of the Parrot)

Rabindranath Tagore (1861–1941) was a Bengali poet, writer, painter, musician, philosopher, and educator who made fundamental contributions to education, including the establishment of a school at Santiniketan and a university at Visva Bharati at Bolepur, near Kolkata (Calcutta). His parable, *Totakahini: The Tale of the Parrot* stands as one of the great satirical attacks on repressive education (see Fasheh, Chapter 3, this volume). A play based on *Totakahini* is performed by children in India. The translation presented here, from her native Bengali, is by Swapna Mukhopadhyay.

Like Freire, Tagore projected a vision of education as emancipatory. As Sen (2005, p. 98) put it “For Tagore it was of the highest importance that people be able to live, and reason, in freedom”, quoting a poem from Tagore’s work *Gitanjali*, for which he won the Nobel Prize for Literature in 1913:

> Where the mind is without fear and the head is held high;
> Where knowledge is free;
> Where the world has not been broken up into fragments by narrow domestic walls; ...
> Where the clear stream of reason has not lost its way into the dreary desert sand of dead habit; ...
> Into that heaven of freedom, my Father, let my country awake.

(This poem first appeared in Naivedya, July 1901, in Bengali. The English translation, by Tagore himself, first appeared in 1912 in *Gitanjali).*

Part III: Mathematics and power

It is arguable that one of the greatest shortcomings of mathematics education is that – at all levels, including tertiary – scant attention is paid to the societal effects of activities in which mathematics is instrumental. To a very limited extent, this deficiency has been ameliorated by increasing emphasis in some curricular programs on mathematical modelling and data handling. However, much of the work that is done on mathematical modelling fails to go beyond what Verschaffel and Greer (2007) termed “explicit modelling”, namely the case-by-case modelling of specific situations, to “critical modelling”, which includes both the critique of particular models, but also more general consideration of the roles played by mathematics in action in our societies, and the limitations of applying technical solutions to human problems. In particular, it is a scandal that such issues are minimally treated at the university level.

Technology has greatly facilitated the extension of mathematical modelling to format more and more aspects of our lives; in D’Ambrosio’s phrase, it has produced both “miracles” and “horrors”. A particularly pernicious effect is the dehumanisation that can take many forms ranging from the glaringly obvious to the extremely subtle. Moreover, above and beyond particular mechanisms and devices there is often a disposition to dehumanise, strikingly and literally exemplified by the time and motion studies of Taylor (Frankenstein, 2009; Skovsmose, 2011). Viewing the human body as a machine (or the brain as a computer) can be useful and appropriate, but it can also be profoundly dehumanising. The same comment applies to viewing (mathematics) teaching as a technical enterprise or as a business.

In Chapter 10, Brian Greer and Swapna Mukhopadhyay examine aspects of the relationship between mathematics and power in relation to three contexts of hegemony. The first context of hegemony discussed is the role of mathematics in cultural imperialism (Bishop, 1990), which includes the dominance until relatively recently of a Eurocentric narrative for the history of academic mathematics. In this context, there are striking parallels between the hegemony of mathematics and the description by Macedo, Dendrinos, and Gounari (2003) of the hegemony of English (Greer & Mukhopadhyay, in press). Secondly, the role of mathematics in action in society is discussed in relation to how mathematics can be used to intimidate, a phenomenon that is exacerbated by the lack of societal support for critical mathematical agency. Thirdly, hegemonic aspects of mathematics education include the use of mathematics education within education in general as a means of cultural oppression. The "Math Wars" in the USA are considered as a prime example of hegemonic struggle, and the slogan "mathematics for all" is picked apart, since it is meaningless without clarification of the basic question "What is mathematics education *for*?" As with the final chapter of Said's (1994) *Culture and Imperialism*, forms of resistance that are developing in each of the three arenas are considered.

In Chapter 11, Keiko Yasukawa and Tony Brown "dig where they stand", that is to say, as mathematics educators they excavate buried models within their own context as university staff in Australia. They begin by pointing out that, while adult numeracy is considered highly important in relation to employment, it is directed towards the increase of human capital rather than towards the kind of critical mathematics that would empower workers to critique their economic circumstances. They present a case study in which they worked in their immediate milieu to uncover and explicate opaque and flawed models governing pay of untenured university staff. As they illustrate, the strength of the exercise derived from combining "statistics and stories", i.e. the integration of the analysis of mathematical models and data with the collective lived experiences of the staff. This integration proved necessary in order to mobilize for action. The example also shows how the agency that derives from having mathematical tools to critique may exist at the level of the collective, through individuals mediating as statistical or modelling experts. In the spirit of Pais (Chapter 2, this volume) it could be argued that this local action should be followed by a more global analysis of the circumstances that gave rise to the situation, namely the increasing commodification and privatisation of university education, a global phenomenon.

In Chapter 12 in this section, Keiko Yasukawa, Ole Skovsmose, and Ole Ravn consider mathematics interpreted as a form of technology, and suggest that insights and theoretical tools from the field of Science and Technology Studies may be usefully adapted in the quest to get a better understanding of, and control over, mathematics in action. They discuss two examples, one being the case study of Yasukawa and Brown described in the previous chapter, and the other from the field of cryptography. Mathematics is ubiquitously used as "a technology of rationality", a tool for making and justifying decisions. It is, indeed, a powerful tool, that meshes easily with a range of particular economic and political interests. What is essential is to retain control over that tool (i.e. preserve critical agency), otherwise technologies can become autonomous and beyond control. Finally, it can be observed that social theorizing, in general, does not pay any particular attention to mathematics. Thus, the discussion of the social formation of technology, including mathematics, needs to be combined with a discussion of how the social becomes formed to a substantial extent through mathematics as a tool for rationality.

Part IV: Searching for possibilities

By "socially relevant mathematics education" we mean mathematics education that connects with the present and future forms of life of the students, and with the issues of importance to them, their families and communities, and humankind in general. It is an education that recognizes the important of critical agency in general. The search for a socially relevant mathematics education concerns all forms of mathematics education: in schools and out of schools, in all possible socio-political and economic contexts, and at all levels of education. Thus, it also concerns the education of future experts, including researchers in mathematics and technical disciplines. It is essential that expertise be combined with social responsibility (e.g., see Skovsmose, Valero, & Ravn Christensen, 2009). To search for a socially relevant mathematics education means to consider what could be in the interest of humankind in general. Clearly, such ideas align with the humanization of mathematics education as described above, with the aspirations of a "mathematics education for social justice", with many forms of ethnomathematical approaches, and with critical mathematics education in general.

In this section, a number of approaches that search for such an education are described and evaluated. (Gutstein's work is a very important example, and his contribution in Part I could equally well have been included here.) In Chapter 13, Eva Jablonka and Uwe Gellert, within a framework of how mainstream and resistant curricular positions are formed, review a number of curricular conceptions in relation to access to tertiary education. They raise the tension between trying to make mathematics education socially relevant while not closing gates to educational and economic opportunities, a discussion that recurs at many points in this book. This tension is embodied in the framework that they introduce, in particular the contrast between tactical resistance (playing the game) and deconstruction (changing the game) in positions that challenge the mainstream. Having commented on four orientations, namely Inquiry-based Mathematics Education, Ethnomathematics, Mathematical Modelling, and Critical Mathematics Literacy, they put forward another perspective, that they term "Radical Conservative Pedagogy" informed by the work of Basil Bernstein and his followers, that might be characterized as "making explicit for the students the rules of the game", in particular the differences between the discourses of formal mathematical knowledge and practices and of non-academic mathematical knowledge and practices.

As discussed by Jablonka and Gellert, Bernstein and his successors have been influential in problematising contextualisations in mathematics education (Greer, Verschaffel, Van Dooren, & Mukhopadhyay, 2009), the focus of Chapter 14 by Annica Andersson and Ole Ravn. In the Swedish context that they address, they report that mathematics is rarely taught/learnt with social relevance. They argue that this state of affairs represents an unbalanced ideological conception of the nature of mathematics as abstract, existing independently of the human concerns of its users and those who are impacted by mathematics in action. Such a position is aligned with the later Wittgenstein's characterization of mathematics as a network of language games. In their first case study, Andersson and Ravn pick apart a number of textbook problems and their use in an upper secondary class. Thereby, they demonstrate the clear gap between a mission statement from the Ministry of Education about socially relevant mathematics education and what happens in schools in the form of a mathematics language game that is dysfunctional in a multitude of ways (Verschaffel, Greer, & De Corte, 2000). In the second case study, they present examples of students studying more complex and realistic issues, with more personal input, over a longer period of time. As with Jablonka and Gellert, they argue for transparency towards the students as to the nature of formal mathematics, applications of mathematics, and the relationships between them, within which framework ethical values can be naturally situated and addressed.

In Chapter 15, Bill Atweh, who has a wealth of experience teaching mathematics within a socially relevant approach, takes up the theme of ethics, in particular in relation to the roles of mathematics in preparing students for democratic participation. He cites Skovsmose and Valero (2001) on the need to subject that highly complex and contested concept to unceasing critique. His analysis of the impossibility of finding a unique set of principles to define democracy echoes the position argued by Sen (2009), in relation to justice, that it is necessary to accommodate a plurality of divergent, but reasonable, points of view, and to transcend narrow views of nationality and identity. Agreeing with Skovsmose on the need for "critique without foundation" he appeals to the idea of responsibility, and echoes the position of the philosopher Anderson in his paper *Ethical rights as ethical fictions* presented in the play *Professional Foul* by Tom Stoppard (1978) that: "There is a sense of right and wrong which precedes utterance. It is individually experienced and it concerns one person's dealings with another person". The notion of social justice is equally contested, full of tensions and even contradictions; nevertheless, the lack of an absolute foundation for ethical behaviour in no way absolves us from action, within and through mathematics education as everywhere else. Atweh hails the reappearance of ethics at the centre of philosophy, particularly due to the work of Emmanuel Levinas, and proposes that a conception of ethics is implicated in any critique of the relationship between mathematics and democratic participation. The conception of ethics that he advocates is based on responsibility to/for the other. From feminist critical theory, he takes the distinction between responsibility and "response-ability", the ability to respond with agency on one's own behalf and with responsibility on behalf of the other. From these considerations follow recommendations for the curriculum, and for pedagogy.

In Chapter 16, Ole Skovsmose points towards a critical mathematics education research programme by addressing three salient features of critical mathematics education. The first concerns the diversity of the situations in which mathematics is taught and learned. In previous chapters, authors have discussed what it is like to teach and learn mathematics for diverse people in diverse settings, such as a school in Chicago explicitly identified with teaching/learning for social justice, a Pakistani student in Barcelona, landless peasants in Brazil, and many more. None of these fit the "prototypical mathematics classroom" that dominates published research in mathematics education. To challenge this bias forms part of critical mathematics education. The second feature concerns the forms of mathematics in action. Mathematics operates as part of very many different work practices and technical settings. Often mathematics is integrated in work practices in a form that is not transparent to people involved in the professional practice. It is, however, important to address such variety of practices to provide a critical investigation of how mathematics might function, and it is important to investigate possible relationships between out-of-school mathematical practices and how mathematics might be contextualised in a school setting. The third feature concerns the exploration of educational possibilities, which can be mediated by notions like empowerment, social justice, and mathemacy. All of these features appeal to explosive and contested concepts, as they are not confinable within strict definitions. Any critical mathematics education research program, accordingly, comes to reflect a deep uncertainty.

Finally, in Chapter 17, we return to the notion of critique and to the impossibility of founding critique on an unassailable logical, epistemological, or political foundation. Emphatically, this position does not imply that action is impossible, but the grounds for that action are value systems and judgments.

Collectively, the authors of this book do not at all expect that readers will agree with us on everything. Indeed, we do not always agree with each other, and may even face internal personal contradictions, doubts, and changes of mind. We regard such reactions as healthy, better by far than the illusion of certainty and the dogmatism that accompanies it. What we do ask for is intellectual openness to examining unexamined assumptions, and to the consideration of alternatives.

The cover of the book shows Ole's painting of a parrot, as an illustration of Tagore's powerful parable that stands as a condemnation of education as the construction of an intellectual cage.

NOTE

1 The definitions cited are from the Concise Oxford English Dictionary. This cluster of words and their complex ramifications in philosophy and critical theory, including critical theory of mathematics education, has been extensively analysed by Skovsmose (e.g., 2005, pp. 38–47), and further reflections on this pivotal concern will be found in Chapter 17.

REFERENCES

Adler, J. (1988) Towards democratic mathematics education in South Africa. *Lengwitch*, March, 83–98.

Apple, M. W. (2000). Mathematics reform through conservative modernization? Standards, markets, and inequality in education. In J. Boaler (Ed.), *Multiple perspectives on mathematics teaching and learning* (pp. 243–259). Westport, CT: Ablex.

Apple, M. W., Au, W., & Gandin, L. A. (2011a). Mapping critical education. In M. W. Apple, W. Au & L. A. Gandin (Eds.), *The Routledge international handbook of critical education* (pp. 3–19). New York and London: Routledge.

Apple, M. W., Au, W., & Gandin, L. A. (2011b). *The Routledge international handbook of critical education*. New York and London: Routledge.

Atweh, B., Graven, M., Secada, W., & Valero, P. (Eds.). (2011). *Mapping equity and quality in mathematics education*. Dordrecht, The Netherlands: Springer.

Bishop, A. J. (1988). *Mathematical enculturation: A cultural perspective on mathematics education*. Dordrecht, The Netherlands: Reidel.

Bishop, A. J. (1990). Western mathematics: the secret weapon of cultural imperialism. *Race and Class*, *32*(2), 51–65.

Breen, C. (1986). Alternative mathematics programmes. In *Proceedings of the Eighth National Congress on Mathematics Education*, 178–188.

D'Ambrosio, U. (1984). Socio-cultural bases for mathematical education. In M. Carss (Ed.), *Proceedings of the Fifth International Congress on Mathematical Education* (pp. 1–6). Boston: Birkhausen.

D'Ambrosio, U. (1985). Ethnomathematics and its place in the history and pedagogy of mathematics. *For the Learning of Mathematics*, *5*(1), 44–48.

D'Ambrosio, U. (2010). Mathematics education and survival with dignity. In H. Alro, O. Ravn & P. Valero (Eds.), *Critical mathematics education: Past, present, and future* (pp. 51–63). Rotterdam: Sense Publishers.

Damerow, P., Elwitz, U., Keitel, C. & Zimmer, J. (1974). *Elementarmathematik: Lernen für die praxis? Ein exemplarischer versuch zur bestimmung fachüberschreitender curriculumziele*. Stuttgart: Klett.

De Corte, E., Greer, B., & Verschaffel, L. (1996). Mathematics teaching and learning. In D. Berliner & R. Calfee (Eds.), *Handbook of Educational Psychology* (pp. 491–549). New York: Macmillan.

Ernest, P. (2009). New philosophy of mathematics: Implications for mathematics education. In B. Greer, S. Mukhopadhyay, S. Nelson-Barber & A. B. Powell (Eds.), *Culturally responsive mathematics education* (pp. 43–64). New York: Routledge.

Fasheh, M. (1982). Mathematics, culture, and authority. *For the Learning of Mathematics*, *3*(2), 2–8.

Fasheh, M. (1997). Is mathematics in the classroom neutral – or dead? *For the Learning of Mathematics*, *17*(2), 24–27.

Frankenstein, M. (1983). Critical mathematics education: An application of Paolo Freire's epistemology. *Journal of Education*, *165*, 315–339.

Frankenstein, M. (1989). *Relearning mathematics – a different third R – radical maths*. London: Free Association Books.

Frankenstein, M. (2009). Developing a criticalmathematical numeracy through *real* real-world word problems. In L. Verschaffel, B. Greer, W. Van Dooren, & S. Mukhopadhyay (Eds), *Words and worlds: Modelling verbal descriptions of situations* (pp. 111–130). Rotterdam: Sense Publishers.

Freire, P. (1997). *Pedagogy of the oppressed* [Revised 20th anniversary edition]. New York: Continuum.

Freire, P., & Macedo, D. (1987). *Literacy: Reading the word and the world*. South Hadley, MA: Bergen & Garvey.

Freire, P., D'Ambrosio, U., & Mendonça, M. C. (1997). A conversation with Paul Freire. *For the Learning of Mathematics*, *17*(3), 7–10.

Greer, B., & Mukhopadhyay, S. (in press). The hegemony of English mathematics. In P. Ernest & B. Sriraman (Eds.), *Critical mathematics education: Theory and praxis*. Charlotte, NC: Information Age Publishing.

Greer, B., & Verschaffel, L. (2007). Modelling competencies – overview. In W. Blum, P. L. Galbraith, H.-W. Henne, & M. Niss (Eds), *Modelling and applications in mathematics education (ICMI Study 14)* (pp. 219–224). New York: Springer.

Gutstein, E. (2006). *Reading and writing the world with mathematics: Toward a pedagogy for social justice*. New York and London: Routledge.

Gutstein, E. (2009). The politics of mathematics education in the US: Dominant and counter agendas. In B. Greer, S. Mukhopadhyay, S. Nelson-Barber, & A. B. Powell (Eds.), *Culturally responsive mathematics education* (pp. 137–164). New York and London: Routledge.

Hardt, M., & Negri, A. (2003). *Empire*. Cambridge, MA: Harvard University Press.

Kilpatrick, J. (1981). The reasonable ineffectiveness of research in mathematics education. *For the Learning of Mathematics, 2*(2), 22–29.

Klein, N. (2007). *The shock doctrine*. New York: Picador.

Macedo, D., Dendrinos, B., & Gounari, P. (2003). *The hegemony of English*. Boulder, CA: Paradigm Publishers.

Mellin-Olsen, S. (1977). *Indlæring som social process*. Copenhagen: Rhodos.

Mellin-Olsen, S. (1987). *The politics of mathematics education*. Dordrecht, The Netherlands: Reidel.

Mukhopadhyay, S., & Greer, B. (2001). Modelling with purpose: Mathematics as a critical tool. In B. Atweh, H. Forgasz, & B. Nebres (Eds.). *Socio-cultural aspects in mathematics education: An international perspective* (pp. 295–311). Mahwah, NJ: Lawrence Erlbaum.

Noss, R., Brown, A., Dowling, P., Drake, P., Harris, M., Hoyles, C., & Mellin-Olsen, S. (Eds.). (1990). *Political dimensions of mathematics education: Action and critique*. London: Institute of Education, University of London.

Powell, A. B., & Frankenstein, M. (Eds.) (1997). *Ethnomathematics: Challenging Eurocentrism in mathematics education*. Albany, NY: SUNY Press.

Russell, B. (1992). *Human knowledge: Its scope and limits*. London: Routledge [first published 1948].

Said, E. W. (1994). *Culture and imperialism*. London: Vintage Books.

Sen, A. (2005). *The argumentative Indian*. London: Allen Lane.

Sen, A. (2009). *The idea of justice*. London: Penguin.

Skovsmose, O. (1980). *Forandringer i matematikundervisningen*. Copenhagen: Gyldendal.

Skovsmose, O. (1981a). *Alternativer og matematikundervisning*. Copenhagen: Gyldendal.

Skovsmose, O. (1981b). *Matematikundervisning og kritisk pædagogik*. Copenhagen: Gyldendal.

Skovsmose, O. (2005). *Travelling through education: Uncertainty, mathematics, responsibility*. Rotterdam: Sense Publishers.

Skovsmose, O. (2010). Mathematics: A critical rationality. *Philosophy of Mathematics Education Journal. Special issue on Critical Mathematics Education, No. 25.*

Skovsmose, O., & Valero, P. (2001). Breaking political neutrality: The critical engagement of mathematics education with democracy. In B. Atweh, H. Forgasz & B. Nebres (Eds.). *Socio-cultural aspects in mathematics education: An international perspective* (pp. 37–55). Mahwah, NJ: Lawrence Erlbaum.

Skovsmose, O., Valero, P., & Ravn Christensen, O. (Eds.) (2009). *University sciences and mathematics education in transition*. New York: Springer.

Stemhagen, K. (2009). Social justice and mathematics: Rethinking the nature and purposes of school mathematics. In P. Ernest, B. Greer & B. Sriraman (Eds.), *Critical issues in mathematics education* (pp. 337–349). Charlotte, NC: Information Age Publishing.

Stoppard, T. (1978). *Every good boy deserves favor/Professional foul*. New York: Grove Press.

Urton, G. (1997). *The social life of numbers: a Quechua ontology of numbers and philosophy of arithmetic*. Austin, TX: University of Texas Press.

Verschaffel, L, & Greer, B. (2007). Characterising modelling competencies. In W. Blum, P. Galbraith, H.-W. Henne, & M. Niss (Eds.), *Modelling and applications in mathematics education (14th ICMI Study)* (pp. 219–224). New York: Springer.

Verschaffel, L., Greer, B., & De Corte, E. (2000). *Making sense of word problems*. Lisse, The Netherlands: Swets & Zeitlinger.

Vithal, R. (2003). *In search of a pedagogy of conflict and dialogue for mathematics education*. Dordrecht, The Netherlands: Kluwer.

Volk, D. (1975). Plädoyer für einen problemorientierten mathematikunterricht in emanzipatorischer absicht. In M. Ewers (Ed.), *Naturwissenschaftliche didaktik zwischen kritik und konstruction*. Weinheim: Belz Verlag.

Volk, D. (Ed.). (1979). *Kritische stichwörter zum mathematikunterricht*. München: Wilhelm Fink Verlag.
Zaslavsky, C. (1973). *Africa counts*. Boston: Prindle, Weber & Schmidt.

Brian Greer
Portland State University
Oregon, USA

Ole Skovsmose
State University of São Paulo, Brazil
Aalborg University, Denmark

PART I

MATHEMATICS EDUCATION IS POLITICS

CHAPTER 1

MATHEMATICS AS A WEAPON IN THE STRUGGLE

ERIC (RICO) GUTSTEIN

In one of several "spoken-word" books that Paulo Freire wrote with friends and collaborators – this one with Chilean philosopher Antonio Faundez – the two discussed models of reality, the role of intellectuals in popular struggles, and the reinvention of others' work. The following is from their conversation:

> Antonio: … what is wrongly called your "method" … a lot of people think that your method is basically a model. I don't think you ever considered your method to be a model.
>
> Paulo: No, never.
>
> Antonio: A method for you is a series of principles which must be constantly reformulated, in that different, constantly changing situations demand that the principles be interpreted in a different way. And thereby enriched. And thus basically your method is a sort of challenge to intellectuals and to reality to reformulate that method in order to translate its principles as the situation demands and thus be a response to different concrete situations. What do you think of that?
>
> Paulo: I am in complete agreement. That is exactly why I always say that the only way anyone has of applying in their situation any of the propositions I have made is precisely by redoing what I have done, that is, by not following me. *In order to follow me it is essential not to follow me!* [emphasis added] (Freire & Faundez, 1992, p. 30).

Instead of "following", Freire (Freire & Macedo, 1987) urged *reinvention,* which he wrote, "demands the historical, cultural, political, social, and economic comprehension of the practice and proposals to be reinvented" (p. 133). He continued "I have to challenge other educators…to take my practice and my reflections as the object of their own reflections and analyse their context so they can begin to reinvent them in practice" (p. 135). To reinvent and apply a set of principles from another's context to one's own, and to move social struggles forward, one needs to comprehend how others see their own lives and the objective conditions of their realities – as well as the dialectical connections of these two – all in relation to understanding the same within one's own contexts.

This spirit of reinvention guides my approach to what Freire (1994) called *reading* [understanding] *and writing* [acting on and changing] *the world* – in a mathematics class in a Chicago public high school. I call this using *mathematics as a weapon in*

O. Skovsmose and B. Greer (eds.), Opening the Cage: Critique and Politics of Mathematics Education, 23–48.

the struggle. Although I still describe my work as teaching "mathematics for social justice" or "critical mathematics", I have adopted this more radical phrasing because, in 2007, then-President Bush travelled to South America and claimed that the U.S. was promoting "social justice" there. When I heard that, I remembered that others can appropriate our language, and so I starting using words that resist easy co-optation. Briefly, I mean by this that, "students need to be prepared through their mathematics education to investigate and critique injustice, and to challenge, in words and actions, oppressive structures and acts" (Gutstein, 2006, p. 4).

In this chapter, I examine, theorize, and problematise how Freire's concept of *generative themes* (defined below) figured into an attempt to enact liberatory education in a neighbourhood[1] Chicago public school. The school is in a culturally and spiritually strong, but economically battered, Black (African American) and Brown (Latina/o) community.[2] I analyse practices within a 12th-grade mathematics class (2008–09) that focused entirely on students investigating their lived realities using mathematics, through which they learned college-preparatory math. Students and teacher (myself) together decided which contexts to study, and when, either based on students' generative themes that they proposed, or on themes I suggested that they accepted. Together, we co-constructed a classroom embracing a pedagogy of questions, connections, critique, and challenge. Students engaged in the complexities of reading their world with mathematics and, to write the world with mathematics, they shared what they learned with their community and others in public presentations and through actions. As a contribution to this present volume on the politics of mathematics education, I analyse how we reinvented and lived out in the praxis of our class – with its many challenges and contradictions – Freire's notion that "education *is* politics" (Shor & Freire, 1987, p. 46).

CONTEXT MATTERS

The history of social movements attempting to upend unequal power relations shows that transformational processes can be neither exported nor imported. The internal conditions within the particular locale must exist for genuine change to occur, from the bottom up and inside out. As Amilcar Cabral (1979), leader of the victorious independence struggles of Cape Verde and Guineau-Bissau against almost 500 years of Portuguese colonialism, said:

> We also know that on the political level – however fine and attractive the reality of others may be – we can only truly transform our own reality, on the basis of detailed knowledge of it and our own efforts and sacrifices... however great the similarity between our cases and however identical our enemies, unfortunately or fortunately, national liberation and social revolution are not exportable commodities – more or less influenced by (favourable and unfavourable) external factors, but essentially determined and conditioned by the historical reality of each people. (p. 122)

This is entirely consistent with Freire's contention that others reinvent his praxis to (re)make their own history.

In other words, context matters – foundationally. Freire's contexts were literacy and post-literacy campaigns, explicitly and radically political, which often took place through *culture circles* – qualitatively different from schools: "Instead of a teacher, we had a coordinator; instead of lectures, dialogue; instead of pupils, group participants; instead of alienating syllabi, compact programs that were 'broken down' and 'codified' into learning units" (Freire, 1973, p. 42). He worked primarily with adult farmers, fisherfolk, peasants, and workers. They wanted to learn to read, on their own volition. Furthermore, there were no tests (let alone high-stakes assessments), passing grades, in-grade retention, or more to the point – educational failure. In Freire's settings, "performance standards" would have been nonsense, and neoliberal ideas of privatising public education, parents as educational consumers in a marketplace, and profits being extracted from school would have been downright retrograde. He worked mainly in rural areas, and the economically developing countries with which he collaborated were in the Global South, including Brazil, Chile, Guinea-Bissau, Cape Verde, Nicaragua, Grenada, and El Salvador. Except for his two and a half years as Secretary of Education in São Paulo (Brazil's largest city), he had little power over those with whom he worked on educational campaigns. Culture circle facilitators created curriculum (in the broad sense) from learners' lived realities and held after-work sessions in the community. No bells rang to signal the period's end. The *school to prison pipeline*[3] was unknown – and the purpose of education was for liberation and humanization.

Contrast this to urban, U.S. public schools, particularly middle/high school mathematics classrooms, my focus. Similar to Freire's contexts, most students are low-income (in Chicago public schools, 85%) and of colour (92%). But the similarities end there. Here, education is compulsory, to age 16 or 17, so whether or not learners want to attend school (of course many do, although as children age, their enthusiasm fades for many reasons), they have little choice. The mathematics "literacy campaign", so to speak, of U.S. public education emphasizes technological innovation for U.S. economic competitiveness, global market supremacy, capital accumulation, and increased productivity – all to benefit the wealthiest (Gutstein, 2009). This campaign's politics are reactionary and racist in that they serve and enrich U.S. capital and corporate/financial elites primarily at the expense of low-income and working-class people and people of colour within the U.S. (Gutstein, 2009), while, externally, the policies bolster U.S. supremacy over the Global South through "the control of new technologies, natural resources, flows of financial capital, communications and information, and weapons of mass destruction" (Amin, 2008, p. 15). Unlike Freirean culture circles, U.S. mathematics classes do have "teachers", "lectures" (often didactic, despite the mathematics reforms), "pupils", and "alienating syllabi" (in that they rarely touch upon or come from young people's lives). Students' school experiences, and those of teachers and administrators, are driven by high-stakes, punitive, (often) multiple-choice, exams, with profound consequences – for students wanting access to college and scholarships, and for communities whose schools can be closed or *turned-around*[4] and made ready for "new" residents in gentrified spaces (Lipman, 2011). In the most recent model of top-down educational power in the U.S., mayors appoint district officials, who have power over principals, who have power over teachers, who have power over students. District, state, and national standards serve the larger agenda and influence curricula and textbooks, with little (or no) responsiveness to local needs or desires (Gutstein, 2010b). Schooling's regimentation instils labour discipline – school bells, discipline codes, rule books, tardy passes, detentions, suspensions, expulsions – and greases the skids from school to prison, the military, gangs, street life, and early graves (all of which youth often critique). Macedo (1994) aptly described this as "literacy for stupidification".

In short, one would have trouble finding two more diametrically opposed educational contexts – Freirean culture circles supporting radical political democracy and U.S. mathematics classrooms in urban schools. How then to reinvent Freire in this foreign space of a Chicago public high school math class?

GENERATIVE THEMES

A key way to address this question is to use generative themes, which are essentially key social contradictions in people's lives. A theme can become "...the starting point for a political-pedagogical project [which] must be precisely at the level of the people's aspirations and dreams, their understanding of reality, and their forms of action and struggle" (Freire & Faundez, 1992, p. 27). Based on generative themes of students and their communities – as *they* understand and articulate them – one can attempt to create liberatory education. Curriculum, then, is not imported, but is particularized to students' lived realities, corresponding to Cabral's (1979) assertions that "we can only truly transform our own reality, on the basis of detailed knowledge of it" and that the path is "essentially determined and conditioned by the historical reality of each people" (p. 122). This reality is embedded within the generative themes that people experience and express, and it forms the basis for transformative curriculum (Camangian, 2006).

However, generative themes encompass more than social problems, and Freire considered them in relation to how people perceive their existence and act in response. While a social contradiction might constitute the basis of the theme itself, one cannot fully comprehend it apart from how people understand and interact with it. As he wrote, "I must re-emphasize that the generative theme cannot be found in people, divorced from reality; nor yet in reality, divorced from people.... It can only be apprehended in the human-world relationship" (1970/1998, p. 87). He continued and critiqued the position that:

> ...presupposes that themes exist, in their original objective purity, outside people – as if themes were *things*. Actually, themes exist in people in the relations with the world, with reference to concrete facts. The same objective fact could evoke different complexes of generative themes in different epochal sub-units. There is, therefore, a relation between the given objective fact, the perception women and men have of this fact, and the generative themes. (p. 87)

Freire viewed themes as interacting with and influencing each other in various ways. For example, he saw them in relationship to their negation, or opposite: "I consider the fundamental theme of our epoch to be that of *domination*" (p. 84), and he added that this theme simultaneously "implies its opposite, the theme of *liberation*, as the objective to be achieved" (p. 84). He also thought about the generality and specificity of themes: "Generative themes can be located in concentric circles, moving from the general to the particular" (p. 84). For him, the overall consideration was that people develop an understanding of, and act to change, the conditions of their lives as expressed in their generative themes.

Complexities in using generative themes

Previously, I examined various difficulties in building on generative themes in mathematics classes (Gutstein, 2006, 2007). My perspective was that three types of interrelated, yet distinct, knowledges relate to "reading and writing the world with mathematics" (that is, using mathematics to understand or "read" reality (Frankenstein, 1998), and to change or "write" it). These ideas are not novel, but in general, have been under-elaborated in mathematics education. The three types are *community*, *classical*, and *critical* knowledges, which all have mathematical components. Briefly, community knowledge refers to what some call popular or informal knowledge – knowledge of one's life circumstances and perspectives on reality. Classical knowledge refers to "traditional" academic knowledge, and critical knowledge means critiques and analyses of relations of power and issues of (in)justice. Developing (and acting upon) critical mathematical knowledge is akin to reading (and writing) the world with mathematics. I use Freire's (Freire & Macedo, 1987) notion of "reading the *world*" and extend it to mathematics to be synonymous to developing classical mathematical knowledge. Knowledge of generative themes might be considered part of community knowledge, but can include aspects of critical and classical knowledge as well.

My argument then was that one way to enact critical mathematics pedagogy was to build on students' community knowledge to support the simultaneous, interwoven development of their classical and critical mathematical knowledges. That is, build on generative themes and move outward from there. But the mathematics education literature contains few sustained examples of these practices. Certainly, educators have tried and studied these ideas. For example, the *Algebra Project* (Moses & Cobb, 2001) and the *Funds of Knowledge* work in mathematics education (Civil, 2006) connected community and classical mathematical knowledge, and culturally relevant mathematics pedagogy attempts this also (Gutstein, Lipman, Hernández, & de los Reyes, 1997; Ladson-Billings 1995; Tate, 1995). But, in practice, these efforts have under-emphasized critical mathematical knowledge. Other educators have connected critical and classical knowledges in mathematics (Brantlinger, 2006; Frankenstein, 1983; Gutstein, 2006), but did not fully proceed from student-articulated generative themes (community knowledge). Turner's (2003) dissertation based in a 6th grade mathematics class and Varley Gutiérrez's (2009) work in an after-school setting are notable exceptions of attempts to build upon urban students' community knowledge to develop both critical and classical mathematical knowledge. In Brazil (not coincidently, Freire's home), urban school systems develop curriculum based on students' and adults' community knowledge (O'Cadiz, Wong, & Torres, 1998; Gandin, 2002), but I know of no detailed research on sustained efforts to create and teach mathematics curriculum based on generative themes. In sum, this territory is insufficiently explored, practiced, and theorized. Below, I examine an attempt to use generative themes to teach critical mathematics and describe and theorize some difficulties, complexities, and contradictions.

GENERATIVE THEMES IN READING AND WRITING THE WORLD WITH MATHEMATICS

In the 2008–2009 school year, I taught a 12th grade mathematics class at Chicago's Greater Lawndale/Little Village School for Social Justice (known as *Sojo*). (I am a university faculty member and occasionally teach in Chicago public schools.) Sojo was born out of struggle through a 19-day hunger strike by residents of Chicago's Little Village community, a Mexican immigrant neighbourhood. Community members had demanded a new neighbourhood high school for years (the existing one was overcrowded), and when the district allocated funds (in 1998) but then reneged and built two selective, magnet high schools in wealthier, whiter communities, residents intensified the battle. This campaign culminated in the victorious hunger strike in 2001. A new building was built in which four small, neighbourhood schools opened in 2005 (Russo, 2003; Stovall, 2005), of which Sojo was one.[5] I joined Sojo's Design Team in 2003 and have worked with the school since its opening, participating in mathematics classes and collaborating with students and teachers to develop, teach, and study social justice mathematics.

The schools are roughly 70% Latina/o (mainly Mexican from Little Village, also known as South Lawndale) and 30% African-American students, who live in adjacent North Lawndale. Almost all students are low-income (96%), and any Lawndale student can attend any of the four schools. Each school is independent of the others, but they share common space (auditorium, gyms, lunchroom). Each has about 375 students, and as of this writing (June 2011), they have just finished their sixth year. Sojo's mean 2009 ACT score was 16.8, about the district average for neighbourhood schools and well below the national mean of 21.0 (American College Testing Program, 2010), but the school's first two classes (June 2009) graduated about 70% of entering 9th graders, well above Chicago's average of roughly 50% (Swanson, 2008).

Sojo uses IMP, the *Interactive Mathematics Program* (Fendel, Resek, Alper, & Fraser, 1998), as its basic curriculum, although we also incorporate several critical mathematics projects (Gutstein, 2007, 2010a) each year. IMP is considered one of the mathematics "reform" curricula and emphasizes student conceptual understanding. Students take four years of mathematics, and, in spring 2008, the 11th grade class selected their senior math class from the "math for social justice" class, year four of IMP, or pre-calculus. This was the only time juniors had this option, as

I only taught the math for social justice class in 2008–09 to 21 seniors (6 Black and 15 Latina/o; 6 male and 15 female). Previously, Sojo teachers and I surveyed and asked students about topics they wanted to study. They mentioned HIV/AIDS, neighbourhood displacement (gentrification), and immigration. That spring (2008), I met with the students twice and had several informal conversations to discuss our 12th grade units. During the conversations, a student proposed studying the criminalisation of youth of colour/people of colour, and I proposed two topics. One was sexism, because my reading of the literature suggests that K-12 classes rarely study it, and because the class was almost three fourths female, I wanted it to build on, support, and be relevant to students' identities. I also suggested that we study some statistical anomalies, suggesting possible fraud, in the 2004 presidential election that I believed (correctly, it turned out) students would not want to reoccur in the upcoming 2008 presidential election. Through dialogue, we collectively agreed on five topics: elections, displacement (gentrification, foreclosures, and immigration/deportation), HIV/AIDS, criminalisation, and sexism. Students expressed why they wanted to study these (personal connections, experiences, specific knowledge, observations, general concern, localness of the contexts, and interest). The three themes students proposed – displacement, HIV-AIDS, and criminalisation – were particularly meaningful and concerned many in the class. Several students had experienced themselves, family members, or friends struggling to save their homes from foreclosure or losing them altogether. HIV-AIDS in North Lawndale was a serious problem and some students knew people living with AIDS. And criminalisation of people of colour affected virtually the whole class as many students had family members involved, one way or the other, in the criminal justice system, Indeed, North Lawndale has an extremely high rate of criminalisation, with one estimate of up to 57% of its adults having had contact with the prison system (McKean & Raphael, 2002). Through the discussions and throughout the whole next year, I better understood how the themes related to students' lives and what was happening in their neighbourhoods.

I believe it is important to clarify some issues about generative themes as I understand them and as they played out in our class. First, teaching critical mathematics based on generative themes expressed by low-income, urban students of colour should *not* be interpreted to mean that none of these youth love or want to study mathematics. There were students in my class (and throughout Sojo) who were mathematically strong and gravitated to and enjoyed the abstractions, patterns, and intellectual stimulation of mathematics. One of my students was even occasionally impatient when we stayed too long (for him) on sociopolitical contexts and spent insufficient time on mathematics.

Second, using generative themes is not a motivational gimmick to entice students to learn more mathematics. Rather, the purpose is to develop what Freire (1970/1998) called a *problem posing pedagogy* that starts from "…the present, existential, concrete situation, reflecting the aspirations of the people" (p. 76) and that…"does not and cannot serve the interests of the oppressor" (p. 67). Such pedagogy is inherently dangerous to the power structure because questioning, critiquing, challenging, and ultimately taking action to transform an unjust society are the goals of using generative themes. Of course, learning mathematics for marginalized students is definitely important for their personal life and education opportunities, and for economic survival for their families, communities, and themselves. Moreover, learning to read the mathematical word (developing classical mathematical knowledge) is necessary to read and write the world with mathematics (developing and acting upon critical mathematical knowledge).

Third, using generative themes is not a panacea – students' lives are too complicated and the pain from harsh conditions of oppression does not stay outside the schoolhouse walls. Furthermore, the profoundly racist *miseducation* (Woodson, 1933/1990) of low-income youth of colour in the U.S. ensures that many have under-developed mathematical knowledge. Though some of my students were both conceptually and procedurally strong in mathematics, others – deeply intelligent, inquisitive, insightful, and resilient youth – had very weak comprehension of mathematics and rarely had had the opportunity to make genuine meaning, for themselves, of symbols, procedures, and concepts. Although documentation exists that student engagement in K-12 critical mathematics tends to outstrip that in reform or traditional mathematics (Brantlinger, 2006; Gutstein, 2006; Turner, 2003) – and engagement is clearly related to learning – one should not infer that critical mathematics, with or without using generative themes, necessarily or miraculously transforms mathematics learning.

Complexities in using generative themes: HIV/AIDS in our communities

Beyond these broad observations about generative themes lie other issues, because having a meaningful theme is just a starting point. This statement may be obvious, but the theme, by itself, does not tell how to teach it. When teaching students to read and write the world with mathematics, the challenges are both mathematical and political. I examine the HIV/AIDS unit to illustrate specific complications in teaching with generative themes, how we tried to resolve them, and what we learned.

Mathematical complexities There were two key challenges related to mathematics in this unit – one, to create curriculum that helped students better understand the HIV/AIDS epidemic in their communities and Chicago, and two, for students to learn the difficult mathematics that this required. To create the unit meant comprehending how scientists, public health workers, community organizers, epidemiologists, and AIDS activists studied and used mathematics to understand HIV-AIDS. I, and two of the graduate students helping develop curriculum frameworks, contacted health professionals and researched medical journals to understand the spread, prevention, and treatment of the disease. We learned that epidemiologists study disease transmission with differential equations (beyond my students' knowledge) and that the mathematical models for HIV/AIDS are complex.

However, my initial (naïve and erroneous) thinking was that we could develop a working HIV/AIDS model of Greater Lawndale and then project the impact of certain policies on the disease's spread. As I wrote in my journal:

> …I want students to have/create math models of AIDS transmission, and then think about tweaking them based on more (for example) gender equity so as to

> reduce "survival sex"[6], or better/more accessible HIV testing or full free access to condoms everywhere. The idea would be for us to make a mathematical argument that we'd have less deaths if we did one or more of these things. Of course, this is a bit contrived because we are just playing with numbers here and we cannot know how to change the parameters to affect the number of deaths…and that should be part of the discussion as well. (3/19/09).

As we progressed through the unit, it became obvious that creating the model for Lawndale was impossible. First, developing a model for HIV/AIDS was more difficult than I had anticipated. Hyman, Li, and Stanley (2003) capture some reasons for this: "Because the transmission dynamics [of HIV/AIDS] form a complex non-linear dynamical system, the behaviour of the epidemic is a highly non-linear function of the parameter values and levels of intervention strategies" (p. 18). Second, good data are scarce. We could not get reliable figures on HIV/AIDS in the community. And third, while mathematically modelling diseases has life-saving potential (but only if political will exists!), the models rely on many difficult-to-make assumptions. How does one account for individual choices? The number of partners one has? Access to prevention programs? The probability that unprotected sex results in disease transmission? Whether HIV-positive individuals inform potential sexual partners? Such uncertainties pervade the modelling of complex social phenomena. Typically little is done in school mathematics to give students a sense of the unavoidable uncertainty in the interpretation of the output from such models. Indeed, the overall effect of school mathematics may be to inculcate an implicit belief that mathematical models map aspects of the real world unproblematically on to mathematical equations that can then be manipulated to yield precise results.

Beyond the challenges for me to create the unit, students had to do quite difficult mathematics. To model HIV-AIDS, we used *discrete dynamical systems* (DDS) to represent change over time (Sandefur, 1993). We built on the work from the previous unit (displacement) in which students used DDS to study mortgages (Gutstein, 2012). A sociologist/biostatistician who teaches doctors about epidemic modelling graciously sent me her materials (Morris, Goodreau, & Cassels, 2008), which I adapted for class. Students built simulations, first drawing cubes from a bag, and then graphed the results on their calculators. To give a sense of their mathematical challenges, I reproduce a graph and an assignment from the unit. What made the mathematics for the HIV-AIDS unit so much harder than for the displacement unit is that one needs a DDS of two interacting equations to model a mortgage but four or six to model disease transmission. This assignment required students to derive and program into their calculators the four equations below (I did not give students the equations below). After the assignment, we analysed why the graph was shaped as it was. These equations describe much simpler disease transmission dynamics than those of HIV-AIDS.

$u(n) = u(n-1) + .0001v(n-1)u(n-1)$ [$u(n)$ means number of infected at time n]
$u(1) = 1$
$v(n) = v(n-1) - .0001u(n-1)v(n-1)$ [$v(n)$ means number of susceptible at time n]
$v(1) = 999$

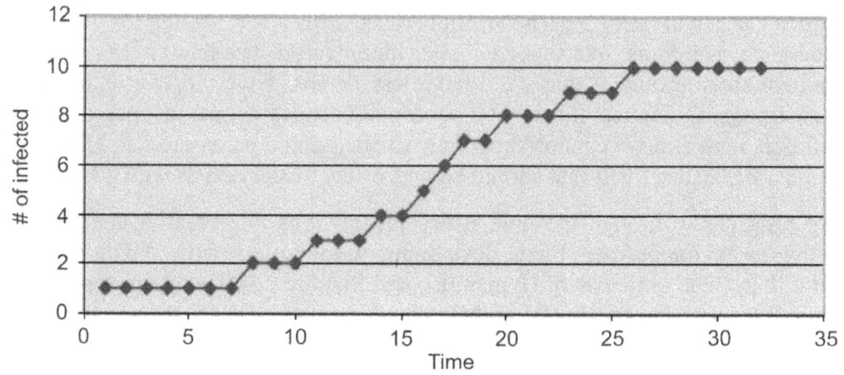

1. Suppose you had 1 infected and 999 susceptible people. Answer a–f below based on these numbers. This is too hard to do with cubes, so we can use our yellows [our TI-84 calculators].
    a. How could you find the number of interactions in a day using I and S for the number of infecteds and susceptibles? (Think about what you did to answer this w/ the chart).
    b. What is the minimum number of interactions in one day that you could have?
    c. What is the maximum number of interactions in one day that you could have?
    d. Suppose there is a probability of .01% (or .0001) that an interaction occurs between an infected and susceptible person and the susceptible is infected. What is the maximum number of infections that can occur in any one day?
    e. A dynamical system would tell us that "the number of infected at the start of day *n* is equal to the number at the start of day *n*–1 plus the newly infected that day". What is that as a difference equation? And what is the starting equation for the system?
    f. Create the dynamical system (the difference equation and starting equation) for the number of susceptibles.
2. Graph both systems on your yellow (think about your window variables!). When does your system hit equilibrium and what are the equilibrium values?
3. Experiment with different values for the probability, changing it slightly. Record your results for the probability value and number of days until equilibrium. What happens?

Political complexities The mathematical issues of the HIV/AIDS unit were complex – and the political ones were no less so. A principal challenge was to support students in developing sociopolitical explanations for the data. Shortly before the unit began, I asked students how we should spend the rest of our year (roughly three months). We had planned to complete three more units (HIV/AIDS, criminalisation, and sexism) but that seemed impossible. Students collectively decided to embed sexism in the other two units, and we started on HIV/AIDS. Carmen, a Black female student, had proposed the unit, and we were focusing on the impact on women, with Black women being the most disproportionately impacted demographic group in Chicago. I assembled and showed a short PowerPoint on Chicago data. As I wrote in my journal:

> We then moved to the slide that showed that 80% of the female diagnoses in 2006 [in Chicago] were of African Americans. That really blew people away. Carmen sat straight up and said, "damn!" Since we knew that African Americans made up only about 35% of the population, this was really out of line. (3/25/09)

This raises another key issue in using generative themes. Teachers and curriculum creators need to develop what Freire (1998) called "political clarity" on the themes. If a teacher uses a controversial topic with such powerful potential to stigmatise those most affected (e.g., Black women with HIV/AIDS), then the teacher herself has to read the world and deeply understand the sociopolitical genesis of the injustice. Then one can provide students access to various perspectives that differ from dominant narratives so they can critique and develop their own. Though this might involve mathematical issues, it is fundamentally a political question. I wanted to ensure that students did not leave class demonising Black women for their HIV/AIDS rate. Initially, I was not totally clear how to explain these data, but I knew that Black women were not the cause of the problem.

I found a book deconstructing various HIV/AIDS myths including that of "dangerous behavior" (Irwin, Millen, & Fallows, 2003). The authors argued that sociopolitical forces strongly constrained individuals' choices. The chapter on dangerous behaviour recounted several tales: a young woman in India pressured to have unprotected sex by social mores and by her HIV-positive, much older, and unfaithful, husband who subsequently infected her; a young man in Puerto Rico pressured by his family to earn more money, who turned to dealing drugs and eventually contacted AIDS; and other stories. The chapter's point was not to absolve individual responsibility or excuse poor choices, but to help readers contextualise and understand why people sometimes acted self-destructively.

We spent three (50-minute) periods in groups reading the chapter and making large posters to present the ideas. A central one was of "survival sex", which meant that women (and men) sometimes had sex because of economic pressures. This led into engaged debate about women, HIV/AIDS, sexism, gender, racism, power, prostitution, "sugar daddies", (a term the book used) and more. From my journal:

> Even though students know that there are real issues here, they have trouble getting past the discourse, the very powerful discourse of individual

> responsibility. Jenny [a Black female], in particular, wants to look at the individual and said that she felt the book was making excuses for people. (3/23/09).

Later in that journal, I wrote about more complexities:

> I drew Calvin and Antoine into the conversation about survival sex because their portion of the chapter was on gender discrimination. I think they have different views. Calvin seems to think that females who have sugar daddies do so because they want things, whereas I heard Antoine talk about needs. How do we approach this and distinguish? Jenny also talked about a woman she knows (I assume a Black woman) who knows her partner has HIV but sleeps with him anyhow, unprotected. We didn't ask, but one question might be if she has the power to demand condoms. Or not. Such a complicated situation. And if a woman acquiesces, even knowing the risk, is she weak? Roxanne pointed out that women stay in abusive relationships even when they should know better, often times because of economic insecurity. Is that also survival sex? Lots of really good questions emerged today… (3/23/09)

We never totally settled this matter at the time, which is not necessarily a problem in my view. One should not expect that youth (or adults!) easily resolve so complicated an issue, and critical (mathematics) pedagogy should allow for ambiguity, open questions, and contradiction. As Freire (Freire & Macedo, 1987) wrote, "A pedagogy will be that much more critical and radical the more investigative and less certain of 'certainties' it is. The more 'unquiet' a pedagogy, the more critical it will become" (p. 54). As I wrote in my journal:

> So something that is really interesting to me is this: in the elections unit, as we've talked about (me, Anita, Patty [doctoral-student co-researchers studying the class]), we've said that was a unit in which you had to do a mathematical analysis of a political situation to understand whether or not the election was stolen. Here, you have to do a political analysis of a mathematical situation (i.e., the data) to understand the data, the question of "why is there the disproportionality?" It's the flip side. (4/21/09)

I return to the "dangerous behavior" myth below when I discuss the related themes of race, class, and gender.

Mathematical-political complexities The above leads into what could be called the mathematical-political complexity of using generative themes in critical mathematics. One needs clarity on constraints and possibilities within both the mathematical and the political spheres – *and* their dialectical interactions – when creating and teaching curriculum based on the themes students and their communities bring. Since my general framework was to have context, not content, drive the curriculum, I tried to ensure that students wanted to learn mathematics to answer their own questions. This occurred in the elections unit during which students themselves reached the point of wanting to understand *if* the 2004 presidential election was "stolen" (a term used by Freeman and Bleifuss, 2006), because of the then-upcoming 2008 Obama election. After all, we were in Chicago (Obama's town), several students were 18 and preparing to vote for the first time, and some had helped monitor elections or participated in voter registration drives. The only way to answer the central question and be prepared for possible fraud in 2008 was to study binomial and normal distributions so students could make appropriate mathematical arguments about whether – or not – the irregularities were just coincidence (Gutstein, 2012). But the HIV/AIDS unit was different. I had trouble linking the mathematics of the DDS to students' questions about HIV/AIDS. My journal:

> After class, Anita contrasted what we're doing in the AIDS unit with what we did in the Elections unit. Students are not clear about why we're doing what we're doing! And that's my responsibility. I'm not even super sure…but she pointed out that we need to one, frame it more clearly in terms of the sociopolitical and economic context, and two, make it clear what we're doing mathematically that relates to the sociopolitical context. So I will work on this! (3/19/09)

That is, students did not need to *model* HIV/AIDS to understand the disease, grasp the disproportionate impact on certain populations, or not blame Black women. There was no focus question that motivated learning mathematics in the same way as in the elections unit ("what really happened in the 2004 election?") or displacement unit ("what were the sociopolitical forces displacing people from Lawndale and whose community was it?"). I felt committed to the challenging mathematics of DDS and was reluctant to give it up because I wanted students to have broad mathematics experiences and to learn serious, college-preparatory mathematics. I did not want students only to analyse data, as important as quantitative literacy is. But the difficulties of creating an HIV/AIDS model for Lawndale that students could learn from, along with my not-quite-rational proposition to students that they needed one, demonstrate the complexities of interweaving mathematical and sociopolitical aspects when using students' generative themes to teach critical mathematics. My journal:

> At times, we use mathematics to explain social things (like the election being stolen), at other times, we use social analysis to explain mathematics (like high AIDS rates). The point here is, I think, that you cannot easily explain one without the other. (3/23/09)

GENERATIVE, "RELATED" THEMES: INTERCONNECTIONS OF RACE, CLASS, AND GENDER

I have tried to illustrate some constraints and affordances of using generative themes in critical mathematics as a way to reinvent, but not follow, Freire. But the challenges of using themes from students' lives also extend to themes that I call "related". These were bound to, and both encompassed and were encompassed within, the situational themes that drove the units (e.g., displacement, HIV-AIDS). Given who my students were – low-income youth of colour from Lawndale – these related themes included race, class, and gender. These were specifically involved in four of our five units – displacement, HIV/AIDS, criminalisation of youth of colour/people of colour, and sexism. These themes interconnected and were central to the class as a whole. They also related to the mathematics students learned. Below, I explain how these related themes manifested themselves differently in two of our units and some possibilities and challenges that emerged. I also provide more context for how they connected to what Freire (1970/1998) called the fundamental themes of our epoch: domination and liberation.

Race, class, and gender – and the displacement unit

The displacement of low-income communities of colour in Chicago demonstrates these themes' interconnections. Displacement takes multiple forms, but in Lawndale, it takes shape in three main ways – gentrification (North Lawndale), deportation (Little Village), and foreclosures (both communities). North Lawndale's gentrification is part of Chicago's attempt to become a "global city" with "command centers" of the global economy (Demissie, 2006, Lipman, 2004). To become global, cities need to attract knowledge workers and develop the commensurate amenities to satisfy them. This includes schools for their children and up-scale housing (and culture and social opportunities, low-skill services, etc.) that this stratum demands. The confluence of housing and education is particularly interlinked in Chicago, and a key component is urban gentrification along with school closings and openings – rebranding of the "new" (transformed) neighbourhoods, along with the euphemistically termed "rebirthing" of "new" schools (Lipman, 2011).[7]

Gentrification in Chicago overwhelmingly affects Black, Brown, and low-income areas. Whole swaths of the city's Black community have been wiped off the map, partially due to the destruction of public housing and its subsequent replacement with "mixed-income" communities that effectively exclude most original residents (Lipman & Haines, 2007). Since 2004, the school board shut about 70 neighbourhood schools and opened approximately 100 new "renaissance" schools that are publicly funded, but mostly privately run by management corporations (i.e., charter or contract schools) (Chicago Public Schools, 2010). Administrators of these schools use implicit and often subtle mechanisms to restrict enrolment to students whom they consider "desirable" and to remove "undesirable" ones (Karp, 2010; Lutton, 2010). In the process, low-income students of colour and their families are forced out.

Women are particularly impacted by displacement. In general, they disproportionately head low-income and working-class families – North Lawndale's 2000 rate of female-headed households was about two and a half times greater than Chicago's as a whole (Census 2000) – and this is particularly true for public housing residents. In 2000, North Lawndale was about 60% female (Census 2000). When I asked a Black male Sojo student to explain this, he answered, "Cuz all the brothers [Black males] are locked up or in the ground", reflecting the confluence of racism (with respect to criminalisation), sexism (women bearing the burdens of life and family by themselves), and class (lack of meaningful job and educational opportunities for low-income Black males). This exclusion and removal devastates people's communities and histories. A mural in one gentrifying neighbourhood captures this well: "Gentrification = Ethnic Cleaning!"

While gentrification assaults North Lawndale, Little Village is relatively unscathed (so far). This is probably due both to geography (North Lawndale is closer to downtown with better transportation) and housing stock (North Lawndale has the most architecturally desirable *greystones*[8] of any Chicago community). Little Village's displacement nemesis is deportation. It is common knowledge in the neighbourhood that thousands of undocumented migrants (including many public school students) live in Little Village – the largest Mexican migrant community in the U.S. outside of East Los Angeles. Many are economic refugees from Mexico where poverty forces people off rural lands to Mexican cities and *maquiladora* areas along the U.S.-Mexico border, and often, eventually, north to the U.S. and Chicago where they live in fear of discovery and deportation (Bacon, 2008; Bigelow, 2006; Oxfam, 2003). Their jobs are disproportionately low-wage, non-union, difficult and dangerous, in the agriculture, low-skill service (e.g., gardening, maintenance, custodial), fast food, and construction (e.g., day labour) sectors (Bacon, 2008; Mehta, Theodore, Mora, & Wade, 2002).

Besides class issues, the experiences of Mexicans in Little Village and throughout the U.S. are highly racialised. A recently passed (April 2010) State of Arizona law exemplifies anti-Mexican racism. The legislation states that any Arizona law enforcement officer can demand identification documents from anyone they think is unauthorized in the U.S., effectively sanctioning racial profiling (although mass protests and resistance immediately erupted). Furthermore, the predicament of economic migrants is often gendered as well. Most from Mexico are males seeking jobs in the new land while their wives and children stay behind. These women often head families by themselves, hoping for remittances from their husbands up north. Men have to deal with alienation, loneliness, and exploitation, but women often experience these *and* the extra burdens of being single parents.

Like gentrification and deportation, home foreclosures have racialised, gendered, and classed aspects. Home foreclosures strongly correlate with race in Chicago, and the communities with the most numerically and the highest rates are overwhelmingly African American (National People's Action, 2010). I started the displacement unit by telling a story about Carmen's family, after receiving permission to share it. Her retired grandmother lived in North Lawndale for decades and paid off her modest home where she raised Carmen's mother and aunt. When her property taxes increased (common in gentrifying communities when property values go up), she took out an adjustable-rate home equity loan and borrowed extra cash. When the loan reset and increased, she could not make the payments, lost her house, and moved in with Carmen's aunt – a typical story in Chicago and across the nation. And although low-income and working-class Black communities had the highest rate of foreclosures in Chicago, Latina/o neighbourhoods were close behind. The families of two Latinas in our class were also struggling to keep their homes. In both Lawndales, foreclosures more than tripled from 2005 to 2008 (Woodstock Institute, 2010).

Here, too, women often bear the brunt because women and female-headed families are overall poorer than men and male-headed families, and poverty generally correlates with foreclosure rates. In recent years, predatory lenders targeted low-income women of colour (Fishbein & Woodall, 2006), showing the interconnections of race, class, and gender. Even though women in 2005 had slightly better average credit scores than men, they were more likely to receive subprime (financially unsound and typically more expensive) loans than men within their income range. This disparity increased as incomes increased – upper-income Black women were five times as likely to have a subprime loan as upper-income white men – and overall, Black women and Latinas have the highest rates of subprime borrowing (Fishbein & Woodall).

Thus, the generative theme of displacement and its related themes of race, class, and gender were salient in our class. In particular, this unit provided the best opportunity for students to use mathematics to understand that despite racial and cultural differences (and sometimes antagonisms) between the two Lawndales, the political commonalities outweighed the differences. I provide student perspectives on this point when I return to it below.

Race, class, and gender – and the HIV/AIDS unit

Similarly to the displacement unit, the HIV/AIDS unit also embodied the related themes of race, class, and gender, but raised different challenges. Since students decided to embed sexism within this unit (and criminalisation), the theme of gender was particularly present. As I mention above, this took shape as we analysed HIV/AIDS rates for Black and Latina women and the myth of dangerous behaviour. But the data surprised us, in more ways than one. We discovered that in the U.S., people of colour disproportionately have HIV/AIDS, but Chicago had some anomalous data. Eighty percent of its newly diagnosed female HIV cases in 2006 were of Black women, while the city was only 37% African American. During the same time, Latinas/os made up 26% of the population, but Latinas accounted for only 16% of the new diagnoses (AIDS Foundation of Chicago, 2008). Whites were also under-represented.

Students drew on their out-of-school knowledge in discussing HIV/AIDS, and implicitly and explicitly linked racism, poverty, and sexism. When we read the chapter on dangerous behaviour, Jenny and Gregory reported on the story of the young Indian woman whose husband infected her. Jenny said that "she [the woman in the story] is obligated to have sex with him…she doesn't have a choice, she's a woman". Marisol, in a very matter-of-fact voice, added that "when you get married, you have an obligation to have sex with your husband". I asked "So where does the power reside in the relationship?" Several students simultaneously responded, "with the man". Students were clear that the text was making the point that women often lack power to demand protected sex.

Students then discussed survival sex, and one read aloud a poster, titled "Women and AIDS". A line read, "Survival sex is necessary for women in poor [economically] situations." When Renee asked, "does survival sex mean prostitution", as I mention above, some students defined it as being about economic *need* but distinguished it from women having sugar daddies. That, they claimed, was about low-income women meeting their *wants*. Roxanne complicated this by stating that some economically dependent women who were not prostitutes stay in abusive relationships to feed their kids, and Ann added, "there are different forms of survival sex".

As we continued to discuss the chapter, contradictions emerged. Sometimes students contradicted themselves. Several accepted the book's argument that economics influenced people's decisions, and they used this to explain why women of colour had disproportionately high rates of HIV/AIDS. However, the class was perplexed when we confronted the Chicago data of Latinas having disproportionately low rates of infection in 2006. From my journal:

> So we established that Blacks were disproportionately [over-] affected by HIV/AIDS. But then Roxanne or Gema or Ann asked "but what about Latinas/os?" since they were 16% of the new [female] cases in 2006. I told them that Latinas/os were about 26% or so of the Chicago population. I also told them that whites were also about 35%, like African Americans. That led to a really interesting discussion. Roxanne said, "then we're not right, because we said that poverty was really related to AIDS, but then how can Latinas/os have a lower rate than they're in Chicago?" This prompted Ann, who said, "Then Jenny and I were right." Jenny said, "What do you mean?" And Ann responded, "then it is about promiscuous behavior, not poverty". What a complicated issue! I asked Roxanne what she made of all that, but she could not answer this. But what we did establish is that in Chicago, *both* Latinas/os and whites are under-represented in the 2006 HIV diagnoses, while African Americans are clearly over-represented. (3/25/2009)

Again, we did not try to definitively resolve this. In retrospect, I believe I made some mistakes at this point. I do not recall the details, and we have no audio or video of that day. Also, my journal makes no mention that I tried to support students in trying to make sense of the strange data. While I believe, in general, that students have to learn to handle ambiguity as part of learning how to remake the world, this oversight was a missed opportunity. Embracing uncertainty is part of the epistemological, ideological, and political standpoint of problem-posing pedagogies, as Freire's (Freire & Macedo, 1987) earlier quote about "uncertain" and "unquiet" education makes clear. But although that is my general pedagogical stance, in this instance my analysis is that I should have done things differently. At the very least, a teacher in this situation could have had students examine whether this was a trend/pattern and asked them how might they investigate this further and what else did they need to know.

All the themes and their interconnections, complexities, and contradictions

Race, class, and gender were inextricably linked in both the HIV/AIDS and displacement units but differed in how they connected to the generative themes framing each unit. In the HIV/AIDS unit, the intersections of race, class, and gender shed less light on political commonalities and more on analytical ambiguities. The unanswered question of why Latinas were under-represented in Chicago's new HIV cases in 2006 complicated students' understandings. They argued that because poverty was associated with higher levels of HIV/AIDS (though they could not fully explain why) and people of colour tended to be poorer, that the overall poverty of people of colour was a factor in their disproportionately high rates of HIV/AIDS. But Latinas' *under*-representation confounded that analysis.

My goal was that students develop a relatively coherent political understanding of specific mathematical data – Black women's drastically high rate of HIV/AIDS. I do not think that happened to the extent I wanted. Some Black female students argued that social factors did not absolve individual choices, but they could not explain why other Black women made what they considered to be bad decisions. In general, several students remained ambivalent in their views about the relationship of sociopolitical forces to individual choices. Some argued that promiscuity was the leading factor, then changed their minds to say it was poverty, but then could not explain why poor Latinas were under-represented in Chicago, and Black women over-represented. Overall, my analysis of classroom data[9] suggests that most students left the unit understanding that racism, sexism, and poverty all contributed to Black women's high rates – even if the exact mechanisms and connections remained unclear. Students left with open questions, to continue to think about in the future, but with more unclarity than I had wanted on this particular issue.

However, in contrast to the HIV-AIDS unit, in the displacement unit, race, class, and gender's interrelationships provided contexts for students to more clearly examine how their lives were touched by gentrification, foreclosure, and deportation and to better understand the political connections between the two communities. In late May 2009, near the end of the school year, the whole class presented what they had learned to their two communities (on consecutive nights). The group of students speaking on gentrification provided a mathematical analysis showing that families with median wages in both Little Village and North Lawndale could not afford new homes in the neighbourhood. And to end the section on the displacement unit, and the whole 81-slide power point presentation (created entirely by students), Erika[10] presented the following slide that drew together both communities and captured well our work for the year:

Why Should We Care?

- Both communities face the same problems but different situations.
- There are many lies and stereotypes about both Mexicans and African Americans
- "Mexicans steal the jobs of U.S. citizens."
- "African Americans are lazy."
- Don't let them pit us against each other!

Monica echoed these sentiments of unity in her end-of-unit project for the displacement unit. Her response was typical of what students wrote and said about this unit:

> Some connections that I see between these two parts [gentrification/ foreclosure and immigration/deportation] of the unit are that in both communities, people are being forced out their homes. Of course, it's different situations, but similar causes. African Americans are being forced out their homes because they can't pay for their homes. The taxes go up so much that they can't afford to keep living in those communities, so they are forced to look for another place to live. For Mexican people, the problem is that they don't have jobs in Mexico because corn isn't being sold, because it's cheaper to import subsidized US corn than to grow their own.[11] That forces Mexicans to leave their family and homes to come to the US to look for a job. This is how the unit connects.... People are being forced out of their community through gentrification. Latinos (especially Mexicans) are being forced out their countries by not having a good paying job. Also, the house mortgages don't only affect one community, but both. They are sometimes the target of bad loans that only make banks richer!
>
> I want the people in my community to know that we are really similar with these situations. That there is more that makes us similar, less that makes us different. If we want to fight the bigger people out there, the best way is to unite. Fighting each other is not going to take us anywhere. I think this is something very important our community should know.

Finally, there is the question of a white teacher and low-income students of colour. What did it mean to the students (and to me) that I, a white male professional, taught them to use mathematics to study these issues? Students and I talked openly about the differences between us, and I shared some of my experiences that were similar to theirs. I asked them their views in an open-ended survey at year's end (June 2009). Question #11 was:

> I am 56-year old white male, not from your community and don't experience what you do – and I'm the person teaching you to use math to understand racism, sexism, criminalization of youth of color, etc. What do you think?

Students' responses were interesting and varied. Some demonstrated a broad perspective on race and difference. Julie wrote: "You don't have to be from our community or our color to understand that things that are going on in our world are not balanced", and Calvin added: "Even if you don't experience it you hear it from the students…". Others seemed relatively unfazed. Gregory wrote: "I think nothing of it because you ask questions so that you don't put your opinion too much into the units", and Roxanne added: "When I read this question I just realized that I don't think it's a problem."

Others mentioned personal relationships, like Ann, who was broadminded and teased me as well: "I like it. I feel like you are one of us. Being born a certain thing isn't only what makes you, you. You are like a more experienced slightly older friend. Lol ☺." Antoine, in his usual succinct way, wrote, "Understand us. So you fit well in this class." Although none expressed a problem with my whiteness, Ellen wrote that "…it's weird how a white male is teaching us this instead of someone that looks like us", and Carmen, commenting on my having grown up in an inner-city Black and Brown community, also thought it weird: "About you being white, I have always thought that this was weird but it is accepted because you come from a similar background as some of us. You have struggled with us and the stuff you went through is encouragement for us to do things." And Daphne's response was particularly interesting because it reconfirms that students pay close attention to who their teachers are as people: "If I didn't know you so well I wouldn't listen to you. I would think of you as just someone trying to tell me what's right and wrong and criticize…"

I take from these responses and others, as well as other data, that students made relatively little of my race, age, etc. Sojo staff are fairly diverse in race, ethnicity, gender, and sexuality, so students were used to teachers from different backgrounds. I also think that, overall, my sharing power and having students co-determine what and when we would study; my continual pushing for questions, critique, challenge, collaboration, and communication; and our explicitly political relationships and the political framing of everything we did all mattered. Using generative themes demands these attributes from teachers, and while I made many mistakes and was not always able to be such a teacher, that was a conscious, explicit goal that I communicated regularly to students. This disposition may have mitigated the real power differentials of race, class, age, culture, language, and experience between my students and me.

CONCLUSION

My argument here is that teachers can use student/community generative themes to create and teach mathematics curriculum that supports young people in learning to read and write the world with mathematics – that is, to use mathematics as a weapon in the struggle for justice. This approach was an attempt to reinvent and apply Freire's principles to an urban U.S. context. I conclude with some key ideas, as I understand them.

The details of classroom interactions here should not obscure some larger points. First, I am clear on the difficulty of this work within the high-stakes, accountability-driven madness enveloping U.S. schools, exacerbated by the Duncan/Obama *Race to the Top* and neoliberal, education privatisation efforts. I recognize that district, school, and state mandates constrain teachers. This reality contributes to why relatively few teachers use students' generative themes to develop critical mathematics education. But understanding the roots of existing constraints helps us overcome barriers through collective agency. There are always spaces in which to act, even in modest ways. Teachers wanting to teach mathematics for social justice can start small, politicise simple situations, and share and collaborate with others.

Second, navigating what we call the "dance" between teaching social justice contexts and teaching mathematics is complicated, fluid, and challenging. At times (infrequently), my mathematics class resembled a social studies class devoid of mathematics; at times (also infrequently), it looked like a more-or-less reform mathematics class lacking sociopolitical contexts; and at other, more fortuitous times, my classroom genuinely and visibly integrated reading and writing the world with mathematics and reading the mathematical word. And everything in between. There is a continuum here, and one should not expect it to look neat and clean. Teachers need the patience to realize that this is a developmental process. It takes time and experience – for both teachers and students – to create a pedagogy of questioning and to co-create a classroom that supports teaching and learning critical mathematics. At its most fundamental level, the task asks teachers to have a political stance and the commitment to stand in solidarity with students in the struggles of their communities for justice (Freire, 1998; Gutstein, 2008; Ladson-Billings, 1994).

Third, there is the obvious distinction between developing curriculum and teaching it. The literature is clear that even veteran teachers have to learn to teach reform curriculum like IMP (Stein, Remillard, & Smith, 2007). Social justice curriculum is no different, and, in fact, has added challenges as I illustrate here. And there is the inherent contradiction between developing curriculum based on generative themes in one locale and then others using it in different settings. Yes, curricula may be adaptable, but teachers have to make that happen, partnering with their students to concretise and study their own specific contexts. There are no shortcuts here.

Fourth, there are no blueprints either. We make this road while walking and will not find answers in a book. We can build on existing learnings, but we also have to collaborate to produce new knowledge in this field. We can draw on the research on mathematics teaching and learning, which I do – the math reforms definitely inform my critical mathematics teaching and curriculum development – but while this is necessary, it is in no way sufficient. Generic "critical thinking" in mathematics is fundamentally different from critical literacy in mathematics (Apple, 1992).

Furthermore, the mainstream mathematics education community overlooks these questions, and the reforms, by themselves, can even exacerbate inequality (Martin, 2003; Secada, 1996). The political nature of this work demands reinvention.

Fifth, using generative themes in the classroom not only demands genuine democracy but also helps create it. This was central in Freire's writings about resolving what he called the "teacher-student contradiction". When teachers elicit generative themes from students and use them to develop liberatory curriculum, they instigate a dialectical process that opens a space for deep student engagement. And when students take this up, as they did in our class, they help make democracy be a cornerstone of the setting and participate in remaking education as "the practice of freedom" (Freire, 1973).

Finally, the purpose of reading and writing the world (with or without mathematics) is for sociopolitical change, to end oppression, exploitation, and exclusion, and for full humanization and liberation. As I write this (January 2012), the revolutions in Tunisia and Egypt are one year old, the Arab Spring has sprung, the Occupy Movement has 2600 encampments around the world, and the 99 percent/1 percent divide is on ordinary people's lips and minds, as well as on those of activists. The gap between rich and poor has never been wider and continues to grow, and vast global resources serve to enrich a tiny minority. The present path is not ecologically, economically, or spiritually sustainable.

In my view, the struggle of the world's people needs to be against our present political and economic system that is the root of racism, sexism, homophobia, and other discriminations that cause so much pain to my students and countless others. These young people need to take their rightful place in history as *subjects, not objects*, in the words of Paulo Freire (1970/1998). In the last book he wrote before he died, Freire (1998) wrote:

> I am a teacher who favors the permanent struggle against every form of bigotry and against the economic domination of individuals and social classes. I am a teacher who rejects the present system of capitalism, responsible for the aberration of misery in the midst of plenty. I am a teacher full of the spirit of hope, in spite of all signs to the contrary. (p. 94)

This spirit of hope, Freire (1994) reminded us, comes with the fight for a brighter future based on new relations between people. He wrote that without hope, the struggle dissipates and withers – but without the struggle, hope is meaningless and does not change reality. They need each other. To build on what others have done, to reinvent, and to use the knowledge of generative themes that students and their communities bring as a source for developing and teaching critical mathematics curriculum is a contribution we can make – that is, to make K-12, urban, U.S. mathematics be a weapon in the struggle for social justice, peace, and a better world. As Antoine wrote when I asked him, impromptu, to write what reading and writing the world with mathematics meant to him, and why our class did it:

> Reading and writing the world with mathematics for me is interpreting and making our judgment of the social and political reality of our community and

the world we live in. At the same [time] I learned how to make connections with mathematics and the real world. We do it because it helps us understand and combat against oppression and injustice in our communities and in the world. But more importantly, we do it to be educated in knowing what we are fighting against.

ACKNOWLEDGMENTS

A version of this article was presented at the XVIIth International Association of Sociology's World Congress of Sociology in Gothenberg, Sweden, 2010.

This research was partially funded by a Great Cities Institute Faculty Fellowship from the University of Illinois – Chicago.

Correspondence concerning this article should be addressed to the author at 1040 W. Harrison St., M/C 147, Chicago, IL 60607 or to gutstein@uic.edu

Acknowledgements to: student co-researchers/co-presenters in the "Crew" (Vero González, Rut Rodríguez, Amparo Ramos, George Carr, Channing Redditt, Alex Hernández, Amy Maldonado, Nikki Blunt, Darnisha Hill, & Rogelio Rivera); Anita Balasubramanian and Patricia Buenrostro (doctoral students/research colleagues who shared in this analysis and the curriculum framework development, and commented on an earlier draft); Joel Amidon, Phillip Caldwell, Paula DeAnda, Naama Lewis, Carlos Lopez Leiva, Craig Willey, and Caro Williams (graduate students who helped develop curriculum frameworks); and Marta Civil, Pauline Lipman, Erin Turner, and K. Wayne Yang for comments and critique.

NOTES

1 A neighborhood school in Chicago is one that any student living within the school's attendance boundaries can attend.

2 In Chicago, many use Black and Brown to refer to African Americans and Latinas/os respectively, and interchangeably, which I follow here.

3 This refers to the way students are forced out of school and eventually wind up in prison.

4 Chicago school turnaround means firing *all* adults and bringing in new administration and staff, though a small portion of teachers return.

5 Each school on the campus has its own theme. One is a multicultural arts school, one a math/science/technology school, and one a "world languages" school. Sojo is the only social justice school.

6 We read in class from a book that defined *survival sex* as people having to have sex for various forms of survival.

7 Gentrification has also become part of neoliberal economic development as cities as metropolitan regions compete on a global scale for capital investment. This activity is to compensate, in part, for federal budget cuts to urban areas in the U.S. as part of neoliberal restructuring that began in the 1970s and 1980s (Harvey, 2003).

8 Buildings characteristic of many Chicago communities.

9 Data for the year includes video from 41 classes, audio of classes, field notes and teacher journals from most classes, student surveys, audiotaped focus-group interviews, video of public student presentations, most of the student work, and more.

10 Her real name, because it was a public presentation. All other student names are pseudonyms.

[11] Although I do not have room to describe it, we also studied the impact of NAFTA (North American Free Trade Agreement) on displacement – from Mexico, as Monica alludes to here, and from Lawndale through its deindustrialization via the global race to the bottom.

REFERENCES

AIDS Foundation of Chicago (2008). *HIV/AIDS in Chicago*. Chicago: Author.

American College Testing Program (2010). *The condition of college and career readiness, 2010.* Retrieved from http://www.act.org/research/policymakers/cccr10/pdf/ConditionofCollegeandCareer Readiness 2010.pdf.

Amin, S. (2008). *The world we wish to see: Revolutionary objectives in the twenty-first century.* (J. Membrez, Trans.). New York: Monthly Review Press.

Apple, M. W. (1992). Do the Standards go far enough? Power, policy, and practice in mathematics education. *Journal for Research in Mathematics Education, 23,* 412–431.

Bacon, D. (2008). *Illegal people: How globalization creates migration and criminalizes immigration.* Boston: Beacon Press.

Bigelow, B. (2006). *The line between us: Teaching about the border and Mexican immigration.* Milwaukee, WI: Rethinking Schools, Ltd.

Brantlinger, A. (2006). *Geometries of inequality: Teaching and researching critical mathematics in a low-income urban high school.* Unpublished doctoral dissertation. Evanston, IL: Northwestern University.

Cabral, A. (1979). *Unity and struggle: Speeches and writings of Amilcar.* New York: Monthly Review Press.

Camangian, P. (2006, March 30). *Transformative teaching and youth resistance.* Talk given at DePaul University, Chicago, IL.

Chicago Public Schools (2010). *Office of New Schools.* Retrieved from http://www.cps.edu

Civil, M. (2006). Building on community knowledge: An avenue to equity in mathematics education. In N. S. Nasir & P. Cobb (Eds.), *Diversity, equity, and access to mathematical ideas* (pp. 105–117). New York: Teachers College Press.

Demissie, F. (2006). Globalization and the remaking of Chicago. In J. P. Koval, L, Bennett, M. I. J. Bennett, F. Demissie, R. Garner, & K. Kim (Eds.), *The new Chicago: A social and cultural analysis* (pp. 19–31). Philadelphia, PA: Temple University Press.

Fendel, D., Resek, D., Alper, L., & Fraser, S. (1998). *Interactive mathematics program.* Berkeley, CA: Key Curriculum Press.

Fishbein, A. J., & Woodall, P. (2006). *Women are prime targets for subprime lending.* Washington, DC: Consumer Federation of America.

Frankenstein, M. (1983). Critical mathematics education: An application of Paulo Freire's epistemology. *Journal of Education, 165,* 315–339.

Frankenstein, M. (1998). Reading the world with math: Goals for a criticalmathematical literacy curriculum. In E. Lee, D. Menkart, & M. Okazawa-Rey (Eds.), *Beyond heroes and holidays: A practical guide to K-12 anti-racist, multicultural education and staff development* (pp. 306–313). Washington DC: Network of Educators on the Americas.

Freeman, S. F., & Bleifuss, J. (2006). *Was the 2004 presidential election stolen? Exit polls, election fraud, and the official count.* New York: Seven Stories Press.

Freire, P. (1970/1998). *Pedagogy of the oppressed.* (M. B. Ramos, Trans.). New York: Continuum.

Freire, P. (1973). *Education for critical consciousness.* (M. B. Ramos, Trans.). New York. The Seabury Press.

Freire, P. (1994). *Pedagogy of hope: Reliving "Pedagogy of the oppressed".* (R. R. Barr, Trans.). New York: Continuum.

Freire, P. (1998). *Pedagogy of freedom: Ethics, democracy, and civic courage.* (P. Clarke, Trans.). Boston: Rowman & Littlefield.

Freire, P., & Faundez, A. (1992). *Learning to question: A pedagogy of liberation.* (T. Coates, Trans.). New York: Continuum.

Freire, P. & Macedo, D. (1987). *Literacy: Reading the word and the world.* Westport, CN: Bergin & Garvey.

Gandin, L. A. (2002). *Democratizing access, governance, and knowledge: The struggle for educational alternatives in Porto Alegre, Brazil.* Unpublished doctoral dissertation. Madison, WI: University of Wisconsin.

Gutstein, E. (2006). *Reading and writing the world with mathematics: Toward a pedagogy for social justice.* New York: Routledge.

Gutstein, E. (2007). Connecting community, critical, and classical knowledge in teaching mathematics for social justice. *The Montana Mathematics Enthusiast, Monograph 1*, 109–118.

Gutstein, E. (2008). Building political relationships with students: What social justice mathematics pedagogy requires of teachers. In E. de Freitas & K. Nolan (Eds.), *Opening the research text: Critical insights and in(ter)ventions into mathematics education* (pp. 189–204). New York: Springer.

Gutstein, E. (2009). The politics of mathematics education in the United States: Dominant and counter agendas. In B. Greer, S. Mukhopadhyay, S. Nelson-Barber, & A. B. Powell (Eds.), *Culturally responsive mathematics education* (pp. 137–164), New York: Routledge.

Gutstein, E. (2010a). Critical approaches to mathematics education in urban, K-12 Classrooms. In S. May & C. Sleeter (Eds.), *Critical multiculturalism: Theory and praxis* (pp. 121–131). New York: Routledge.

Gutstein, E. (2010b). The Common Core State Standards Initiative – A critical response. *Journal of Urban Mathematics Education, 3*(1), 9–18.

Gutstein, E. (2012). Using critical mathematics to understand the conditions of our lives. In T. Wrigley, P. Thomson, & B. Lingard (Eds.), *Changing schools: Alternative ways to make a world of difference* (pp. 181–193). New York: Routledge.

Gutstein, E., Lipman, P., Hernández, P., & de los Reyes, R. (1997). Culturally relevant mathematics teaching in a Mexican American context. *Journal for Research in Mathematics Education, 28*, 709–737.

Harvey, D. (2003). *The new imperialism.* New York: Oxford University Press.

Hyman, J. M., Li, J., & Stanley, E. A. (2003). Modeling the impact of random screening and contact tracing in reducing the spread of HIV. *Mathematical Biosciences, 181*, 17–54.

Irwin, A., Millen, J., & Fallows, D. (2003). *Global AIDS: Myths and facts.* Cambridge, MA: South End Press.

Karp, S. (2010). One in 10 charter school students transfers out. *Catalyst Chicago.* Retrieved from http://www.catalyst-chicago.org/print/index.php?item=2653&cat=5

Ladson-Billings, G. (1994). *The dreamkeepers.* San Francisco: Jossey Bass.

Ladson-Billings, G. (1995). Making mathematics meaningful in multicultural contexts. In W. G. Secada, E. Fennema, & L. B. Adajian (Eds.), *New directions for equity in mathematics education* (pp. 126–145). Cambridge: Cambridge University Press.

Lipman, P. (2004). *High stakes education: Inequality, globalization, and urban school reform.* New York: Routledge.

Lipman, P. (2011). *The new political economy of urban education: Neoliberalism, race, and the right to the city.* New York: Routledge.

Lipman, P., & Haines, N. (2007). From accountability to privatization and African American exclusion: Chicago's "Renaissance 2010". *Educational Policy, 21*, 471–502.

Lutton, L. (2010). *Chicago charter schools struggle to hold onto weakest students.* Retrieved from http://www.wbez.org/story/news/education/charters-struggle-hold-their-weakest-students

Macedo, D. (1994). *Literacies of power: What Americans are not allowed to know.* Boulder, CO: Westview.

Martin, D. (2003). Hidden assumptions and unaddressed questions in mathematics for all rhetoric. *The Mathematics Educato*r, *13*(2), 7–21.

McKean, L., & Raphael, J. (2002). *Drugs, crime, and consequences: Arrests and incarceration in North Lawndale.* Chicago: North Lawndale Employment Network.

Mehta, C., Theodore, N., Mora, I., & Wade, J. (2002). *Chicago's undocumented immigrants: An analysis of wages, working conditions, and economic contributions.* Chicago: University of Illinois – Chicago, Center for Urban Economic Development.

Morris, M., Goodreau, S., & Cassels, S. (2008, August). Epidemic modeling and simulation. Presentation given at the meeting of the Advanced Research Methodology Workshop, Chiromo, Kenya.

Moses, R. P., & Cobb, C. E. Jr. (2001). *Radical equations: Math literacy and civil rights.* Boston: Beacon Press.

National People's Action (2010). *The home foreclosure crisis in Chicago: An assessment of foreclosures and their impacts in 2009.* Retrieved from http://showdowninamerica.org/files/npa_2009_chicago_foreclosure_report_0.pdf

O'Cadiz, M., Wong, P., Torres, C. (1998). *Education and democracy: Paulo Freire, social movements, and educational reform in São Paulo.* Boulder, CO: Westview Press.

Oxfam (2003, August). *Dumping without borders: How US agriculture policies are destroying the livelihoods of Mexican corn farmers.* Oxfam Briefing Paper. Oxford, UK: Author.

Russo, A. (2003, June). Constructing a new school. *Catalyst.* Retrieved from http://www.catalyst-chicago.org/06–03/0603littlevillage.htm

Sandefur, J. T. (1993). *Discrete dynamical modeling.* New York: Oxford University Press.

Secada, W. G. (1996). Urban students acquiring English and learning mathematics in the context of reform. *Urban Education, 30,* 422–448.

Shor, I., & Freire, P. (1987). *A pedagogy for liberation: Dialogues for transforming education.* New York: Bergin and Garvey.

Stein, M. K., Remillard, J., & Smith, M. S. (2007). How curriculum influences student learning. In F. Lester, Jr. (Ed.), *Second handbook of research on mathematics teaching and learning* (pp. 319–370). Charlotte, NC: Information Age Publishing and the National Council of Teachers of Mathematics.

Stovall, D. (2005). Communities struggle to make small serve all. *Rethinking Schools, 19*(4). Retrieved from http://www.rethinkingschools.org/archive/19_04/stru194.shtml

Swanson, C. B. (2008). *Cities in crisis: A special analytic report on high school graduation.* Bethesda, MD: Editorial Projects in Education, Inc.

Tate, W. F. (1995). Returning to the root: A culturally relevant approach to mathematics pedagogy. *Theory into Practice, 34,* 166–173.

Turner, E. E. (2003). *Critical mathematical agency: Urban middle school students engage in significant mathematics to understand, critique, and act upon their world.* Unpublished doctoral dissertation. Austin, TX: University of Texas – Austin.

Varley Gutiérrez, M. (2009). *"I thought this U.S. place was supposed to be about freedom": Young Latinas engage in mathematics and social change to save their school.* Unpublished doctoral dissertation. Tucson, AZ: University of Arizona – Tucson.

Woodson, C. G. (1933/1990). *The mis-education of the Negro.* Trenton, NJ: Africa World Press, Inc.

Woodstock Institute (2010). *Chicago city and regional foreclosure activity.* Retrieved from http://woodstockinst.org/publications/applied-research-reports/foreclosure-updates/

Eric (Rico) Gutstein
University of Illinois at Chicago
USA

CHAPTER 2

A CRITICAL APPROACH TO EQUITY

ALEXANDRE PAIS

INTRODUCTION

Equity has been on the agenda of mathematics education research for at least two decades. The (first) *Handbook of Research on Mathematics Teaching and Learning* (Grouws, 1992) contains two chapters dealing with issues of equity and access: one focusing on gender (Leder, 1992), and the other on race, ethnicity, social class, and language (Secada, 1992). In 1995, a collection of pioneering contributions concerning the research on equity within mathematics education was published (edited by Secada, Fennema, and Adajian). In the same year, Rogers and Kaiser (1995) edited a book compiling research on the relation between equity and gender. This interest in equity was followed by a proliferation of theories in mathematics education research that progressively de-emphasized cognitive psychology as an interpretative framework for the learning of mathematics in favour of more socio-culturally oriented frameworks – what Lerman (2000) called the "social turn" in mathematics education research.

Acknowledging this trend, Valero (2004) coined the term "socio-political perspectives on mathematics education" to describe the branch of research emerging when the practice of mathematics education is investigated as part of the larger social, political, economical, cultural, and historical frameworks for education, schools, and classrooms. The scope and diversity of the research around these issues is vast. Authors such as Ole Skovsmose, Leone Burton, Marilyn Frankenstein, Paul Ernest, Ubiratan D'Ambrosio, Alan Bishop, Paola Valero, Robyn Zevenbergen, Mogens Niss, Paul Cobb, Gelsa Knijnik, among many others, have been developing research concerning issues that in some way handle social and political aspects of mathematics education. These subjects include equity, democratic access, social justice, critical mathematics education, social exclusion, and ethnomathematics.

Regarding the issues of equity and social justice, the last decade was prolific in publications. In 2002, a special edition of the journal *Mathematical Thinking and Learning*, edited by Na'ilah Nasir and Paul Cobb, was dedicated to the issue of *Diversity, Equity, and Mathematical Learning*. Five years later, the same authors edited a book (Nasir & Cobb, 2007) collecting contributions from several well-known mathematics education researchers on the issue of diversity and equity in the classroom. The theme of the *28th International Conference of the International Group for the Psychology of Mathematics Education* (proceedings edited by Johnsen Haines and Anne Fuglestad) was *Inclusion and Diversity*. The *Handbook of International Research in Mathematics Education* (English, 2008a) was

O. Skovsmose and B. Greer (eds.), Opening the Cage: Critique and Politics of Mathematics Education, 49–92.

particularly concerned with social, cultural, economical, and political influences in mathematics education. It collected several articles dealing with the issue of equity, under the label of "democratic access to powerful mathematical ideas". Also in 2008, three articles, by Sarah Lubienski and Rochelle Gutiérrez, were published in the *Journal for Research in Mathematics Education.* The authors engage in a discussion about the most fruitful way to address issues of equity in mathematics education research, concerning the existence of an achievement gap in the USA between middle-class white (and certain Asian) students and groups of people considered to be disadvantaged. Finally, in 2009, this same journal dedicated two special editions to the issue of equity, considered to be one of the current top priorities in mathematics education by the NCTM Board of Directors. Therefore, there is no risk in saying that the issue of equity is currently on the agenda, and the moment seems propitious to develop a theoretical analysis on the way equity is being addressed in mathematics education research.

Although notions of what it means to achieve equity diverge – and some authors prefer to use other terms such as "social justice" (e.g. Gutstein, 2003), "democratic access" (e.g. Skovsmose & Valero, 2008), or "inclusion/exclusion" (e.g. Knijnik, 1993) – it is usually acknowledged that research on equity requires social and political approaches that situate the problem in a broader context than the classroom or school (Anderson & Tate, 2008; Gates & Zevengergen, 2009; Valero, 2004, 2007). Although studies dealing with equity in mathematics education acknowledge the social and political dimensions of the problem, I shall argue that such studies insist on addressing the problem of inequity as if it could be understood and solved within mathematics education. It is as if we admit that the problem has an economical and political nature, going way beyond the classroom, but, since we are mathematics educators, we must investigate it in the classroom.

This approach – which consists in reducing a political problem to a didactical one, thus possible to be solved through the development and implementation of better stratagems to teach and learn mathematics – cannot be said to have produced the desired results, namely the commonly shared desire of "mathematics for all". Indeed, authors such as Roberto Baldino and Tânia Cabral have called our attention to the discrepancy between the huge amount of research targeted at diminishing the worldwide problem of failure in mathematics and the fact that failure persists and, as they highlight, is getting worse:

> So the situation is this: in spite of all research efforts along the lines such as the few ones above mentioned that are more directly concerned with exclusion issues, the social gap continues to increase. Authors ask for more research, better understanding of exclusion processes, more initiatives for change, more detailed and in depth studies, more comprehensive and interactive studies and data, new teaching strategies, and, recently, more studies addressing the social and cultural aspects of school mathematics. In one word, our problem is: *why do so many people insist in asking for more of that which cannot be said to have produced results for change so far*? (Baldino & Cabral, 2006, p. 21, my emphasis)

The question posed by Baldino and Cabral makes us wonder if mathematics education is on the right track in relation to research on equity.[1] The discrepancy between the increasing sophistication of research and the persistence of failure reminds me of an old joke about a man who lost a needle. Someone asks him: "Why are you searching for your needle in the kitchen if you lost it in the toilet?" He responds: "Well, the light is better in the kitchen." I will return to an exploration of this joke latter. For the moment, I suggest that most mathematics education research on equity may be leaving unaddressed crucial dimensions of the problem.

The main purpose of this article is to provide a deeper theoretical understanding of how equity is addressed in mathematics education research. The appeal for stronger theoretical conceptualisations has been endorsed by some authors (e.g. Gutiérrez, 2008; Nasir & Cobb, 2002), and I strongly believe that solution-oriented research is not enough when addressing problems that by their very nature are political and economical. One can suggest that, in the case of equity, perhaps it is a good idea to stop and ponder rather than repeating existing research. Nevertheless, a number of renowned researchers have been calling for the importance of developing a more reflective research: "[I]t appears to be one of the weaknesses of our profession that many of us, myself included, tend to write and speak too much and read and contemplate too little" (Niss, 2007, p. 1311). Despite Niss's appeal, mathematics education research has not been appreciative of research that is not immediately concerned with action, in the sense of providing solutions or strategies for improving the teaching and learning of mathematics (Pais, Stentoft, & Valero, 2010). However, I believe that without deep criticism of our own actions we run the risk of acting blindly.

My premise is that exclusion and inequity within mathematics education, and education in general, are integrative parts of schooling and cannot be conceptualised without understanding the relation between scholarised education and capitalism as the dominant mode of social formation. As mentioned, it is common to find research that presupposes the idea that the problem of inequity transcends school and mathematics education. However, little research has been done that explicitly tries to understand exclusion as an integral part of schooling; that is to say, as something consubstantial to schooling itself.

The article is constructed as follows. After presenting the theoretical and methodological background, which leans heavily on Slavoj Žižek's philosophical interpretation of the works of Marx, Lacan, and Hegel, I proceed to an analysis of recent research on equity, trying to identify: a) the problematic issues affecting research on equity within mathematics education; and b) common shared assumptions in the discourse on equity. The analysis leads to two conclusions. On the one hand, the discourse that vindicates the importance of mathematics to becoming a full citizen, and therefore requires mathematics for all, can carry the germ of exclusion. On the other hand, the way mathematics education research addresses the problem of equity is mostly a technical one, leaving aside a social and political comprehension of it. This imbalance creates an inconsistency since it is acknowledged that the problem of equity is a political and economical one, that goes beyond mathematics and its education, yet the strategies for "solving it" presuppose that it can be solved within mathematics education. The last part of the article is an attempt to address the problem of inequity in its materiality. I argue that inequity is not a marginal problem of a "good" system; rather, it is what gives consistency to the system itself. These conclusions will be related with capitalism wherein, from a Hegelian point of view, equity amounts to universal exclusion.

CRITIQUING RESEARCH ON EQUITY IN MATHEMATICS EDUCATION

Positing equity in the Political

As a field of research, mathematics education has been historically concerned with the improvement of the process of teaching and learning mathematics in schools (Cobb, 2007; Sierpinska & Kilpatrick, 1998; Silver & Herbst, 2007). As a consequence, the theories used in the field, even within a sociocultural framework (which sees meaning, thinking, and reasoning as products of social activity), are, in the majority of cases, theories of learning, designed to improve students' performance in mathematics (Pais, Stentoft, & Valero, 2010). In this sense, the work of researchers is "to identify important teaching and learning problems, consider different existing theories, and to try to understand the potential and limitations of the tools provided by these theories" (Silver & Herbst, 2007, p. 45).

As explored by Paola Valero and me (Pais & Valero, 2010), the tendency to focus on issues of learning – and thereby of teaching – is not exclusive to the field of mathematics education research, but has over the last two decades also proliferated in educational research in general. The language of education has largely been replaced by a technical language of learning. The contradictions of the role of schooling in society and the goals of education, that fuelled part of the educational debate during the last century,[2] seem to have been superseded. We seem to have reached a consensus on the benefits of schooling. Therefore, the central concern now is to make it more effective. The problems of schooling and school subjects are no longer political or ideological, but have become primarily technical or didactical. In most cases, solutions to educational problems are reduced to better methods and techniques to teach and learn; to improve the use of technology, to assess student's performance, and so on. Education has progressively been reduced to be a controllable, designable, engineerable, and operational framework for the individual's socialization.

Such a tendency is what the philosopher of education Gert Biesta calls the *learnification* of education (Biesta, 2005), and can be understood as part of a larger societal propensity to address fundamental social problems (such as the worldwide pressure to succeed in school mathematics) as if they were the object of expert management and administration (Agamben, 1998; Foucault, 1991, 1997).[3] On the one hand, if we take a look at two of the recent articles on the role of theory in mathematics education research[4] (Cobb, 2007; Silver & Herbst, 2007), we can notice how "theory" is perceived as providing "tools for action", where action is normally the practice of school mathematics, thus reducing research to a matter of providing the solutions for the problems of practice. On the other hand, the discursive construction of students as cognitive subjects and "schizomathematicslearners" (Valero, 2004) is a good example of the way mathematics education research reduces full political and historical human beings to "bare learners"[5], whose cognition can be scrutinized in the interest of devising appropriate and effective techniques for learning mathematics. All the complexity of the social and political life of the student is wiped out of the research focus. The student is reduced to a biological entity, likely to be investigated in a clinical way. For example, some researchers find it useful to draw an analogy between mathematics education research and medicine. Mathematics education research is formulated as a science of treatment that, by understanding the symptoms that characterize students' learning difficulties in mathematics, aims at designing and applying proper treatments, with the hope of curing what is a defect in students' learning: "The evolving understanding of the logic of errors has helped support the design of better instructional treatments, in much the same way that the evolving understanding of the logic of diseases has helped the design of better medical treatments" (Silver & Herbst, 2007, p. 63).

To some extent, philosophy offers us an antidote against this tendency to "technicise" educational problems. By taking a step back, by resisting the temptation to engage immediately in some form of action, one can propose that sometimes the best way to act is to stop acting, in the sense of doing more of the same research which proved not to have the solutions for the core problems the community faces (Baldino & Cabral, 2006). When addressing the problems of the field – most notably the persistent failure in school mathematics worldwide, the senseless mathematics instruction which reduces mathematics education to the apprehension of a set of routine procedures to be reproduced in exams, and the problem of inequity and how school mathematics is associated with processes of social exclusion – I concur with Brown (2008, p. 229) that "greater attention to policy domains, rather than focusing primarily on developing apparatus for working with individuals' minds" is necessary.

I argue that this judgment is especially true when the topic under research is equity, because achieving equity transcends the field of mathematics education. The problem of guaranteeing an equitable and just society, where all people have access to the material and cultural goods they need to become fully integrated citizens has been at the same time the flag of Modern Liberal-democratic politics, and its prevalent disenchantment. In line with Marx's (1857) critique of capitalist societies, modernity carries the spirit of universalising bourgeois ideology wherein equality, freedom, civil rights, humanitarianism, free speech, and open media stand for the Common Good. Of course, as noted by Jameson (1991), Marx sees nothing wrong with these principles in themselves, only the assumption that they can be achieved under capitalism.[6]

In this chapter, and taking advantage of the recent revitalization of capitalism's critique carried out by philosophers such as Slavoj Žižek and Frederic Jameson, I take the Marxian standpoint that equity within capitalism (both as an economy and an ideology), is doomed to fail. Indeed, there should be no need for sophisticated theories to recognize this, since we are living today at the pinnacle of world economical inequality[7] at a moment where capitalism has become what Žižek (2004) calls the "concrete universal" of our historical époque: "what this means is that while it remains a particular formation, it *overdetermines* all alternative formations, as well as all noneconomic strata of social life" (p. 3). In this way, contrary to the assumption defended by many economists (most notably Daniel Bell's *post-industrial society*) that we have arrived at a new type of society, where the new social formation in question no longer obeys the laws of classical capitalism, "late capitalism" (Jameson, 1991) signals instead that this "new society" is a *purer* stage of capitalism than any of the moments that preceded it. In this sense, every position on postmodernism in culture "is also at one and the same time, and *necessarily*, an implicitly or explicitly political stance on the nature of multinational capitalism today" (Jameson, 1991, p. 3). And this implication also goes for mathematics education research: what is the political stance towards capitalism informing the studies around the issue of equity? As will be argued in this text, the way the community engages in the research on equity is based on the assumption that despite the recognized problems in achieving equity in school mathematics, these can be overcome through an amelioration of the system (better teacher education, valorising students' learning diversity, more equitable school practices, use of technology, etc.). The system itself is not questioned. Change is conceptualised as a change within the system.

This view of change assumes that although the modern project based on capitalist economics and liberal ideologies has not been without its problems, these are contingent, conceived as "deviations" from an otherwise good system. The message conveyed is that modernity gave us the Ideals for which we should strive (equality, freedom, etc.), and our task should be to eliminate all the impediments standing in the way of their full actualisation. In *The Ethics of Psychoanalysis*, Jacques Lacan (2008) describes this ideology – *evolutionism* – as implying a belief in a Supreme Good, in a final goal of evolution that guides its course from the very beginning.

This perspective is notably evident in the influential works of John Rawls and Jurgen Habermas. Despite their differences, they share the assumption that a theory of the social should be primarily concerned with the delineation of a set of "universal" principles that should guide our action towards a better society: Habermas's society of fully communicative beings and Rawls's society of distributive justice. According to Chantal Mouffe (2005), who developed a groundbreaking critique of both, Rawls and Habermas don't deny that there will be obstacles to the realization of the ideal discourse, but those obstacles are conceived as *empirical* ones. That is, when dealing with a universal structuring principle, it is assumed that it is possible to apply this principle to all its potential elements, so that the principle's empirical non-realisation is merely a matter of contingent circumstances. What in this thesis are subsidiary problems of a "good" system are, in Marxian theory, the points at which the "truth", the immanent antagonistic character of the system, erupts – what Žižek calls, after Lacan, the symptom:

> A symptom, however, is an element which – although the non-realization of the universal principle in it appears to hinge on contingent circumstances – *has* to remain an exception, that is, the point of suspension of the universal principle: if the universal principle were to apply also to this point, the universal system itself would disintegrate.[8] (Žižek, 1997, p. 161)

Capitalism and its ideology posit progress, equality, and freedom as natural ideals shared by all humankind. These are presented as the goals we have to strive for – we know what we want, so the question is how to achieve it. The fact that we are today (still) living in an unequal society, where democracy has been reduced to elections and our freedom to freedom to select among a set of pre-given conditions[9], is seen by today's liberal-democracy as degenerations of the normal functioning of society, and as such capable of being abolished through amelioration of the system. By inventing the symptom (as Lacan (2008) put it), Marx called our attention to the fact that such "empirical obstacles" are the necessary conditions for the maintenance of the system that generates them, and it is through them that we can perceive the antagonistic structure of society:

> … for the standard capitalist view, crises are "temporary, correctable glitches" in the functioning of the system, while from the Marxist point, they are its moment of truth, the "exception" which only allows us to grasp the functioning of the system. (Žižek, 2007, p. 6)

In this light, the problem of inequity appears not as a contingency of a good system, but as a *necessity* of the same system that posits equity as a goal to strive for. Inequality is a necessity of capitalist economics, while equality functions as the necessary ideological supplement concealing the obscenity of what is going on. That is, the values of equity and freedom are generated by the market itself, as the necessary "double" concealing the lack of freedom and inequality of the system. Inequity is not foreign to the same system that struggles to eliminate it. Rather, it is its own motor. As Marx and Engels put it:

> Exchange value, or, more precisely, the money system, is indeed the system of freedom and equality, and what disturbs in the more recent development of the system are disturbances immanent to the system, i.e., the very realization of equality and freedom, which turn out to be inequality and unfreedom. It is an aspiration as pious as it is stupid to wish that exchange value would not develop into capital, or that labor which produces exchange value would not develop into wage labor. What distinguishes these gentlemen [the Proudhonists of the time, or the Habermasians of today] from the bourgeois apologists is, on the one hand, their awareness of the contradictions inherent in the system, and, on the other, their utopianism, manifest in their failure to grasp the inevitable difference between the real and the ideal shape of bourgeois society, and the consequent desire to undertake the superfluous task of changing the ideal expression back into reality, whereas it is in fact merely the photographic image of this reality. (Marx & Engels quoted in Jameson, 1991, pp. 261–262)

Ideals such as freedom and equity are indissolubly linked to capitalist economics that, in practice, turn out to be unfreedom and inequality. In this sense, as stated by Jameson (1991), the only thing real about the ideal is its unrealisability: "everybody wants to want them [freedom and equality]; but they cannot be realized. The only thing that can happen to them is for the system that generates them to disappear, thereby abolishing the 'ideals' along with the reality itself" (pp. 262–263). This is a courageous statement to make. It posits change in its *radicality*, that is, not aiming for a multitude of particular changes that, with time and faith, will allow us to correct the errors standing in the way of the Ideal, but to change the Ideal itself, that is, the whole system that coordinates our desire for equality and freedom. At stake here is the importance of conceiving, in a proper dialectical way, the relation between ideal and real: not just that the ideal cannot be actualised (an equitable economic organisation of society), this unrealisation is disavowed by the presence of the ideal. That is, the ideal discourse functions as the proper staged discourse – what we usually call ideology – that makes the real sustainable, accepted, and reproduced. We need to know that the goal for which we all strive is equality and freedom (that the presupposition of the system is a "good" one), so that we can accept the unequal reality in which we live.

Ideology and dialectics of necessity and contingency

The kind of analytical exercise carried out in this chapter conceives critique as something more than the critique of particular features of a system. Rather, it seeks to criticize the system as a whole, in this case, the whole of a system that posits mathematics and its education as crucial factors in becoming a citizen. According to Žižek (1993, p. 2), this is the most radical meaning of critique: an exercise of suspending what exists so that an experience of wondering, of conceiving possible alternatives to the system as a whole, becomes possible. However, the need to avoid evaluations of the system as a whole is now an integral part of its own internal organization as well as its various ideologies (Jameson, 1991, p. 350). Suffice to recall the recent measures deployed by governments to deal with the well-known "economical crisis". Despite some disruptive voices calling our attention to the need for a complete re-evaluation of the current economical system as a whole, the measures taken to solve the crisis are what Paulo Freire (1998, p. 508) called "superficial transformations", that is, transformations designed precisely to prevent any real change in the core features of Capital. In the same way that, for capitalism, achieving social equity is a matter of correcting market mechanisms (increasing taxation of private profit, governmental limitations on capital speculation, the criminalisation of individual magnates, etc.), similarly the vast majority of research on equity to achieve equity in school mathematics is a matter of developing better teaching and learning strategies. In both cases, a philosophical and political critique of the system as a whole, whether economical or schooling, is foreclosed.

In this chapter, in the form of a critical essay, I seek to re-map the way the community of mathematics education research gives meaning to the problem of equity. This reconceptualisation implies not accepting what exists as given, but, rather, raising the question of how what we encounter as actual is also possible. That is, it implies that we should conceive as contingent what is usually addressed as necessary. This interplay between necessity and contingency brings up one of the central concepts of Žižek's philosophy: Ideology. According to Žižek (1994), ideology can emerge from considering what is no more than a historical contingency as a necessity. For instance, in our times the existence of schools is perceived as a necessity, yet is, in fact, the result of a set of historical contingencies (Foucault, 2003; Rose, 1999). But also the converse: ideology can emerge from considering contingent what was always a necessity. For instance, school exclusion is dealt with as a contingent occurrence of a necessary system, whereas exclusion is what is necessary to maintain school. The task of a critique of ideology is precisely to "discern the hidden necessity in what appears as a mere contingency" (Žižek, 1994, p. 4). This task is also the central purpose of a dialectical approach – to articulate necessity with contingency.

On the one hand, necessity is only understood as such (as "necessary") retroactively, after a set of contingent episodes crystallizes and gives place to a sense of necessity. In this sense, necessity is a retroactive effect of a contingent process:

> "Dialectics" is ultimately a teaching on how necessity emerges out of contingency: on how a contingent *bricolage* produces a result which "transcodes" its final conditions into internal necessary moments of its self-reproduction. It is therefore Necessity itself which depends on contingency: the very gesture which changes necessity into contingency is radically contingent. (Žižek, 1991, p. 129)

On the other hand, to complete the dialectics, what appears as a mere contingency of a necessary system is to be posited as a necessary constitutive – and thereafter foreclosed as a mere contingency – of the same system. Take the slogan "mathematics for all" as an example. "Mathematics for all" has to be posited as a necessary goal if researchers, teachers and politicians are to find some meaning in their task of providing an equitable mathematics education. The fact that people continue to fail in school mathematics is seen as an "excess" introduced from the outside; its elimination would enable us to obtain a stable, inclusive school mathematics. All obstacles impeding the full actualisation of the Ideal are aliens to the Ideal, thus susceptible of being overcome through a correction of "empirical" intruders. However, although the goal should be easily within our grasp, it appears as if, to paraphrase Žižek (1997, p. 164), the entire universe has somehow been adjusted to produce, again and again, the unfathomable contingency of failure blocking the full actualisation of "mathematics for all". The kind of dialectical twist I am suggesting here is one that posits this "unfathomable contingency" as a *necessity.* Žižek articulates the argument this way:

> The other side of this necessity, which realizes itself in the guise of a series of contingent intrusions, which again and again prevent the universal notion of the project from realizing itself (…), is the necessity, the absolute certainty that within the field of a universal Lie the "repressed" truth will emerge in the guise of a particular contingent event. (1997, p. 165)

In our case, the "universal Lie" is no more than the slogan "mathematics for all", the "repressed truth" being the crude reality of those who year after year continue to fail in school mathematics. The systematic failure of people in school mathematics points towards the system's antagonistic character: the condition of impossibility of realizing the common goal (mathematics for all) is simultaneously its condition of possibility. That is, what makes schooling such an efficient modern practice is precisely its capability of excluding people by means of promotion. Thus, schooling is possible (as an institution *overdetermined* (Althusser, 1994) by capitalism) only to the extent that universal schooling, where everybody will be successful, remains impossible.

The motto "mathematics for all" functions as the necessary ideological double concealing the crude reality that, as any mathematics teacher knows, mathematics is not for all. This "social fantasy", as Žižek (1997) calls it, keeps us on the "right track" by avoiding putting in question the system as a whole: who will dare to challenge a system that seeks the Common Good, in this case, mathematics for all?

Ideology simultaneously conceals its "motives" whilst making them actual and effective. It is in this sense that Žižek says that ideology always appears in its sublated form, that is, its injunctions make effective what it "officially" conceals. In the example of "mathematics for all", this official claim conceals the obscenity of a school system that year after year throws thousands of people into the garbage bin of society under the official discourse of an inclusionary and democratic school. It is in this discrepancy between the official discourse and its (failed) actualisation that ideology is made operational. Within the official discourse, what is *necessary* is the abstract motto of "mathematics for all", all the exceptions to this rule (the ones who fail) being seen as contingencies. However, from the critical/dialectical discourse I am deploying here, what is *necessary* is precisely the existence of those who fail, the abstract proclamation being a purely contingent result of the frenetic activity of individuals (researchers, teachers, politicians) who believe in it. The antagonistic character of social reality – the crude reality that in order for some to succeed others have to fail – is the *necessary* real which needs to be concealed so that the illusion of social cohesion can be kept.

Mapping the "hot topics" in mathematics education research on equity

One of the most extensive reviews on the issue of equity in mathematics education is the article by Bishop and Forgasz (2007) published in the *Second Handbook of Research on Mathematics Teaching and Learning* (Lester, 2007). The authors try to give us an overview of the different research approaches on the issues of access and equity in mathematics education. Right from the beginning, they call our attention to the artificiality present in the construction of groups of people as being in disadvantage (girls, ethnic minorities, indigenous minorities, western "ex-colonial" groups, non-Judeo-Christian religious groups, rural learners, learners with physical and mental impairments, and children from lower classes) and how such constructions can in themselves convey discriminatory actions. This problem has been recently labelled by Gutiérrez (2008) as the "gap-gazing fetish in mathematics education", which triggered an interesting discussion between her and Sarah Lubienski, published in the *Journal for Research in Mathematics Education*, in which the two authors confronted their ideas and different strategies for approaching the issue of equity. Roughly speaking, the dilemma is how to know if research based on an achievement gap focus can benefit or not the purpose of equity. The position of Gutiérrez (2008) is that there are dangers in conceiving and performing research focused on an "achievement gap". These dangers include:

> [O]ffering little more than a static picture of inequities, supporting deficit thinking and negative narratives about students of color and working-class students, perpetuating the myth that the problem (and therefore solution) is a technical one, and promoting a narrow definition of learning and society. (p. 358)

She argues that such research, which usually leans on quantitative methods of data collection, does no more than provide a description of the problem without presenting understandings that allow us to change it. She argues that less research should be done focusing on the "gap", and more on qualitative analyses of successful experiences among groups of people considered to be in disadvantage.

On the other hand, Lubienski (2008) assumes the position that research on the achievement gap is necessary. She is against the suggestion of Gutiérrez of lowering the intensity of gap analysis, arguing, rather, for moving towards a more skilled and nuanced analysis, since: "analysis of gaps also inform mathematics education research and practice, illuminating which groups and curricular areas are most in need of intervention and additional study" (Lubienski, 2008, p. 351).

While Lubienski is concerned with the question "Is there a gap?", which leads to all the studies that analyse when those gaps begin, under what conditions they grow or shrink, and what consequences underserved students ultimately suffer because of the gap, Gutiérrez is concerned with the question "How to diminish the gap?", leading to studies orientated towards effective teaching and learning, making research more accessible to practitioners, and more intervention by the researcher. Despite the difference regarding the ways to conduct research on equity, both authors share a common assumption: "For both of us, the goal is for students to gain access to dominant and critical ways of viewing the world so that they might become empowered citizens" (Lubienski & Gutiérrez, 2008, p. 367). We already start to notice how this common assumption – that mathematics empowers people by giving them the tools for "viewing" the world – functions as a *quilting-point*[10] in mathematics education research.

Although we can discuss better ways to do research on equity, there is a fundamental question that cannot be left unaddressed: Why is there a gap at all? That is, why does school (mathematics) systematically exclude and include people in the network of social positioning? Why do schools perform this selective role that inevitably creates inequity? As Bishop and Forgasz (2007) note, "in every country in the world mathematics now holds a special position, and those who excel at it or its applications also hold significant positions in their societies" (p. 1149). Why does our society need to have such an institution that guarantees from very early ages an accumulation of credit? This question is rarely posed by the community of researchers in mathematics education when addressing equity. Gates and Zevenbergen (2009) state that "mathematics and social justice has been the focus of much research – however this has largely focussed on such issues as the process of learning, the content of the curriculum and its assessment" (p. 162). They also make a very suggestive point. They argue that it is common in mathematics education research to discard such "political" questions, since it is not the responsibility of mathematics education to address them (p. 165). One can argue that such a position removes the possibility of subversiveness in research on equity, rendering it harmless.

In the same article, Gates and Zevenbergen (2009) take advantage of Bourdieu's theory on social reproduction to theorize about social justice in mathematics education in relation to teacher education. They categorize research on social justice into three types. The first type, which they call "moderate forms of social justice" (p. 166), does not relate the failure in mathematics to structural inequalities in society. These perspectives do not challenge the status quo and show a tendency to see the social inequalities as something natural, as a result of people's different capabilities and merit. In this sense, the exclusion provoked by the school system is seen as a natural selective process of our societies. The second, "liberal forms of social justice" (p. 167), recognises structural inequalities in the way mathematics is taught in schools, and proposes as a solution more detailed and accurate research on the process of mathematical learning, teacher education, assessment, curriculum, and so on. Most of the research compiled by Nasir & Cobb (2007) assumes this perspective, the idea that social inequalities will be solved through better classroom practices. Finally, the "radical forms of social justice" (p. 167) also recognize structural inequality, as well as social class and ideology. This perspective avoids engaging in salvation discourses, and assumes that the only way for mathematics education to become more equitable is through a deep change in the class structure of society, to which we can contribute by ways of developing with students a "class consciousness".[11]

Perhaps we should extend the radicalism of the third type. What still appears problematic is the notion of empowerment through mathematics, or why it is important to learn mathematics. Using the theoretical framework developed by Skovsmose and Valero (2008) to deal with the issue of power in mathematics education, we can say that mathematics education can empower people through the intrinsic characteristics of mathematics itself (logical thinking, abstraction); by providing students with psychologically meaningful experiences (solving problems, metacognition); by enhancing the relation between cultural background and foreground, therefore allowing students to learn "in context" (connection between everyday practices and school mathematics; providing opportunities to envision a desirable range of future possibilities); and finally students can get empowered through school mathematics by exploration of situations of "mathematics in action", which make visible the way mathematics formats reality (exploring real mathematical models in a critical way). What is missing in these four conceptualisations of the way mathematics empowers people is a fundamental element – mathematics empowers people not so much because it provides some kind of knowledge or competence to them, but because it gives people a value. It allows students to accumulate credit in the school system that will allow them to continue studying and later to achieve a place in the sun. Mathematics empowers people because it is posited as a socially valuable resource.

This powerful dimension of mathematics is often absent in mathematics education research. One feels tempted to conceive such obliteration as the repression of a trauma. The discourses about the way mathematics empowers people through knowledge and competence disguise the traumatic role of school mathematics as a social gatekeeper. So, if one wishes to extend the radicalism of the last approach mentioned above, it must be added that it is not just that the problem of failure is a structural problem, a solution to which mathematics education can contribute by developing a class awareness, but that mathematics education research itself is part of the problem as long as it continues to neglect the importance of school mathematics outside of knowledge and competence.

Nolan (2009) developed a critique on how the issue of social justice has been researched in mathematics education. She calls our attention to the risk of considering issues of equity and social justice just as a current fashion, and explores what, in her view, is missing in the current push to marry off mathematics with social justice. She starts by criticizing the usual way in which issues of social justice are translated into practices of teaching and learning, which are basically in terms of contents: "[t]he most common approach to realizing a mathematics education in and through social justice is by integrating the facts and figures of poverty, exploitation, and discrimination into ready-to-use problem-based lesson plans" (p. 207). According to her, this is the easiest approach to realizing a simplified consensus on the nature of the complex union between mathematics and social justice. Unsatisfied with this conceptualisation and practice of social justice within mathematics education, she suggests that for a lasting relationship between social justice and mathematics we need to focus our attention on teacher education, by preparing teachers to develop in their classrooms practices of social justice. She then continues the article by exploring her own experience in teacher education for social justice, and all the problems involved in it.

The author makes a good point by saying that, if we wish to take social justice seriously, it is not enough to work with students' issues where mathematics appears as a tool to explore socially unjust situations. The mathematical content approach "seems to leave the dominant characteristics and personality of mathematics intact, while molding and shaping the concerns of social justice to fit into life-as-usual of mathematics" (Nolan, 2009, p. 207). She suggests, in a similar way as ethnomathematics (e.g. Powell & Frankenstein, 1997), that we must not just deconstruct social reality using mathematics as a tool, but to deconstruct mathematics itself – its image of neutrality, social independence, universalism, and so on.

Despite her good will, producing "teachers for social justice" suffers from repeated failure. Notwithstanding the fact that prospective teachers revealed some "theoretical" interest in issues of social justice during the classes with Nolan, when they go to schools they feel that it is nonsensical to explore such issues with their students. At most they feel compelled to explore some ready-to-apply examples of mathematics as a tool to understand reality. This gap between knowing the research findings (in the case of social justice: the nature of mathematics, issues of critical mathematics education, issues of democracy, etc.) and the practice of teaching in schools is common. Although teachers can be aware of the social and political dimensions of mathematics, they tend to reproduce discourses that posit mathematics as an apolitical subject (de Freitas, 2004). There are many possible explanations for the fact that teachers do not simply transfer to their practices the theoretical ideas learned in the academy. Nolan suggests that to reduce this gap is a teacher education problem. In her view and in her work, the way to decrease the gap is to alter teachers' conceptions about mathematics and the role of school mathematics in social justice.

We already have research within mathematics education that shows how teacher education cannot be taken for granted. Although some say that teachers are not prepared to conveniently implement reforms, Klette (2004) takes another approach. She conceptualises the lack of change in mathematics education reforms as an embedded part of research itself. The author suggests that researchers are an important element in the non-concretisation of school reform. This idea goes against the commonsensical one, which posits the problem on the side of the practitioner. The idea is that in research everything goes well, we know the best methods, theories and strategies; the problem lies in its application. Klette criticizes this view, and argues that the denial of change is being constructed from the beginning, in the theoretical, methodological and conceptual ways in which research is done. The author suggests that more is needed than merely investigating how reforms change schools; we should investigate how schools change reforms. This is precisely what is missing in Nolan's work: not so much concern with how teachers will change classroom routines, but how classroom routines change the (weak) theoretical awareness of teachers.

Finally, I wish to argue that although the issue of equity is usually conceived as having to do with specific groups of people (women, indigenous, poor, etc.), other issues are at stake. Valero (2007) analysed a paradigmatic episode of discrimination in a regular Danish mathematics classroom. The case of the "lonely girl" (p. 227) is about Gitte, a Danish teenager who was positioned by the school administration and by her teacher as a student with learning difficulties in mathematics. The girl was isolated in the classroom, performing banal tasks like sharpening the pencils for her colleagues or picking up dropped rulers. She was allowed to be in school, but everybody unassumingly knew she wasn't learning any mathematics. The exploration of this case led Valero to assert that "disadvantage is being built even at the heart of an educational system that has inclusion and democracy as an organizational principle" (pp. 228–229). The problem of equity is not exclusive to people who are positioned as being in disadvantage due to their association to some category (ethnicity, gender, linguistic, socio-economical, etc.). Indeed, I wish to argue for a displacement of the problem of equity that conceptualises inequity not so much as a problem affecting particular groups of people, but a generalized problem of the school system, that affects everyone by the way schooling is involved in social stratification. The paradox is that such systematic "social selection" is happening at the core of a school organized around democratic and inclusionary principles.

CRITIQUING THE IMPORTANCE OF MATHEMATICS

Common shared assumptions: the importance of mathematics in becoming a worker and a citizen

I shall argue that beneath all the different understandings on how to deal with the issue of equity in mathematics education research, there are some common shared assumptions that function as neutral camps where everybody should agree – in Lacanese, as quilting-points. These common shared assumptions, it will be argued, convey the germ of exclusion.

The literature on equity and mathematics education is full of statements that posit mathematics as a powerful knowledge and competence, required to become a full citizen and worker. These two educational functions, that Biesta (2009) calls qualification – having to do with the need for people with the knowledge, skills, and understanding that allow them to "do something" on a professional basis – and socialization – having to do with the role of education in allowing people to become members of a particular society, by the insertion of the "newcomers" into existing social and cultural orders – comprise the two main goals of mathematics education:

> Mathematics education in schools is thus seen to have a dual function: to prepare students to be mathematically functional as citizens of their society arguably provided equitably for all – and to prepare some students to be the future professionals in careers in which mathematics is fundamental, with no one precluded from or denied access to participation along this path. (Bishop & Forgasz, 2007, p. 1152)

Mathematics is posited as indispensable knowledge and competence to participate in the world – the idea that through mathematics we become empowered citizens. This idea presupposes another one: that mathematics is everywhere. In our times, it is a commonplace to state that mathematics is the dorsal spine of our high-tech world (D'Ambrosio, 1993; Ernest, 1991; Skovsmose, 1994). This paramount presence and influence of mathematics poses a challenge to mathematics education: since most of the mathematics "ruling" our world is "under the veil", students need to critically deconstruct the way in which mathematics formats reality, so that they can socially participate as informed citizens. Such claims can be read in many articles dealing with the issue of equity. From this point of view, to guarantee equity means to provide to all students learning experiences that allow them to become full participative citizens:

> Students are facing a world that is shaped by increasingly complex, dynamic, and powerful systems of information and new ideas. As future members of

> the work force, students will need to be able to interpret and explain structurally complex systems, to reason in mathematically diverse ways, and to use sophisticated tools and resources. (English, 2008b, p. 11)

> Democratic education – accessible to all students – rests on the assumption that all students can learn, given the right circumstance, provides students with an avenue through which they can learn substantial mathematics, and, at the same time, can help students become productive and active citizens. (Malloy, 2008, p. 23)

> The marginal performance in mathematics of minority students, language-minority students, poor students, and to some extent, girls have led several American scholars to raise concerns about the opportunities for members of these groups to compete in an increasingly technological world. (Nasir & Cobb, 2002, p. 91)

> It is impossible to be a democratic citizen and not be proficient in mathematics. Every decision that a citizen must take requires complicated calculations. (Pearl & Knight, 1999, p. 119 quoted in Malloy, 2008, p. 24)

Indeed, this discourse on the importance of mathematics to become a full citizen is not new. Amit and Fried (2008) locate the emergence of this goal for mathematics education in the middle of the 20th century. They refer to the School Mathematics Study Group (SMSG) founded in 1958 in the United States. In one of the documents of this study group we can read:

> Our society leans more and more heavily on science and technology. The number of our citizens skilled in mathematics must be greatly increased; an understanding of the role of mathematics in our society is now a prerequisite for intelligent citizenship. (SMSG newsletter, quoted in Amit & Fried, 2007, p. 389)

This statement seems more contemporary than ever, and it can be read, not just in research, but also in many curricular documents around the world. Its basic structure consists of two steps. First, we realise that mathematics is everywhere in our technological society. We live in a society reigned over by mathematical modelling, which influences our life decisions, most of the time without us being aware of it. Second, this "under the veil" mathematics needs to be understood if we wish to fully exercise our democratic rights. So, a mathematics education concerned with citizenship must prepare students to use mathematics in a way that allows them to participate in an informed and critical way. This is the premise behind the aim of equity: to achieve equity in mathematics education means to provide meaningful mathematics education for all students, so that they can become full workers or citizens. The ways in which this can be done could diverge, but this "neutral camp" is common to all of them. This discourse seems to be accepted as benign for the majority of mathematics education research. So, why do I feel uneasy with it?

The problem arises when one forgets that such abstract aims for mathematics education are to be implemented in a concrete reality where not everyone will manage to achieve success. If one seriously looks at the Pearl and Knight quotation above, one realises how, behind a plea for the importance of mathematics – mathematics education as a right – we are indeed constructing a barrier in becoming a citizen – mathematics education becomes a duty. In the desire to provide significant mathematics education to all students, we are constructing a discourse that puts anyone who is not proficient in mathematics (the majority of the population?) outside citizenship and democracy. It is not that people need mathematics to become democratic citizens. This necessity is retroactively constructed by this discourse itself.

The dissemination of discourses like this one – mathematics as a prerequisite to becoming a citizen – ends up creating the ideological injunction that you really need mathematics to attain citizenship. Given that mathematics is socially posited as a prerequisite to be a citizen, if you fail in school mathematics you will find difficulties in school and professional advancement. Here lies the importance of learning mathematics. The proper act of enunciation functions retroactively to perform mathematics as a powerful thing. It is not that school mathematics is powerful because people use it in their daily lives; mathematics is powerful because it gives people school and professional credit. That is, mathematics is constructed as a prerequisite to citizenship because school needs mathematics to perform its role as a credit system.

What I am suggesting is an inversion of the cause-effect relation in the discourse on the importance of mathematics. At a first glance it seems that the relation between cause and effect is as follows: people need mathematics in their daily lives, so they must learn it in school (preferably by exploring "real" examples). From this point of view, learning school mathematics is a consequence of the necessity people have to use it. But what if one reverses the argument: since people need to learn mathematics in school, a discourse attesting its importance must be constructed by characterizing daily practices as being mathematical practices. Therefore the importance of mathematics in daily life is the result of a performative act that *a posteriori* performs this importance for the sake of school mathematics. From this perspective, the importance attributed to mathematics is a consequence of the need for teaching school mathematics.

The discourse on the importance of mathematics

The last statement is polemic, and requires justification. I agree that people need mathematics to become citizens, that is, to become included in the social formation. However, I suggest that the reasons why people need it are not related with mathematical knowledge or competences, but with the school valorisation that mathematics gives to people. People need school mathematics not because they will use it directly in democratic participation (as knowledge or competence), but to continue having success in school, undertake a university course and find a stable job, so that they become "normal" social beings. I argue that the importance of mathematics must be discussed not in the field of knowledge but in the field of value. Hence it will be necessary to support the claim that people do not really use school mathematics in the exercise of citizenship or as workers.

On a personal note, I begin with a story about my grandfather. He was a cultivator during the last sixty years of his life, in a rural village near Lisbon. He never went to school. Until now, I never met anyone so knowledgeable and with such an affinity with nature. He knew everything about agriculture, from irrigation, to the calendar routines of planting all kinds of plants, not to mention the mechanical aspects of dealing with a tractor and all the heavy machinery used in agriculture. He was not just a good farmer, but also a reasonable citizen. He had an acute critical sense of politics and of the social discrepancies that pervade society. He voted and, although a solitary man, he complied with his communitarian obligations. From my point of view, my grandfather performed well his duties as a worker and a citizen. He never needed school mathematics to do so, and he was not excluded from participation in society. From the moment someone urges that if you do not know mathematics you are not a full citizen, then my grandfather, who never felt excluded, was to be found in that position. Not because he suddenly realised that he did not know any school mathematics necessary to live his life, but because someone asserted that in order to be a full citizen you need to know mathematics.[12] Then my grandfather was placed in the difficult situation of being sixty years old and being obliged to go to school and learn mathematics if he wished to continue working. This happened because new laws came up saying that people should have some kind of schoolarisation if they wanted to be employed. My grandfather realised that he needed a minimum scholarship to continue doing what he did all his life. So, my grandfather went to school, not to acquire mathematical skills and knowledge, but to have a diploma attesting that he can now be a citizen.

To be a participative citizen does not have to do with how much you know about the world that surrounds you (my grandfather was a master in cultivating the land, dealing with nature in a way that remains original to me) but how the world that surrounds you – Society – recognizes, valorises or not what you know about the world. The problem of participation is not a problem of skills or knowledge, but a matter of what is valorised in our social symbolic order. And the same goes for equity. What if guaranteeing equity were not a problem of providing meaningful mathematical learning to pupils, but guaranteeing that even if someone fails in school mathematics, they will be no less a citizen and no less able to participate in society than those who succeed?

Fortunately, there has been research in mathematics education that criticizes the idea that we use school mathematics in our daily lives. The most well known example is arguably the eye-opening work of Dowling (1998). He identifies three fundamental myths that pervade the field of mathematics education. For our purpose, we will focus on the myth of participation, which has to do with the way mathematics justifies its existence in the school curriculum. Dowling states that the current trend in mathematics education "is orientated more towards the widespread dissemination of mathematical use-values: not more mathematicians, but a more mathematically competent workforce and citizenry" (p. 3). Thus, he continues, "[m]athematics justifies its existence on the school curriculum by virtue of its utility in optimizing the mundane activities of its students. This is the *myth of participation*." (p. 9). Dowling suggests that the way in which mathematics is incorporated in mundane/everyday practices leads to the individual who lacks mathematics being acknowledged and, ultimately, acknowledging himself/herself as being "handicapped" (p. 9). So, in order not to be handicapped, the individual must learn mathematics. But the notion of "handicapped" is already being constructed "inside" the discourse of the importance of mathematics in daily life:

> However, Sewell's assertion that a lack of mathematical skills constitutes a handicap within these contexts is untenable; in fact, mathematical skill is neither necessary nor sufficient for optimum participation within these practices. (p. 10)[13]

> The myth of participation "recognizes" the operation of mathematical tools in diverse practices. It constructs a role for mathematics education in providing the toolbox and a pathological lack on the part of the yet-to-be-tutored. (pp. 11–12)

What Dowling (1998) calls the myth of participation – the idea that mathematics is a necessary feature of everyday practices – ends up creating a school curriculum where mundane activities are *mythologized* in a way that privileges mathematical rather than everyday principles. That is to say, everyday activities, in order to be introduced in school, need to be amputated of all the complex vicissitudes that make them what they are. This amputation – the result of casting the mathematical gaze onto public domains – privileges what Dowling (2001, p. 22) calls the "esoteric domain", while, at the same time, concealing the purely fictional status of the importance attributed to mathematics.

Another useful resource is the study conducted by Riall and Burghes (2000), who gathered together employees from a wide range of industry, commerce, and the public sector. Their intention was to evaluate the extent to which these people use mathematics in their professions. They conclude that "almost the entire population of the study said that they had had to learn at school some maths that they had never then used again" (p. 110). In the voice of one of the workers, who stands for the general opinion: "I think that a lot of maths that is taught is not used in later life. I've forgotten most of what I had to learn and I never use it" (p. 104). Also Hudson (2008) found that the people of his study did not transfer what they learnt in school mathematics to their daily work activities. Rather, they developed their mathematical skills in the workplace. Within a philosophical approach, Ernest (2007), when discussing epistemological issues in the internationalisation and globalisation of mathematics education, generalizes these findings by arguing that the mathematics behind our high-tech society is just a small part of the huge amount of research being done in mathematics. Some kind of applied mathematics that ends being *routinised* and used in a "technical" way. This mathematics is learned by people in practice, outside school:

> It is not academic mathematics which underpins the information revolution. It is instead a collection of technical mathematised subjects and practices

> which are largely institutionalised and taught, or acquired in practice, outside of the academy. (p. 31)

There are two important dimensions here. First, the argument that what people use is just some kind of "applied mathematics" does not in any way diminish the importance of research in "pure" mathematics: we could argue that it is because there is research in "pure" mathematics that we can apply more particular results. But we realise how the mathematics behind the "information revolution" is not directly academic mathematics. The second dimension involves another concern. Ernest argues that the mathematics we use in our daily lives as workers and citizens is acquired not in school but in practice. Indeed, this statement is in line with the research on the situated dimension of learning (Lave, 1988), which has been largely applied in mathematics education research. If we consider the research on the sociocultural aspects of knowledge and learning we can assert that all knowledge is eminently situated in the places where it is used, whether a workplace or an indigenous community. The meaning of some practices and knowledge is deeply involved in the community of practice (Lave & Wenger, 1991) where it is exercised and developed. There is no guarantee that people transfer knowledge from one practice to another without some kind of "misrecognition". School mathematics, although it can explore "real" situations, will always be school knowledge, learned in a specific place called school where students are not necessarily concerned with learning.[14]

Also, the studies carried out in mathematics education under the auspices of Activity Theory reveal that "the activity of situated problem solving in the school context seems to be fundamentally different from decision-making in the real world because of the difference of the activity systems that govern them" (Jurdak, 2005, p. 296). This difference is reflected in the responses of the students[15], who suggest that they operate under different rules when solving a school problem and a real problem, no matter if they are, from a mathematical point of view, the same (e.g., car loans or buying a cell phone): "they define their own problems, operate under different constraints, and mathematics, if used at all, plays a minor role in their decision making" (p. 296). Williams and Wake (2007) also take advantage of Activity Theory to study the relations between workplace mathematics and college mathematics. In a similar way they conclude that:

> … while it is true that some mathematics can readily be identified by academics in workplace practices, we find that workplace mathematics has its own distinct genre, inflected by the local practice and its activity system, its instruments and division of labour and power, as well as the productive goal of the whole activity. (p. 336)

From the point of view of Activity Theory, it seems that school and out-of-school practices are fundamentally different due to differences in the activity, the goal, and the operation, or the conditions under which the action is carried out.

EXCHANGE-VALUE AND CREDITATION

Shifting the importance of mathematics from knowledge to value

Thus a question should be raised. If the value of school mathematics for citizenship is deceptive, what *could* be the value of school mathematics? Dowling (1998) calls our attention to the *exchange-value* of school mathematics:

> [M]athematics justifies its existence on the school curriculum by virtue of its utility in optimizing the mundane activities of its students. This is the myth of participation. It constructs mathematics, not as a system of *exchange-values*, but as a reservoir of use-values. (p. 9, my emphasis)

Exchange-value is a notion explored by Karl Marx in his book *Das Kapital* to refer to one of the two values a commodity can take. While the use-value of a commodity is strictly related to the concrete use someone makes of a commodity – the mathematical know-how necessary to perform a profession, for instance – the exchange-value posits this commodity in relation to all the others, that is, as part of a structure of equivalences where its value can be gauged. Thus exchange-value has a purely relational status: it is not inherent to a commodity. It expresses the way this commodity relates to all the others. When looking at a commodity, say a table, we see its use-value. What we cannot see is its exchange-value, which remains invisible. If we transpose this line of thought to school mathematics, we can speculate how the gesture of positing the value of mathematics in its use hides its exchange-value – that is, the formal place school mathematics occupies within capitalism. The fundamental gap between an object and the structural place it occupies (Žižek, 2006) is at work in school mathematics: while perceiving the importance of school mathematics as use-value, we neglect the importance of this subject in maintaining schools' functioning as credit systems.

To my knowledge, it was Shlomo Vinner (1997) who first called mathematics educators' attention to the fact that issues stemming from the recognition that "the educational system is, above all, a credit system" (p. 68) have not been dealt with in mathematics education. In Vinner (2007), the author makes a crucial distinction between the importance of mathematics in maintaining important aspects of our present life and the importance for everybody to be proficient in mathematics: "No doubt mathematical knowledge is crucial to produce and maintain the most important aspects of our present life. This does not imply that the majority of people should know mathematics" (p. 2). And, going further, Vinner argues that the real importance for students to learn mathematics is because through a good performance in mathematics they will be able to achieve a higher social position:

> I suggest that the students have very good reasons to study mathematics. It is not the necessity of mathematics in their future professional life or their everyday life. It is because of the selection role mathematics has in all stages of our educational system. (p. 3)

Despite Vinner's plea, we lack research that explicitly connects these social phenomena – shamefully associated with school mathematics – with the broader political and economical spectrum. How, then, shall we understand the relation between school and capitalism? The usual way is to conceive education as an increasingly commodified social space. It is a commonplace in critical educational studies to assert how education has become merchandise and schools some kind of corporation.[16] In this view, education is conceived as something originally pure that has been progressively contaminated by the capitalistic structure of society. Educational industries, from publishing houses producing textbooks to computer firms developing technology, see schools as a profitable market; administrators and politicians use the metaphor of schools as companies to envision ways of managing education; governments attribute primordial importance to results in high-stake tests as a means to do school evaluation and make grades and scores a matter of profit; the labour market and industries demand the production of the highly qualified people needed; all these are few examples of such a view. In this perspective, education has become capitalized, and the "solution" would be a *decapitalisation* of education, to return to its original purity, based on humanistic ideas (the place to learn the cultural heritage, educating the free man, formative assessment instead of summative assessment, etc.). The purpose is to keep the capitalist logic of production/consumption outside the educational enterprise.

Despite being true, this characterization does not exhaust the relationship between capitalism and education, nor is it the crucial aspect. The problem is that school itself, more than just being contaminated by some capitalistic ideas, is the crucial ideological state apparatus in the reproduction of capitalism (Althusser, 1994). Education in its *schoolarised* form has in its kernel the capitalist logic.[17] It is against this background that we should conceive education not as being contaminated by capitalism, or a part of capitalism, but as sustaining the capitalist system itself, by assuring its reproduction. Education is not just a product (education as a piece of a profitable market) but a means of (ideological and material) reproduction.

However, this role of school as an ideological apparatus is concealed by means of a "naturalization" of schooling. Capitalist ideology represents school as a neutral environment purged of ideology:

> [W]here teachers respectful of the "conscience" and "freedom" of the children who are entrusted to them (in complete confidence) by their "parents" (who are free, too, i.e. the owners of their children) open up for them the path to the freedom, morality and responsibility of adults by their own example, by knowledge, literature and their "liberating" virtues. (Althusser, 1994, p. 20)

This concealment is essential to maintain the role of school as an ideological state apparatus. Seeing school as a place free of ideology disables bringing ideological struggle to school. All enterprises undertaken by teachers to unmask the "invisible" ideology are immediately accused of being ideological acts. In this way, the dominant ideology ensures that no ideology is present in school except, of course, the dominant one. The dominant one is precisely the one that presents itself as ideologically free, by positing the importance of mathematics as knowledge and competence.

One of the few exceptions within mathematics education research to acknowledge the importance of mathematics, not as knowledge or competence but as exchange-value, is the work of Roberto Baldino and Tânia Cabral. They have been, to this reader's knowledge, the only persons in mathematics education who have been analysing schools and mathematics education as part of capitalist economics (Baldino, 1998a, 1998b; Baldino & Cabral, 1998, 1999, 2006). Their suggestion is that we should look at school not so much as a place of knowledge but as a place of production. Very briefly, if we consider salary as school credit, and work as the presence of students in school, then, taking into consideration that students who fail do not receive any credit (diploma), the salary/school credit is not equivalent to the student's work in school. There is something that is missed by the student. Even though he spent all the year in school, went to classes, carried out all the regular activities, if after everything he does not get approved, he will not receive anything for all the work he did:

> Only students who get certificates recapture their labour force. This labour force embodies the work done by all, by those who flunked, by those who abandoned the course, by those who could not buy a higher education and remained at the lower levels of the pyramid. Graduates get higher salaries because their labour force embodies more value, more work done by themselves but, mainly, by others who were left behind. (Baldino, 1998a, p. 43)

From this perspective, failure is a school necessity. It is because some of us fail that others can achieve higher positions in social hierarchies. The value of the ones who flunk is appropriated by the ones who pass as surplus value. At school the student learns, above all, to participate in and accept the conditions of production and seizure of surplus value. Failure is posited as a *necessary* condition for schooling: "in order to perpetuate the process of production/seizure of surplus value, a certain amount of failure is necessary" (Baldino, 1998a, p. 5). Therefore, "failure of students means success of the institution" (Baldino & Cabral, 2006, p. 34).

Some people justify inequity by saying that since some perform better than others, they should be compensated for it. This argument presupposes that schools are places where equal students meet freely, and where some kind of "invisible hand" guarantees that the competition of individuals' egotisms works for the common good. What such an approach makes invisible is that such merit is possible only by the demerit of others, i.e., the notion of personal merit is only possible as long as others fail. This is the capitalist ideology at work, by means of making individuals recognize their choices as their own, as free choices that they took – especially when these choices imply failure.[18] All the work a failed student produced, all the time he spent in school, is not his (since he will not receive the diploma at the end of the year) but without producing it would be impossible for him to keep living.[19] Individuals must realise failure as the result of a wealth competition among equals, and repress the traumatic truth that they fail so that others can succeed.

The key element to be noticed here is that schools need this subversive supplement in order to retain their indispensable role in maintaining our democratic and inclusive society. In order for school to be the most important ideological apparatus, to function as a credit system, it is not productive for it to be presented as an exclusionary institution. That would cause criticism from the whole of society, and would be unbearable from an educational or political point of view. In order to perform well in the role of credit systems, schools need to be presented as inclusionary and emancipatory places, places where phenomena such as exclusion and failure are seen not as necessary parts of the same system which purports to be trying to abolish them, but as contingent problems, malfunctions of an otherwise good system.

The materiality of exclusion

In the article *Inclusion and diversity from Hegel-Lacan point of view: Do we desire our desire for change?* Baldino and Cabral (2006) create a parody concerning where one can find exclusion in school. The authors suppose that we enter an elementary school and ask the staff where the so-called "exclusion" is happening. Who will be able to answer such a question? Where to locate exclusion in schools? It seems as if exclusion has no "materiality", no precise site where it is happening. It seems as if it is a name to represent some structural impalpable reality, resulting from several complex factors, having to do with teacher engagement with the students, with the quality of the mathematics learning, with issues of race, gender, and social class, with lack of resources, and so on. Equity is understood as a complex phenomenon involving several dimensions, not identifiable in some place or in some practice. From this perspective, achieving equity means to fight in different battles (for groups of people considered to be in disadvantage, inequity of resources, teacher formation, mathematical content for social justice, etc.).

I argue that such dissemination of the problem of inequity disavows its materiality. Although exclusion may be related with all these different aspects, one should insist that exclusion has a materiality visible in assessment. In the story of Baldino and Cabral (2006) we have the chance to meet a special girl:

> Suppose we enter an elementary school and ask the staff where the so-called "exclusion" is happening. We will get no answer, but if we are lucky to meet the child who told us that the king was naked she will take us directly into Mr. Smith's office where the teacher is grading students' final exams. (p. 33)

Let us take a similar situation, more akin to my own experience. The girl will take us to the final evaluation meeting of the year where all teachers of the class will present their grades and decide who will pass and who will flunk. From my experience as a teacher, I felt the final moment of evaluation as a disheartening one. It is a commonplace among teachers to wail about their vain efforts to promote success among their pupils: "I don't know what else I could have done?"

A sense of disbelief falls on teachers. But the year is at the end, and vacations are imperative now. So, they forget all the unrewarding strategies and, at the beginning of the next year, they appear again optimistic, with lots of new ideas to promote success among their students. Next year, touched off by a kind of compulsive bias for repetition, they start all over again. As Baldino and Cabral (2006) suggest, we can distinguish between two kinds of teachers. The "healthy-but-not-ethical" one, who continues his work trying to improve teaching, alleviating the suffering of the most oppressed, mitigating segregation, but who realises that the problem of inequity will not be solved by such actions: "it does not solve the problem, but, they say, it is the most we can do" (Baldino & Cabral, 2006, p. 31). And the "obsessive" one who emphatically defends that mathematics should be for all, assuming the role of students' saviour from the dark reality of failure in mathematics. This teacher assumes that it is possible to solve the problem of inequity through his practice, by endowing students with sufficient mathematical skills so as to lead to full social equity – he completely endorses the (hysterical) societal claim that mathematics is the key for full citizenship. However, at the end (in the traumatic moment of final evaluation) he fails again in providing success to all students, and repeats the process again. About them, Baldino & Cabral (2006) state that:

> We might, perhaps, be tempted to praise these people as true heroes of mathematics education, the ones who refuse to lose their hope. However, if equity is an economical problem that escapes school, at least in the short run, the society's demand is really the demand of the hysterics and, as such, impossible to satisfy. (p. 31)

This desire for repetition reflects the desire for not knowing the deep roots of failure. The demand of society upon school is a hysterical demand, because it is impossible to satisfy. To such impossible demand the teacher responds with an obsessive behaviour of repetition of failure. This demand of society conceals a specific purpose: "the pleasure of this gaze [society; super-ego's gaze] in looking at the obsessive teacher's failure is that, in so doing, it is able to avoid looking at other issues" (Baldino & Cabral, 2006, p. 31).

These "other issues" are the facts that assessment is exclusion and these evaluation meetings are places of judgement. In them, teachers decide which students will continue to the next year, and who will be left behind, to repeat the same year sometimes three, four times[20]. In such places we witness all the materiality of exclusion. However, this materiality is not experienced as such. It is unbearable for the teacher to conceive himself as entrusted with the smooth functioning of school as an excluding machinery. Teachers cannot live with the guilt feeling of perceiving evaluation as judgement, as pure exclusion/inclusion. Here begins all the process of "disavowing" (we can hear things like: "I would like to pass him but all his test grades were negative" or "he didn't acquire the competences necessary to proceed to the next year"). What these arguments disguise is the fact that in evaluation we are not evaluating content or competences, but people. This displacement is the fetish of evaluation, whereby we have to forget that we are evaluating people and act as if we were evaluating objects (contents, competences) in order to avoid the guilt feeling. This fetishistic attitude[21] enables the teacher to avoid the confrontation with the trauma of acknowledging that assessment is exclusion.

Concerning the absence of research that specifically addresses assessment as exclusion, one can find support in the study of Marshall and Thompson (1994, quoted in Baldino, 1998b) that surveyed six at-the-time recent books on assessment. There was no reference in those books about the implications of assessment for social promotion and selection. Baldino (1998b) conjectures that this absence is due to a belief that social selection is a natural consequence of the various evaluation processes incorporated in society. That is, even though we acknowledge the role of selection perpetrated in school we address it as an anomaly, as something that, with time and research improvements, will be solved by constructing the evaluation instruments that guarantee success to all children:

> They [studies on assessment] seem to feed the hope that trustful evaluation procedures in mathematics could contribute to the edification of a just society: to each according to his/her merit. In fact, an ideology of justice and an implicit validation of instructional objectives is observable at the basis of most research about evaluation. (Baldino, 1998b, p. 1)

It is beyond the scope of this chapter to present an analysis of how studies on assessment do not address assessment as promotion. However, I suggest looking at *Topic Study Group 36* in the *11th International Conference in Mathematics Education* (ICME11), dedicated to *Research and development in assessment and testing in mathematics education.*[22] The reading of the aims and focus of this study group shows that the emphasis of research is not on assessment as a promotional mechanism but on assessment as part of the learning process. The text emphasizes the importance of moving from assessment *of* learning to assessment *for* learning. According to the team members of this study group, in the last fifteen years, assessment and testing have been evolving in the direction of addressing the need of the student, to help him or her learn better, rather than making judgments on the achievement of the student. The challenge for the mathematics education community is how to get social recognition for this new role of assessment.

However, in the last years we have been witnessing an increasing concern with measurement within education. Biesta (2009) argues that we are living in an age of measurement in which pressure is put on teachers, schools, and governments to increase educational results measured by mass-scale comparative studies such as the *Trends in International Mathematics and Science Study* (TIMSS) and the OECD's *Programme for International Student Assessment* (PISA). These international, comparative studies are to an increasing extent brought into the political sphere, placing pressure on national governments to regulate their educational systems according to the standards stipulated by those tests (Biesta, 2009; Wilson, 2007). This is what has been happening in the last eight years in very many developed countries where education tends to be transformed, by the pressure of politicians' demands for accountability, into an evidence-based profession. Consequently, political measures contribute to formatting teaching and learning of mathematics in a clear and crude way. Teachers tend to tailor their instructional practices to the format of the test out of concern that if they design their teaching differently, their students will fail. Although they might know all the didactical novelties and methods to promote learning in a way meaningful to the students, if what counts is to pass the test, that is how they will "educate" their students (Lerman, 1998; Wilson, 2007).

This scenario contrasts with the scenario imagined by the text of the ICME11 study group. On the one hand we have mathematics education research's call to reduce the importance of making judgments on the achievement of the student. On the other hand, we have a societal trend that puts pressure on teachers and schools to reduce the educational process to a promotional process. I argue that studies on assessment should make this antagonism visible, and avoid engaging in discourses that, by harmonizing assessment and promotion, ended up disguising the promotional role that assessment has today.

I agree that assessment is indispensable for learning. A teacher needs to constantly assess student's learning so that both can move forward in the educational process. However, assessment is just one part of evaluation. The other part is promotion, which is related with what we value when we decide the grade that should be attributed to a student. To grade a student is to materialize in a mark the (school) value of that person. That mark will have consequences for the student's future, both in terms of school and professional life. This part of evaluation is not about assessment, but about promotion. When we "blur" assessment and promotion, and start conceiving of promotion as the "fair" consequence of a good assessment, then we no longer have to be worried about exams or other forms of classifications, since students are being evaluated strictly for their mathematical knowledge, which has been developed by means of assessment for learning. As a consequence, school as a credit system is completely buried in oblivion in the large majority of the studies on assessment in mathematics education.

WHERE DOES THIS TAKE US, IN TERMS OF EQUITY?

Reaching equity

De Freitas (2004) has been an important voice in mathematics education research working in the field of teacher education. She does not take for granted the work of teachers in schools, and she is aware of all the structural constraints that can compromise good intentions of promoting mathematical success among students. She is also aware of the connections between school and political ideologies, and how the problem of inequity extends beyond the educational sphere:

> The goal of learning is no longer elite class privilege and intellectual awakening. The goal is market expertise and infinite consumption. (…) Mathematics becomes the handmaiden to yet another industrial agenda, an agenda outlined by aspiring, managerial, corporate needs. (…) Mathematics

> becomes a means of transforming experience into bits of information, and concurrently mutates learners into consumers. (pp. 264–265)

In de Freitas and Zolkower (2009), the authors address the issue of how to prepare prospective mathematics teachers to teach social justice. Although the authors acknowledge all the social and political dimensions of the problem of social justice within education, they engage in an obsessive task, stating that: "our central argument in this article is that a social semiotic theory of learning, in the hands of teachers, contributes to interrupting the cycle of inequity reproduced through education" (p. 191). The problem of inequity, previously described as a political problem, can be solved by changing teacher education. I criticize this minimal political reduction that posits a solution to the problem of equity both in the hands of teachers and in better theories for learning. I must say that I found the research conducted by De Freitas and Zolkower of high value, especially the way in which they deconstruct teacher discourses, showing how they can convey social prejudices. However, there is the risk of falling into salvationist discourses that displace the problem from its political dimension. Especially that is so, if we bear in mind the huge gap that exists between research and practice, and how most of the time this research comes to the teacher already "institutionalised", and is seen by teachers as another research novelty.

In the research carried out by Nolan and previously described, this obsession to promote equity and the repeated failure that follows it becomes visible. As Nolan (2009) insistently admits, her effort to promote a teacher education for social justice suffers from repeated failure. What she seems not to realise is that this repeated failure, cause of anxiety, is originated by a fundamental antagonism that remains unaddressed in her work: the antagonism between the societal demands for equity and the role of schools as exclusionary institutions. In the last paragraph of the article, after describing all the failed attempts to educate students in her own image (as people concerned and willing to implement social justice within their classrooms), something uncanny but highly revealing occurs:

> Finally, a break-through moment. On one final course evaluation, a prospective teacher wrote:
>
> "I finally get that the way [the instructor] taught her class WAS about social justice… that teaching mathematics about, or though, social justice isn't just about poverty statistics and world population figures… it's also in the thoughts and actions of the teacher towards his/her students and in the thoughts and actions of students toward each other. It's about feeling safe to be who I am and, at the same time, to critically question who I want to become and what (and who) I value. And, most of all, I think it's also about opening up the content of mathematics (what and how we teach) to this same kind of critical questioning". (pp. 214–215)

What appears to be a light at the end of the tunnel, a sign of hope that gives strength for the researcher to continue her work towards social justice is … a dream: "Ok, so that's a lie. No one actually wrote that on the course evaluations. This is the fictional part of my story – it's the opportunity I am seizing to convey my dream for mathematics in and through social justice" (p. 215). What if the whole truth of her story is precisely this dream? In these last words, we realize all the phantasmatic support that grounds Nolan´s desire to educate teachers for social justice – the total fantasy that disables us from approaching directly the core of the problem, that is, the traumatic kernel of school inequity.

But the solutions presented by research in equity are not restricted to teacher education. Gates and Zevenbergen (2009) identify a common basis for such measures:

> What might we all agree on then as fundamentals of a socially just mathematics education? Perhaps we can list: access to the curriculum; access to resources and good teachers; conditions to learn; and feeling valued. (p. 165)

The first thing that is evident here is the complete absence of a political conceptualisation of equity. What is recognized as an economical and political problem ends up being addressed in a technical fashion: better ways to teach and learn mathematics for all students. Another thing that looms up is the complete obliteration of the role of assessment in exclusion as was previously described.

We tend to see the problem of inequity as a problem of achievement gaps, of teacher education, of curriculum applicability, and other scientific categories. These ideas are well established in research. According to Schoenfeld (2002, quoted in Langrall, Mooney, Nisbet, & Jones, 2008, p. 127), to achieve equity requires four systematic conditions to be met, namely, 1) high quality curriculum; 2) a stable, knowledgeable, and professional teaching community; 3) high quality assessment that is aligned with curricular goals; and 4) stability and mechanisms for the evolution of curricula, assessment, and professional development. Lubienski (2002) claims that, as far as the issue of equity is concerned, the goal is to learn more about the complexities of successfully implementing meaningful instructional methods equitably with students who differ in terms of social class, ethnicity, and gender. Or, according to Goldin (2008) "to create teaching methods capable of developing mathematical power in the majority of students" (p. 178). The problem of equity is reduced to a problem of developing the best "instructional methods" to allow mathematical success to all students.

Langrall et al., (2008, p. 118) take a more sophisticated approach, which they generalize to the whole mathematics education community:

> Mathematics educators today (...) recognize clear discrepancies among the *desired curriculum* – as it exists in a national goal statement or a ministry of education syllabus, the *implemented curriculum* – as it plays out in classrooms, and the *achieved curriculum* – in terms of what children learn. Ultimately, while these inconsistencies remain, we cannot guarantee that all elementary students will have access to powerful mathematical ideas. (p. 118)

The idea that we will achieve equity when these three phases of curriculum implementation coincide presupposes the idea of society as an organic whole, within which people become what Society stipulates as the "ideal citizens", propagated by the curriculum. The discrepancy between the three levels is seen as a malfunction of the system. When we manage to fix this, all students will have access to powerful mathematical ideas. Equity will be achieved when those three curriculums become one and the same. What I think is highly problematic in this approach[23] is the evolutionistic thesis that it conveys. The idea is that inequity and social exclusion are *still* associated with school mathematics and that, through research and practice, it will be possible to overcome these problems: we already know the right path (make the three levels coincide), and what remains is a particular question of how to do it. In these circumstances, the question of equity is not a political, economical question. It is transformed into a "technical question" – how to reduce the gap between the three levels of curriculum implementation, so that students act and think as stipulated by Society. The fact that equity is a political problem is dismissed.

This refusal to confront the real core of the problem of equity can be seen as the result of an ideological injunction that systematically leads us to repeat the same "abstract" discourses – school as a place for emancipation, mathematics as a powerful knowledge and competence, mathematics for all, etc. In order to critically analyse such discourses we should replace the abstract form of the problem with the concrete scenes of its actualisation within a life-form (Žižek, 1991, p. 145). That is, in order to understand which are the real aims for school mathematics, or the real motives that students have for being in school, we must not repeat ideologically loaded discourses conveyed by the curriculum, by political statements, and even by research, but rather look the "negative" (schools sorting future people for the labour market by means of credit accumulation) in the face and convert it into research problems. Following Žižek's (1989) thought, this implies research should pass from the notion of crisis (in this case, the fact that people fail in school mathematics therefore creating exclusion) as an occasional contingent malfunctioning of the system to the notion of crisis as the symptomal point at which the truth of the system becomes visible. In this case, what is revealed is the inconsistency of a system that, on the one hand, demands mathematics for all and, on the other hand, uses school mathematics as a privileged mechanism of selection and credit.

I argue that research in mathematics education is destined to repeated failure if it continues to avoid facing exclusion in its materiality, and restricting research to a "technical" enterprise. Although issues of equity, social justice, inclusion/ exclusion, can be addressed in a multitude of forms, its fundamental structure has to do with the process of schooling itself. It is the very "nature" of schooling that carries the germ of exclusion. In the first pages of this chapter I told an anecdote about a guy who lost a needle. Why is the guy searching for the needle in the kitchen if the needle is in the bathroom? Because in the bathroom, since there is no light, he can't "work". Therefore (because he has to work) he goes to the kitchen where there is light. Of course no needle will be found there, but does he really want to find the needle, or is the whole purpose of his "work" just to keep him occupied by repeating again and again the same procedure? From what has been said in this chapter, the analogy between this guy and mathematics education research seems obvious. As Baldino and Cabral (2006) suggest, the purpose of our desire is not really to solve the problem of equity (since we know that its causes lie elsewhere) but to keep us occupied, thus guaranteeing that things remain the same: "It appears that the true goal [of research in mathematics education] is repetition: repetition of teaching attempts, repetition of research issues, repetition of explanations; in one word, repetition of failure" (p. 30). The authors raise the question: do we desire our desire for change?

A dialectical materialist approach to the problem of equity

In the remaining pages, a philosophical background to the problem of equity within mathematics education will be suggested. I shall give consistency to the claim that exclusion is not some marginal problem of schooling, but its constitutive element. This approach can be called dialectical materialist, having as reference the philosophical work of Slavoj Žižek.

Mathematics education research acknowledges that schools can represent the opportunity to succeed in life, and can also represent the place that marks you as "disposable" (Skovsmose, 2006). And mathematics is right at the centre of this selective process. Assuming this background, we have two ways of contextualising the problem of exclusion and the aim of equity in mathematics education. For the first one, in line with the evolutionistic world-view addressed at the beginning of this chapter, school is a necessary institution in our societies and mathematics, being one of the biggest achievements of mankind and the basis of our high-tech world, which should be accessible to all people. Assuming this premise, the problem of inequity is seen as a malfunction of the school system. The fact that school mathematics is dishonourably involved in processes of social exclusion is understood as an obstacle for the full actualisation of an equitable society. The focus of research and politics is then, first, to improve the studies on how mathematics excludes people from social life, by studying the peculiarities of some groups of people (based on gender, ethnicity, socioeconomic status, etc.) in their relation with school mathematics. The premise is that when we understand better the mechanisms involved in the exclusion of groups of people we can implement strategies that will allow them to achieve success in mathematics. Second, in a political dimension, we intend to create an international mechanism that could measure the performance of each student, so that we can address the problem of exclusion properly. The idea is that we *still* have inequalities in school mathematics. Nevertheless we are on the right path, which just has to be polished.

I suggest another approach. What if exclusion is not an occasional and contingent phenomenon that will be ameliorated by research and political efforts, but a symptom in which the "truth" of the system becomes visible? That is, what if the exclusion associated with school mathematics is not a particular negativity, a vicissitude of a "good" system, but, on the contrary, represents a glimpse of what the school system really is: a credit system with the main goal of social selection by means of deciding who is capable and who is disposable? This picture of schooling is in line with broader liberal-democratic politics:

> All phenomena which appear to liberal-democratic ideology as mere excess, degenerations, aberrations – in short: signs that the liberal-democratic project is not yet fully realized – are *stricto sensu* its symptoms, points at which its hidden truth emerges. (Žižek, 1991, p. 270)

To better explain this conflict between the *particular* – school exclusion – and the *universal* – school equity – I present some considerations using Žižek's philosophical theorizations, especially his analysis of the Hegelian concept of *negative particularity*. The universal law dictates that school is a benefit: all children should go to school and experience success. The negative particularities are particularities that do not follow this universal rule; they are seen as exceptions, aberrations of the system, such as, for instance, the fact that school provokes exclusion. There are two different ways of dealing with the dialectics of the universal (schooling) and the particular (exclusion). The first one, which Žižek (1991) calls traditional, faces exclusion as a passing moment of a universal process of schooling: "a passing moment of the law's mediated identity-with-itself" (p. 33). It has been argued in this chapter that mathematics education research conceives the problem of equity in this way, as is evident in the research discourses about equity. These discourses convey the idea that we still have inequity, which is understood as a malfunction of a system that aims at universality – mathematics for all. Exclusion is perceived as an "error", something that defies the universal rule and, as such, something that needs to be repressed for the universality to constitute itself. What follows is a multiplication of efforts to insert exclusion (the traumatic kernel of social reality) into our symbolic order by means of conceiving it as achievement gaps, possible to solve by means of better teacher education, as a characteristic of certain groups of people, by improving classroom practices, etc.

The second way, according to Žižek (1991)[24], states that universal schooling is nothing but universalised exclusion, exclusion brought to its extreme, "to the point of self-negation" (p. 38), whereby the distinction between exclusion/schooling collapses into exclusion. Schooling or inclusion "dominates" exclusion when some "absolute exclusion" particularizes all other exclusions, converts them into "mere particular exclusion", in a gesture of universalisation by means of which an entity turns into its opposite. That is, the logic of the universal always carries with it some fundamental exception that is a precondition for its existence. This was first acknowledged by Marx:

> Marx's key theoretical achievement, which allowed him to articulate the constitutive imbalance of capitalist society, was his insight into how the very logic of the Universal, of formal equality, involves material inequality – not as a remainder of the past to be gradually abolished, but as a structural necessity inscribed into the very formal notion of equality. (Žižek, 2005, p. 183)

In other words, (school) inclusion and (school) exclusion are not two opposite poles, "struggling" with each other, in which we take the side of inclusion. Inclusion already presupposes exclusion – it is only by means of exclusion that the notion of inclusion can be made effective: "An entity is negated, passes over into its opposite, as a result of the development of its own potential" (Žižek, 1991, p. 180). In the case of equity and exclusion, we can say that equity is negated, passes over into its opposite – exclusion or inequity – as a result of the development of its own potential. It is the development of a school aimed for inclusion and equity that produces inequity. This dialectical approach brings into the light of day the forgotten reverse of equity – the way equity itself coincides with supreme exclusion: "the Universal is the domain of Falsity *par excellence*, whereas truth emerges as a particular contingent encounter which renders visible its 'repressed'" (Žižek, 1991, p. 196).

Žižek calls this kind of dialectical twist – from contingency (exclusion as an obstacle to equity) to necessity (exclusion as inherent to equity) – an *ideological anamorphosis* (1997, p. 97). In the case of equity, what first appears as an external obstacle to the universal idea of equity – exclusion – is revealed to be an inherent hindrance, i.e., an outside force turns into an inner compulsion. We do not have two different and opposite identities – exclusion and inclusion – so that by eliminating the former we can achieve the later. Rather, both exclusion and inclusion are part of the same identity, they coexist, not because we are doomed to live in a society that excludes people in it, but because to stop (school) exclusion means to put an end to a society wherein school, by means of universalised exclusion, is universal itself. The negative force exclusion has in school (seen as a place for inclusion) is its essence – universalised exclusion. The true antagonism is not between school exclusion and school inclusion, or between equity and inequity, but between this antagonism itself and the end of school as a credit institution.

Against this background, exclusion is not a stumbling block to the emergence of a fully inclusive school but a necessity, insofar as it maintains the illusion that it is exclusion itself that prevents the establishment of a fully equitable school. Exclusion functions not as an obstacle to full schooling, but as a "filler", as something school needs in order to affirm its identity. The traditional way of positing the problem of inequity conceals the fact that achieving equity is not about dealing with some particular forms of exclusion, but to a matter of addressing school as a capitalist state apparatus.

If one assumes the traditional explanation, then our work as mathematics educators worried about the issue of inequity will be to develop strategies that promote a better teaching and learning of mathematics for all people. But if we assume that the role of selection performed by school mathematics is at the core of schooling, as a universal particularity (in which we can see the whole purpose of the system), then we must think more broadly and address social and political structures that are usually taken for granted. Žižek suggests that "one undermines a universal 'thesis' by means of exhibiting the 'stain' of its constitutive exception" (Žižek, 1991, p. 160). This gives us some guidance for addressing the problem of equity. In order to undermine the universal thesis that provides the apologia of a school for all, we must exhibit how exclusion exists at the core of the universal rule. We need to posit exclusion as an intrinsic quality of school.

FINAL COMMENTS

Some will say that such an awareness of the problem takes us to a deadlock. Indeed, by realising that exclusion is something inherent to the school system we realise that to end exclusion means to end schooling as we know it. In the current matrix of world social organization this does not seem possible. Thus, as raised in discussion with the editors of this book during the review process, what should be done?

I would like to think that in our day and age, in which universities are being transformed into service provider companies, there can still be a place in them for contemplation. In our frenetic academic life, it is not easy to find the time and the will to contemplate. Partly because one of the necessary requisites for contemplation is the absence of a concern with the applicability of our thoughts, in these days in which time is money, some will ask: why lose money with all this philosophical/contemplative waste of time? The ethos of scientific research today makes plain that empty words are not enough; we must set to work, do it instead of just talking about it. What we need, some say, is engagement in action, quick solutions ready to be implemented, evaluated and, eventually, discarded, so that the entire process can start again. I argue that this pressure to produce "solution-based" research is part of an ideological injunction to keep us occupied with specific research, while neglecting research that is not immediately concerned with providing solutions but rather to complicate the usual ways we approach problems. Some would say that such an approach to research will lead us into a state of paralysis, lost in an endless discussion from which no practical solutions, no "insights for action" will emerge. My response is that the true act sometimes could be a purely "inactive" one. I strongly believe that sometimes the best way to act is to stop "acting" – in the sense of doing research that immediately implies some kind of action – and ruminate. Žižek (2006) expresses this attitude as follow:

> The threat today is not passivity but pseudo-activity, the urge to "be active", to "participate", to mask the Nothingness of what goes on. People intervene all the time, "do something"; academics participate in meaningless "debates", and so forth, and the truly difficult thing is to step back, to withdraw from all this. Those in power often prefer even a "critical" participation, a dialogue, to silence – just to engage us in "dialogue", to make sure our ominous passivity is broken. (p. 334)

In this chapter, I sought an analytic understanding of how the issue of equity is being addressed in mathematics education research. In order to carry out such an analysis, I mostly deployed the recent revitalization of Hegel and Marx by Slavoj Žižek. My goal was to redefine the theoretical and political coordinates mapping the way mathematics education research engages in the issue of equity and to give a different explanation for the existence of inequity within a system (school) which stands for inclusion and democracy. If the theorization I advanced leads to paralysis it will not be the worst of evils. It would be worse to keep the current state of affairs, wherein huge amounts of resources are disbursed in innocuous research, which has not been proved to have the solutions for the core problems of the field. Indeed, if teachers refuse to participate in school promotion, and if researchers reserve more time for contemplation instead of complying with market demands for *fastresearch*, perhaps paralysis would have a very disruptive effect. As put by Žižek above, the worst threat for the system today is not "activism", but passivity: the refusal to comply with more of the same.

The same capitalist system that needs social exclusion to reproduce itself stimulates the frenetic production of research aimed to eliminate such exclusion. Such research is usually performed as a design of what to do (e.g. Cobb, 2007), planning an entire set of "guidelines for action" both for researchers and for teachers.[25] My call is that if we really desire our desire for change, the task ahead is much more painful than following "What to do?" answers. And the first step is, in line with a long critical tradition from Marx through Adorno and Foucault, thinking of the present as a way of thinking about change. From the moment we critically think about the present – in the sense of suspending what exists so that a sense of strangeness towards the present can appear – we are already changing it. This kind of posture is usually marginal in research. Most of the times we accept the rules of the game (we know the needle is not in the kitchen!), and avoid a questioning of the whole game. My intention was precisely to question the whole game, that is, to realise that all the particular problems we are trying to solve within mathematics education don't change if the school credit system as a whole remains unaddressed. We realise the wilderness of the task ahead!

It appears that mathematics education is confronted with a challenge similar to that of psychoanalysis.[26] The latter challenge is ultimately the theory of why its clinical practice is doomed to fail (Lacan, 2008). And, along the same lines, perhaps mathematics education should be the field of research of why people are doomed to fail in school mathematics, and not so much, as it is today, the field of research on how people should achieve success. If we assume that mathematics education, as a field of research, exists with the purpose of facilitating the universal goal of mathematics for all, then it would become obsolete if school mathematics ceased to be the headache that it is. In this sense, mathematics education research should be concerned not so much with integrating and domesticating the excess (the ones who fail) that resists a society of full democracy and citizenship, but in developing research around the fact that mathematics education systematically fails in achieving the universal goal. That is, to study why inequity persists[27], to take it *not* as a contingency but as a *necessity* of the same system that so eagerly tries to eliminate it.

Against this background, perhaps I can risk a suggestion for research in order to satisfy the demand for action. Instead of leaving the field and become a political activist, or remaining within the field and treating it as one arena within which the political struggle for equity can be waged ("equity through mathematics education"), I suggest that more attention should be given to *failure*. That is, more research efforts to study failed attempts to promote a meaningful mathematics education for all.[28] In a way analogous to the psychoanalytical interpretation, I argue that the community should take the difficulties and impediments in achieving a meaningful mathematics education for all not as particular obstacles to get rid of, but as central issues for educational research. It is my contention, to paraphrase Žižek (1994, p. 7), that such "failures" have the potentiality to point towards the system's antagonistic character, and thus "estrange" us to the self-evidence of its established identity. This "estrangement'" takes place when we no longer take as given the liberal-democratic capitalist view of school as a neutral environment purged of ideology. Instead, we realize the purely manufactured, contingent nature of such discourse. Ultimately, this discourse functions as an ideological mechanism aimed at obliterating the traces of its traumatic origins, that school is a place of social promotion. There is no Supreme Good waiting for us at the end of the tunnel. Inequity is a necessary condition for the consistency of the same system that strives to abolish it. The conclusion to be drawn is that if our purpose is to extinguish school inequality then the whole ideological fantasy structuring schooling as a place of inclusion, freedom, and equality must go altogether. Borrowing an expression from Hegel, which gives the title to one of Žižek's books, we are faced with the challenge of *tarrying with the negative*, that is, of assuming that there is no safe constellation of values (such as equality, freedom, etc.) from which we can stand and preach the Good. We are compelled to tarry with the fundamental social antagonism that perpetrates the very universality of our social formation.

ACKNOWLEDGEMENTS

This article is part of my PhD project, supported by the Foundation for Science and Technology of Portugal, grant SFRH/BD/38231/2007. It is also part of the Project LEARN, funded by the same foundation (contract PTDC/CED/65800/2006). I am grateful to Roberto Baldino who first read this text and pointed out crucial misunderstandings regarding the dialectical materialist approach I sought to deploy.

NOTES

1 At the last ICME Conference, ICME11 in México, during a plenary debate session with Paul Cobb, Mariolina Bussi, Teresa Rojano and Shiqi Li, entitled *What do we need to know? Does research in mathematics education address the concerns of practitioners and policy makers?*, Shiqi Li asked a very insightful (and provocative) question: Why do the students participating in the research always increase their capabilities (to solve problems, to learn with meaning, of communication, to use technology, etc.) but in reality, in the daily routine of worldwide classrooms, failure in mathematics persist, and students are far from reaching those desirable capabilities achieved in research settings?

2 For instance the discussions fuelled by the work of John Dewey, Ivan Illich, Louis Althusser, and Paulo Freire.

3 This characterization of modern, technological, societies is in line with what Heidegger called "enframing" and Adorno the "administered world".

4 Theory is another "hot topic" in recent publications within the field. At the *11th International Congress on Mathematics Education* (ICME11) one of the survey teams was responsible for developing a study of the notion and role of theory in mathematics education research. This survey team had the task of identifying, surveying, and analysing different notions and roles of "theory", as well as providing an account of the origin, nature, uses, and implications of specific theories pertaining to different types of research in mathematics education. The *Second Handbook on Mathematics Teaching and Learning* (Lester, 2007) contains two articles dealing with the issue of theory: "Putting Philosophy to Work: Coping With Multiple Theoretical Perspectives" (Cobb, 2007), and "Theory in Mathematics Education Scholarship" (Silver & Herbst, 2007). At the Congress of European Research in Mathematics Education (CERME) there has been a working group dealing with the problem of linking, contrasting, and comparing the wide variety of theoretical approaches found in the field to tackle the teaching and learning of mathematics (see ZDM, 40(2), published in 2008). Finally, in 2009 the theme of the *33rd Conference of the International Group for the Psychology of Mathematics Education* (PME) was "In search for theories in mathematics education". There seems to be a widespread desire for understanding the role of theory in mathematics education research: "The moment seems propitious for a serious examination of the role that theory plays and could play in the formulations of problems, in the design and methods employed, and in the interpretation of findings in education research" (Silver & Herbst, 2007, p. 41). This chapter is also intended to be a contribution for how theory is understood and used in mathematics education research.

5 For Agamben (1998), who amplified the work of Foucault, the only real question to be decided in contemporary society is which form of organization would be the most suitable for the task of securing the care, control, and use of *bare life*: human life stripped from its entire political dimension, and reduced to its biological entity. Human bare life is that type of existence that can be measured, calculated, and predicted; in other words, the object and result of technical expertise. Recognizing this condition, Žižek (2006) argues that today we live in a *post-political* society; politics has surrendered to specialized social administration, targeting the bare life of the individual by controlling its fluctuations according to global standards of normality.

6 The day I was writing these lines, the US government decided to bar online access to five of the world's most important newspapers to its Air Force personnel. Such censorship, in a country that professes to stand as no other for the modern ideals of equality and freedom, clearly shows the inherent limits of such ideals: they are "universal" insofar as they don't jeopardize the same system which posits them as universals. From the moment they put the system at risk, they are sacrificed. This should make us suspect the values of a system which presents itself as the pinnacle of democracy, freedom, and equality.

7 For evidence that this is the case see, for instance, Cole (2003).

8 We can speculate how, if everyone experienced success in school mathematics, it would disintegrate as part of the school's credit system. In the pages below, I develop further the characterization of schools as credit systems. For the moment, by "credit system" I am referring to the role schools, particularly through school mathematics, perform as gatekeepers.

9 In one of his many pop illustrations of ideology, Žižek explores that uncanny button placed in elevators which enables the user to close the doors, thus apparently speeding up the journey. He argues that this can be seen as a case of a placebo effect, and, as such, a metaphor for freedom in Liberal-democratic capitalist politics. In short, we are led to believe that by our personal, free actions we are making a contribution for the "progress" of society. Philipp Oehmke, in an article in the German magazine *Der Spiegel,* gives this explanation: "One of his famous everyday observations on this subject relates to the buttons used to close the door in elevators. He has discovered that they are placebos. The doors don't close a second faster when one presses the button, but they don't have to. It's sufficient that the person pressing the button has the illusion that he is able to influence something. The political illusion machine that calls itself Western democracy functions in exactly the same way, says Žižek." (Available at http://www.spiegel.de/international/zeitgeist/ 0,1518,705164–2,00.html)

10 As explored by Lacan, an entity that enables us to unify in a single large narrative all the antagonisms – something which "quilts" the social edifice. In Pais & Stentoft (unpublished manuscript) we link this notion with another Lacanian one: the Master-Signifier.

11 See for instance the work of Marilyn Frankenstein.

12 Although my grandfather dealt with what we can call ethnomathematical knowledge, the point is that he didn't need to realize that what he was doing involved any mathematics.

13 Dowling is referring to the recommendations for school mathematics that emerged from the Cockcroft Committee, in which Bridgid Sewell defended the importance of percentages to operate in a shopping environment.

14 See Baldino & Cabral (1998, 1999) for an interpretation of why students in school engage not in learning but in passing.

15 Thirty-one grade 12 students dealing with real-life mathematical problems.

16 For instance, D'Ambrosio (2003) explicitly compares school with factories, where people are components of big machinery that aims for uniformity.

17 From a strictly economical perspective, schools have been performing a crucial role, without which our current mode of living could not be possible. Schools guarantee a place where children could be deposited when their parents go to work. But is not just a matter of "guarding" children. It is also a matter of sorting them, by means of stipulating who is capable of performing specific roles in society. Therefore, at the same time school performs three crucial economical functions in our societies: it guarantees a space where parents can put their children so that they can work, keeps children away from production while sorting them.

18 People accept this inequality because the dominant ideology conceives them as self-conscious subjects (Althusser, 2000). That is, the worker who works all his life and ends up with nothing sees his misfortune as a natural consequence of the way economical relations are built. He can even blame himself for not having worked enough, for lacking initiative, for being an unlucky guy; or blame some "sublime" (Žižek, 1989) group such as immigrants or Jews in Nazism – he naturalizes his poorness. Capitalist ideology must conceive subjects as owners of their own actions, as individuals completely conscious that their misfortune cannot be imputed to anything other than their own lack of capacity to succeed.

19 We should keep in mind that schooling is not free but compulsory.

20 Martin Willis called my attention to the fact that exclusion is also happening when students aren't barred each year. For instance, in the United Kingdom education system there is no opportunity to repeat the same year. Those students who fail still progress to the next year along with those who pass. However, as mentioned by Martin, exclusion is still happening: by allowing students to "move up" to the next year without having succeeded in the previous year it sets them up to fail again and excludes them from the beginning of the new school year. And, we can argue, this exclusion works in a more efficient way since we are dealing here with "exclusion without excluding", a veiled exclusion and, as such, more effective.

21 Namely, forgetting that through assessment we are deciding students' lives; that we are dealing not with students' knowledge but with people; conceiving assessment as a treatment or as a measure for merit and knowledge.

22 The work developed by this group is available at http://tsg.icme11.org/tsg/show/37

23 But also the fact that authors are trying to close the irreducible gap that exists in all social structure. The existence of this fissure – between what society wants us to be, and in what we actually become – is Real in the Lacanian sense, and cannot ever be closed. What it can be is disguised, disavowed. These three levels only coincide in an absolutely neurotic society.

24 The example he uses as illustration is law (universal) and crime (particular).

25 Thomas Popkewitz, in a conversation, called my attention to the modern spirit of individualism behind this posture, reflected in the growing idea in the population that there is someone who does the work of thinking for us, and we just have to apply it.

26 Indeed, Freud considered psychoanalysis, education, and politics the three impossible societal tasks.

[27] Or, to use one of Lacan's (1998) neologisms, *ex-sists*, to signal the paradoxical nature of failure, that is both extrinsic to the system of equality and freedom upon which liberal capitalist society is based and, at the same time, necessary, that is, intrinsic, to the reproduction of this same system.

[28] I leave to the reader the exercise of noticing how mathematics education research is mainly targeted at analysing and reporting successful experiences. Some exceptions are Vithal (2000) and Baldino and Cabral (2005).

REFERENCES

Agamben, G. (1998). *Homo sacer: Sovereign power and bare life.* Palo Alto, CA: Stanford University Press.

Althusser, L. (1994). Ideology and ideological state apparatuses (notes towards an investigation). In S. Žižek (Ed.), *Mapping ideology* (pp. 100–140). [First edition 1970]. New York and London: Verso.

Althusser, L. (2000). *Freud e Lacan/Marx e Freud.* Rio de Janeiro: Edições Graal.

Amit, M., & Fried, M. (2008). The complexities of change: Aspects of reform and reform research in mathematics education. In L. D. English (Ed.), *Handbook of international research in mathematics education (2nd ed.)* (pp. 385–414). New York: Routledge.

Anderson, C., & Tate, W. (2008). Still separate, still unequal: Democratic access to mathematics in U. S. Schools. In L. English (Ed.), *Handbook of international research in mathematics education (2nd ed.)* (pp. 299–318). New York: Routledge.

Baldino, R. (1998a). Assimilação solidária: Escola, mais-valia e consciência cínica. *Educação em Foco, 3*(1), 39–65.

Baldino, R. (1998b). School and surplus-value: Contribution from a Third-World country. In P. Gates & T. Cotton (Eds.), *Proceedings of the First International Conference on Mathematics Education and Society* (pp. 74–81). Nottingham, England: Nottingham University.

Baldino, R., & Cabral, T. (1998). Lacan and the school's credit system. In A. Olivier & K. Newstead (Eds.), *Proceedings of 22nd Conference of the International Group for the Psychology of Mathematics Education* (Vol. 2, pp. 56–63). Stellenbosch, South Africa: University of Stellenbosch.

Baldino, R. & Cabral, T. (1999). Lacan's four discourses and mathematics education. In O. Zaslavsky (Ed.), *Proceedings of the 23rd International Conference of the Psychology of Mathematics Education Group* (Vol. 2, pp. 57–64). Haifa, Israel: Technion Israel Institute of Technology.

Baldino, R. & Cabral, T. (2005). Situations of psychological cognitive no-growth. In H. L. Chick & J. L. Vincent (Eds.), *Proceedings of the 29th Conference of the International Group for the Psychology of Mathematics Education* (Vol. 2, pp. 105–112). Melbourne, Australia: International Group for the Psychology of Mathematics Education.

Baldino, R., & Cabral, T. (2006). Inclusion and diversity from Hegel-Lacan point of view: Do we desire our desire for change? *International Journal of Science and Mathematics Education, 4*, 19–43.

Biesta, G. (2005). Against learning. Reclaiming a language for education in an age of learning. *Nordisk Pædagogik, 25*(1), 54–55.

Biesta, G. (2009). Good education in an age of measurement: On the need to reconnect with the question of purpose in education. *Educational Assessment, Evaluation and Accountability, 21*(1), 33–46.

Bishop, A., & Forgasz, H. (2007). Issues in access and equity in mathematics education. In F. Lester (Ed.), *Second handbook of research on mathematics teaching and learning* (pp. 1145–1168). Charlotte, NC: Information Age Publishing.

Brown, T. (2008). Lacan, subjectivity, and the task of mathematics education research. *Educational Studies in Mathematics, 68*, 227–245.

Cobb, P. (2007). Putting philosophy to work: Coping with multiple theoretical perspectives. In F. Lester (Ed.), *Second handbook of research on mathematics teaching and learning* (pp. 3–38). Charlotte, NC: Information Age Publishing.

Cole, M. (2003). Might it be in practice that it fails to succeed? A Marxist critique of claims for postmodernism and poststructuralism as forces for social change and social justice. *British Journal of Sociology of Education, 24*(4), 487–500.

D'Ambrosio, U. (1993). *Etnomatemática. Arte ou técnica de explicar e conhecer*. São Paulo: Ática.

D'Ambrosio, U. (2003). *Educação matemática: Da teoria à prática*. Campinas, São Paulo: Papirus.

De Freitas, E. (2004). Plotting intersections along the political axis: The interior voice of dissenting mathematics teachers. *Educational Studies in Mathematics, 55*, 259–274.

De Freitas, E., & Zolkower, B. (2009). Using social semiotics to prepare mathematics teachers to teach for social justice. *Journal of Mathematics Teacher Education, 12*, 187–203.

Dowling, P. (1998). *The sociology of mathematics education: Mathematical myths, pedagogic texts*. Washington, DC: Falmer.

Dowling, P. (2001). Mathematics education in late modernity: Beyond myths and fragmentation. In B. Atweh, H. Forgasz and B. Nebres (Eds.), *Sociocultural research on mathematics education* (pp. 19–36). Mahwah, NJ: Lawrence Erlbaum.

English, L. D. (2008a). *Handbook of international research in mathematics education* (2nd Ed.). New York: Routledge.

English, L. D. (2008b). Setting an agenda for international research in mathematics education. In L. D. English (Ed.), *Handbook of international research in mathematics education* (2nd ed.) (pp. 3–19). New York: Routledge.

Ernest, P. (1991). *The philosophy of mathematics education. Studies in mathematics education*. London: Falmer.

Ernest, P. (2007). Epistemological issues in the internationalization and globalization of mathematics education. In B. Atweh, A. Calabrese Barton, M. Borba, N. Gough, C. Keitel, C. Vistro-Yu and R. Vithal (Eds.), *Internationalisation and globalisation in mathematics and science education* (pp. 19–38). New York: Springer.

Foucault, M. (1991). Governmentality. (Lecture at the Collège de France, Feb. 1, 1978). In G. Burchell, C. Gordon & P. Miller (Eds.), *The Foucault effect: Studies in governmentality*. Hemel Hempstead, England: Harvester Wheatsheaf.

Foucault, M. (1997). 'The birth of biopolitics'. In P. Rabinow (Ed.), *Michel Foucault: Ethics, subjectivity and truth*. New York: The New Press.

Foucault, M. (2003). *Vigiar e punir* (27nd Ed.). Petrópolis: Editora Vozes.

Freire, P. (1998). Cultural action for freedom. *Harvard Educational Review, 8*(4), 471–521.

Gates, P., & Zevengergen, R. (2009). Foregrounding social justice in mathematics teacher education. *Journal of Mathematics Teacher Education, 12*, 161–170.

Goldin, G. (2008). Perspectives on representation in mathematical learning and problem solving. In L. D. English (Ed.), *Handbook of international research in mathematics education (2nd Ed.)* (pp. 176–201). New York: Routledge.

Grouws, D. A. (Ed.). (1992). *Handbook of research on mathematics teaching and learning*. New York: Macmillan.

Gutiérrez, R. (2008). A "gap-gazing" fetish in mathematics education? Problematizing research on the achievement gap. *Journal for Research in Mathematics Education., 39*(4), 357–364.

Gutstein, E. (2003). Teaching and learning mathematics for social justice in an urban, Latino school. *Journal for Research in Mathematics Education, 23*(1), 37–73.

Hudson, B. (2008). Learning mathematically as social practice in a workplace setting. In A. Watson and P. Winbourne (Eds.), *New directions for situated cognition in mathematics education* (pp. 287–302). *New York:* Springer.

Jameson, F. (1991). *Postmodernism, or, the cultural logic of late capitalism. Durham, NC:* Duke University Press.

Jurdak, M. (2006). Contrasting perspectives and performance of high school students on problem solving in real world, situated, and school contexts. *Educational Studies in Mathematics, 63*, 283–301.

Klette, K. (2004). Classroom business as usual? (What) do policymakers and researchers learn from classroom research? In M. Høine and A. Fuglestad (Eds.) *Proceedings of the 28th Conference of the International Group for the Psychology of Mathematics Education]* (Vol. 1, pp. 3–16). Bergen, Norway: University College.

Knijnik, G. (1993). An ethnomathematical approach in mathematical education: A matter of political power. *For the Learning of Mathematics, 13*(2), 23–25.
Lacan, J. (1998). *Encore. The seminar of Jacques Lacan: Book XX.* New York: Norton.
Lacan, J. (2008). *The ethics of psychoanalysis. The seminar of Jacques Lacan: Book VII.* New York: Taylor and Francis.
Langrall, C., Mooney, E., Nisbet, S., & Jones, G. (2008). Elementary students' access to powerful mathematical ideas. In L. D. English (Ed.), *Handbook of international research in mathematics education (2nd Ed.)* (pp. 109–135). New York: Routledge.
Lave, J. (1988). *Cognition in practice: Mind, mathematics and culture in everyday life.* Cambridge: Cambridge University Press.
Lave, J. & Wenger, W. (1991). *Situated learning: Legitimate peripheral participation.* Cambridge: Cambridge University Press.
Leder, G. (1992). Mathematics and gender: Changing perspectives. In D. A. Grouws (Ed.), *Handbook of research on mathematics teaching and learning* (pp. 597–622). New York: Macmillan.
Lerman, S. (1998). The intension/intention of teaching mathematics. In C. Kanes, M. Goos and E, Warran (Eds.), *Proceedings of 21st Annual Conference of Mathematics Education Research Group of Australasia* (Vol. 1, pp. 29–44). Brisbane, Australia: Mathematics Education Research Group of Australasia.
Lerman, S. (2000). The social turn in mathematics education research. In J. Boaler (Ed.), *Multiple perspectives on mathematics teaching and learning* (pp. 19–44) Westport, CT: Ablex.
Lester, F. (Ed.). (2007). *Second Handbook of Research on Mathematics Teaching and Learning.* Charlotte, NC: Information Age Publishing.
Lubienski, S. (2002). Research, reform and equity in U. S. mathematics education. *Mathematical Thinking and Learning, 4*(2&3), 103–125.
Lubienski, S. (2008). On "gap gazing" in mathematics education: The need for gap analyses. *Journal for Research in Mathematics Education, 39*(4), 350–356.
Lubienski, S., & Gutiérrez, R. (2008). Bridging the gaps in perspectives on equity in mathematics education. *Journal for Research in Mathematics Education., 39*(4), 365–371.
Malloy, C. (2008). Looking throughout the world for democratic access to mathematics. In L. D. English (Ed.), *Handbook of international research in mathematics education* (2nd ed., pp. 20–31). New York: Routledge.
Marshall, S. P., & Thompson, A. (1994). Assessment: What's new – and not so new – a review of six recent books. *Journal for Research in Mathematics Education, 25*, 209–218.
Marx, K. (1857). *A contribution to the Critique of Political Economy.* Retrieved from http://www.marxists.org/archive/marx/works/1859/critique-pol-economy/appx1.htm
Mouffe, C. (2005). *The democratic paradox.* London and New York: Verso.
Nasir, N. S., & Cobb, P. (2002). Diversity, equity, and mathematical learning. *Mathematical Thinking and Learning, 4*(2&3), 91–102.
Nasir, N. S., & Cobb, P. (2007) (Eds.). *Improving access to mathematics: Diversity and equity in the classroom. New York: Teachers College Press.*
Niss, M. (2007). Reflections on the state of and trends in research on mathematics teaching and learning: From here to Utopia. In F. Lester (Ed.), *Second handbook of research on mathematics teaching and learning* (pp. 1293–1312). Charlotte, NC: Information Age Publishing.
Nolan, K. (2009). Mathematics in and through social justice: Another misunderstood marriage? *Journal of Mathematics Teacher Education, 12*, 205–216.
Pais, A. & Stentoft, D. (unpublished manuscript). *Construction of "importance of mathematics" as setting boundaries for research and education.* Aalbord, Denmark: Aalborg University.
Pais, A., Stentoft, D. & Valero, P. (2010). From questions of how to questions of why in mathematics education research. In U. Gellert, E. Jablonka and C. Morgan (Eds.), *Proceedings of the 6th International Mathematics Education and Society Conference* (pp. 398–407). Berlin: Freie Universitat, Berlin.

Pais, A. & Valero, P. (2010). Beyond disavowing the politics of equity and quality in mathematics education. In B. Atweh, M. Graven, W. Secada, and P. Valero (Eds.), *Mapping equity and quality in mathematics education (pp. 35–48).* Dordrecht, The Netherlands: Springer.

Pearl, A., & Knight, T. (1999). *The democratic classroom: Theory to inform practice.* Cresskil, NJ: Hampton Press.

Powell, A. & Frankenstein, M. (1997). *Ethnomathematics: Challenging eurocentrism in mathematics education.* Albany, NY: State University of New York Press.

Riall, R. & Burghes, D. (2000). Mathematical needs of young employees. *Teaching Mathematics and its Applications, 19* (3), 104–113.

Rogers, P., & Kaiser, G. (1995). *Equity in mathematics education: Influences of feminism and culture.* London: Falmer.

Rose, N. (1999). *Governing the soul: The shaping of the private self* (2nd Ed.). London: Free Association Books.

Schoenfeld, A. (2002). Making mathematics work for all children: Issues of standards, testing, and equity. *Educational Researcher, 31*(1), 13–25.

Secada, W. (1992). Race, ethnicity, social class, language, and achievement in mathematics. In D. A. Grouws (Ed.), *Handbook of research on mathematics teaching and learning* (pp. 623–660) New York: Macmillan.

Secada, W., Fennema, E., & Adajian, L. (Eds.). (1995). *New directions for equity in mathematics education.* Cambridge: Cambridge University Press.

Sierpinska, A. & Kilpatrick, J. (Eds.) (1998), *Mathematics education as a research domain: A search for identity.* Dordrecht, The Netherlands: Kluwer.

Silver, E. A., & Herbst, P. (2007). Theory in mathematics education scholarship. In F. K. Lester Jr. (Ed.) *Second handbook of research on mathematics teaching and learning* (pp. 39–67). Charlotte, NC: Information Age Publishing.

Skovsmose, O. (1994). *Towards a philosophy of critical mathematics education.* Dordrecht, The Netherlands: Kluwer.

Skovsmose, O. (2006). Research, practice, uncertainty and responsibility. *The Journal of Mathematical Behavior, 25*(4), 267–284.

Skovsmose, O., & Valero, P. (2008). Democratic access to powerful mathematical ideas. In L. D. English (Ed.), *Handbook of international research in mathematics education* (2nd Ed.) (pp. 415–438). New York: Routledge.

Valero, P. (2004). Socio-political perspectives on mathematics education. In P. Valero & R. Zevenbergen (Eds.), *Researching the socio-political dimensions of mathematics education: Issues of power in theory and methodology* (pp. 5–24). Boston: Kluwer.

Valero, P. (2007). A socio-political look at equity in the school organization of mathematics education. *Zentralblatt für Didaktik der Mathematik. The International Journal on Mathematics Education, 39*(3), 225–233.

Vinner, S. (1997). From intuition to inhibition – mathematics education and other endangered species. In E. Pehkonen (Ed.), *Proceedings of the 21th conference of the International Group for Psychology of Mathematics Education* (Vol. 1, pp. 63–78). Helsinki: Lahti Research and Training Centre, University of Helsinki.

Vinner, S. (2007). Mathematics education: Procedures, rituals and man's search for meaning. *Journal of Mathematical Behavior, 26*, 1–10.

Vithal, R. (2000). Re-searching mathematics education from a critical perspective. In Matos, J. F. (Ed.), Proceedings *of the 2nd Conference on Mathematics Education and Society.* Retrieved from http://nonio.fc.ul.pt/mes2/renuka.pdf

Williams, J., & Wake, G. (2007). Black boxes in workplace mathematics. *Educational Studies in Mathematics, 64*, 317–343.

Wilson, L. (2007). High-stakes testing in mathematics. In F. K. Lester Jr. (Ed.), *Second handbook of research on mathematics and learning* (pp. 1099–1110). Charlotte, NC: Information Age Publishing.

Žižek, S. (1989). *The sublime object of ideology*. London and New York: Verso.

Žižek, S. (1991). *For they know not what they do: Enjoyment as a political factor*. New York and London: Verso.

Žižek, S. (1993). *Tarrying with the negative: Kant, Hegel, and the critique of ideology*. Durham, NC: Duke University Press.

Žižek, S. (1994). The spectre of ideology. In S. Žižek (Ed.), *Mapping ideology* (pp. 1–33). London and New York: Verso.

Žižek, S. (1997). *The plague of fantasies*. London: Verso.

Žižek, S. (2004). *Organs without bodies: Deleuze and consequences*. London: Routledge.

Žižek, S. (2006). *The parallax view*. Cambridge, MA: MIT Press.

Žižek, S. (2007). Badiou: Notes from an ongoing debate. *International Journal of Zizek Studies 1*(2) [online].

Alexandre Pais
Aalborg University
Denmark

CHAPTER 3

THE ROLE OF MATHEMATICS IN THE DESTRUCTION OF COMMUNITIES, AND WHAT WE CAN DO TO REVERSE THIS PROCESS, INCLUDING USING MATHEMATICS

MUNIR JAMIL FASHEH

PARTING OF PATHS

Since the days of Francis Bacon, the father of European science, who articulated the purpose of science as having "the power to conquer and to subdue [nature]", people (especially the educated) have lived in disharmony with nature. Mathematics – the "queen of sciences" – played a main role in "subduing and conquering" nature. In recent times, conquering and subduing physical nature was extended to human nature and human communities. For 12 years, education seduces students to live in a make-believe world and become active participants in the harm done to themselves, their communities, and to physical nature – believing all the time the claim that all is done for their own good![1] As Wendell Berry (1990) says, harm that was done to the world prior to modern times was done out of ignorance or weakness; today, the "rape and plunder" of the world is done with full awareness and conscious intentions. The role of the sciences and mathematics is central in this process. For Berry, these conscious acts are a "new thing under the sun". According to him, the main division in the world today is between those who work hard to protect life and those who (for greed and control) are consciously destroying it. Mathematics faces this parting of paths: it can contribute to protecting life or destroying it. This article is a personal reflection on one experience in such parting of paths: the Palestinian experience.

PLURALITY OF MATHS AND CONQUEST OF KNOWLEDGE: HOW OFFICIAL MATHEMATICS CONTRIBUTED TO DISMANTLING PALESTINIAN SOCIETY

My first awareness of such parting of paths was when I became aware of my illiterate mother's mathematics (Fasheh, 1990). The kind of mathematics I studied and taught was more in line with the values of greed, power, and control, while my mother's mathematics was embedded in life, inseparable from it. My awareness happened in the mid 1970s when I was in charge of improving mathematics instruction in the schools of the West Bank (in Palestine), and teaching mathematics at Birzeit University. It took me several years to realize that my mathematics and my mother's mathematics do not intersect; they belong to different worlds. That was my first deep understanding of what "plurality of

O. Skovsmose and B. Greer (eds.), Opening the Cage: Critique and Politics of Mathematics Education, 93–106.

knowledges" means. My mathematics could not be reduced to hers and hers could not be reduced to mine; nor was it possible to produce a synthesis higher than both. My mathematics and her mathematics were worlds apart. Such plurality of knowledges[2] is usually not acceptable in universities, which propagate the myth of one and only one kind of mathematics. What made things even more shocking was my realization that even if I study mathematics for another ten years in universities, I still wouldn't be able to do what she was able to do without being taught, and without curriculum, tests, and grades! It was a moment of deep learning for me because it touched and dismantled basic myths I held as given.

That realization was a most profound turning point in my life. Questions that never left me since then include: "Why is my kind of mathematics considered valuable and worthy of being taught in schools and universities (almost all over the world) while my mother's kind of mathematics is totally ignored? Why is my kind of mathematics considered knowledge while hers is not? Is it because it is superior or better? Is it because it is more modern and thus supersedes all other kinds, including hers? Is it because it is more useful, and if so, to whom? What is the cost we pay as a result of my kind of mathematics winning over and wiping out hers?" Slowly I started realizing that the triad that was manifested by Western educational missions in Palestine – despising people and what they have, monopolizing what constitutes knowledge, and being ready to "help" people move along the dominant path of progress – has been fatal to the Palestinian society and its ability to regenerate itself. Gradually, I realized that the mathematics I studied and taught suppressed and won over my mother's mathematics through bullying; by devaluing, ignoring, and belittling her mathematics, and providing instead another mathematics that claimed to be neutral and universal – the only path into the future.[3] It won not because it is superior or better but through being a tool serving interests of dominant political and economic powers, helping them in controlling people, suppressing their knowledges, and robbing them of what they have (including their biological abilities). What frightened me was the fact that I was an active participant in that process and doing it with good intentions, believing all the time that I was helping Palestinians move along the path of progress. I feel that I was given a degree, with all the privileges associated with it, once they were sure that I would carry this subtle weapon, not only into my classes but also into my home, and use it to wipe out people's knowledges, sense of self-worthiness, and biological abilities such as learning. My reflections on the above questions made me perceive official mathematics as one of the Trojan horses that helped conquer us from within. Those reflections also made me rethink meanings of crucial words related to education and life. I started in a spontaneous way (what later became a most important conviction in my life) perceiving people (especially children) as co-authors of meanings of the words they use, hear, or read. I started with words that my interaction with my mother's world triggered, such as learning, knowledge, the worth of a person, pluralism, progress, and humility. My reflections taught me humility. I realized that there are many worlds that inhabit this world and that life is much richer than what the human mind can comprehend and ideas can express. In contrast to my mathematics, which was aloof, my mother's mathematics was embedded in life like salt in food: we can taste it but not see it. It was useful, beautiful, meaningful, and fitting; no woman would have accepted a dress if it lacked any of these qualities. Her work was not only of art but also consisted of main scientific aspects: experience, observation, experimentation, and making sense.

While it was very common for me to use words and concepts that I had no idea what they referred to,[4] my mother never had such a problem; she never used a word that she did not know the source of its meaning, or cut a piece of cloth not knowing where it fits with the other pieces into a whole which is a dress that fits a particular woman. To cite just another example, she never knew the meaning of "good citizen" but she had a deep understanding of what a good human being is.

PLAYING THE ROLE OF THE "CULTURAL IMPERIALIST" IN MY OWN HOME

It was very scary for me to realize that I was playing the role of the "imperialist" in my very home! I was a tool in conquering my mother's mathematics and knowledge; a tool in conquering her world through internalising the myth that there is a single undifferentiated natural path for progress – the European path – that I acquired in schools and universities. Such conquering is an act of violence difficult to recognize by schooled institutionalised minds. I was a tool in this violent conquest where one kind of knowledge conquered and wiped out another – which is probably the deepest and most destructive conquest. That's why I believe that fundamentalism in the modern world did not start in religion or politics but in relation to knowledge.[5] I was spreading the "virus" of believing in a single universal path and source for learning and knowing – contaminating my students the same way a man with AIDS spreads his disease without knowing it. Like him, I was spreading the virus through loving and being loved; wanting to please and be pleased. I enjoyed teaching mathematics and my students enjoyed mathematics as I was teaching it. It was difficult for me as a product of American universities (the American University of Beirut, Florida State University, and Harvard University) to think otherwise. The belief in a universal path to learning and gaining knowledge was very strong. My encounter with my mother's world helped me see that the struggle in the world has always been between people-in-communities[6] and those who actively want to tear apart the inner world in each person and the social-cultural-economic fabric of communities.

The power I had over my students can be likened to a person who has a loudspeaker in a crowd. People would listen to him not necessarily because his voice is nicer or what he says is wiser but simply because he carries a tool that gives him arbitrary power and authority. My "loudspeaker" was constituted by the degree I had and a job that gave me the power to inflict harm on those who refused to listen to me and follow my instructions.

ONE PALESTINIAN MAN'S RESPONSE

This conquest at the knowledge level is absent – in general – from Palestinians' consciousness. One exception was Khalil Sakakini who seemed to have seen the damage that was happening to Palestinian children at a deep level as a result of establishing American and European schools in Palestine at the end of the

19th century. He saw it as early as 1896 and wrote his first book: *Wearing Someone Else's Shoes*. It seems he saw how children were turning into copies of an alien culture that had no roots in the community and no relevance to Palestinian life.

What Sakakini seems to have noticed was that the use of grades to measure children's worth (by comparing them along a vertical line) is inherently a violent act against them. Shifting the source of the worth of a person from the person and community to numbers that claim to be objective and universal, and whose legitimacy came from London, led to conquering Palestinians from within. London matriculation became (in the 1920s, 1930s and 1940s) the main measure of the worth of Palestinian students. Curriculum and grades became idols at whose feet people worshiped and kneeled – Trojan horses the British left behind, which helped dismantle and subdue our communities. Using numbers to measure students' worth has been a destructive act, more dangerous than other forms because it is invisible to most and tries to measure what cannot be measured, which ends in degrading people (ironically called grading), tearing apart communities, harming nature.

It is worth mentioning that Sakakini did not criticize foreign education in words only but also in action. He built his first school in Jerusalem in 1909, basing it on principles radically different from western schools: he did not start with goals and objectives but with values, the main one of which was dignity. He translated that by refusing to have grades, prizes, or punishment in his school. He was able to do that a hundred years ago, under the Ottoman rule. Today, a Palestinian principal who would even dare think of not using grades in his school will be expelled immediately. The British occupation of Palestine in 1917 wiped out Sakakini's spirit. In short, the two "crimes" mentioned above – exemplified by my mathematics wiping out my mother's (i.e., suppressing diversity in learning and knowing) and by using numbers to measure students' worth – were the first manifestations of how school mathematics was used to destroy communities in Palestine.

SCIENCE VS. WISDOM: HOW SCIENTIFIC AND MODERN TOOLS THAT BRITAIN BROUGHT INTO PALESTINE DESTROYED THE PILLARS ON WHICH PALESTINIAN COMMUNITIES RESTED

Community as I use the term here rests on three pillars: local soil, local culture, and local economy. They are the source of community's inner strength and the basis of its regeneration – yet, they are absent from education! Every child is nurtured by the land soil and the cultural soil. However, these two local soils would be mere slogans without local economy. The three form the fabric of community. Once children become rooted within these local soils, they can be enriched and nurtured further by other cultures.[7] Without these pillars, children would lose their roots, and their source of nurturance and inner strength, and community would gradually wither away. The British brought into Palestine three tools (presented as inventions for progress) which destroyed these three pillars: *education* destroyed local culture, *flush toilets* destroyed local soil and wasted local water, and the *state* destroyed local economy. Mathematics is a main ingredient in all three: grading in schools, designing a device (that needs science and mathematics but lacks wisdom) to transfer our excrement, and using productivity and national income to measure economic activity. In all three, no questions are asked about consequences – an act that manifests lack of wisdom. For the mind to move without hindrance, wisdom had to be imprisoned.

Living in harmony with physical and human nature and refusing to do anything that would harm them and tear apart the social fabric of communities are part of living wisely. Wisdom has been ignored, actually imprisoned, when the mind (with its tools of mathematics, science, and technology and its values of competition, control, and winning) was elevated to the power of the throne. The state of the world today compels us to de-throne the mind, set wisdom free, and make science and mathematics go hand in hand with wisdom. One current challenge, thus, is how to conceive and practice science and mathematics in a way that is more in harmony with wisdom, i.e. with the values of respecting creation and protecting nature to continue regenerating itself. This necessarily requires inventing ways that help remedy the harm we do (by our ways in living) to our selves, our communities, and nature.

This is what I have been trying to do since the early 1970s. It is a slow process simply because it is against interests of dominant powers that require current ways of teaching mathematics. However, we don't have any choice other than do what we can in a world that faces real threats to life. We need to ask about the cost to communities and nature that we pay as a result of stressing formal skills and technical knowledge. It is here where I believe we need to rethink the meaning and role of logic in the mathematics curriculum and how to incorporate it in schools and universities. We stress formal logics and ignore how we can use logic in mathematics classes in terms of its relation to life. We rarely ask, for example: What is the logic that governs our actions and relations in education? What are the values that currently govern our behaviour in schools and universities, and what values do we want to live in harmony with? A formal logical system starts with axioms that we do not violate. Similarly, we should ask about the values (axioms) that we currently do not violate in education. If we reflect on our actions in schools and universities, we notice that the values we do not violate are control and winning. Any claim by an institution to the contrary is an act of delusion and deception (no matter whether that is done consciously or not). These claims include slogans such as *Veritas* and *Lux et Veritas*.

Using mathematics to see patterns, relations, systems, and connectedness in life (physical and human) and to discover the values that govern our actions, can help us rethink our ways of living and gain, instead, self-knowledge and self-rule[8]. It is crucial in making sense of experiences. Trying to understand the logic that underlies the behaviour of a person, group, or institution should be part of the mathematics curriculum. In 1979, I introduced a course for entering first-year students at Birzeit University, part of which was observing and seeing patterns in all aspects of life. One student in my class, who was jailed by the Israelis, told the class – after he was released – how he collected from fellow prisoners questions they were asked during interrogations. He wanted to know the logic of the Israeli official mind. When I introduced the course, I never thought that it would go that far!

The role of mathematics in dismantling communities and degrading cultures was similar to its role in agriculture, where it considers productivity as the only measure of agricultural activity. Just like we don't ask about what happens to children as a result of education, we don't ask what happens to the soil as a result of stressing productivity. Such stress led to the corruption of local soil and destruction of the local economy. This corruption took a sharp increase when the Palestinian Authority was installed and the World Bank allowed in. There was more ecological sustainability and social equity before 1993. Development and investment that accompanied the Oslo accords (especially during the past 5 years) have been disastrous at the community level. One manifestation is transforming land and people into commodities. Land that was for 4000 years a source of living and dignity for Palestinian peasants became in recent years a pure commodity. This means that the shift we need in teaching mathematics requires a shift in what we consider as measures. Teaching mathematics would go through a real transformation if the values that govern our actions shift from control and winning to the well-being of children, communities, and nature.

Communities and the ability of nature to regenerate itself cannot be created through plans and minds. Communities are formed over hundreds (if not thousands) of years, and the spirit of regeneration has been part of creation. Civil society perceived as consisting of NGOs has been effective in tearing apart communities; and development as was conceived in 1949 has been a main tool in killing the ability and spirit of regeneration.[9]

GRADING IS DEGRADING: THE ROLE OF MEASUREMENT IN CONQUERING COMMUNITIES

I once read in a mathematics book, "If you can't measure what you are talking about, you don't know what you are saying". I was fascinated by it and enthusiastically taught it to my students. I can't believe I was so blind not to see that most of what is valuable in life cannot be measured.[10] When we talk about agriculture, with productivity as the only measure, not caring about the cost we pay at other levels, we become agents in our own destruction. Using numbers to measure human and community aspects helps distract us from fundamental aspects in life.[11] The Arab Human Development Reports sponsored by the United Nations Development Programme are excellent examples of how mathematics is used to distract us from seeing the source of strength in our communities and the threats to life on earth due to modern patterns in living. The purpose of such reports seems to be making us perceive ourselves as "less", which compels us to move along the path of consumption so we would measure higher on the development scale; i.e. along what tears communities and nature apart. Such reports are a main source of deception using mathematics as means.

Using numbers to measure the worth of people, cultures, and countries marked – in my opinion – the beginning of decadence whose manifestations we can see worldwide. It underlies the falling apart of most societies in the Two-Thirds world (a phrase used by Gustavo Esteva and Madhu Prakash) in the 20th century. Our real enemy (as peoples in all countries) has become what we embrace and pay dearly for with our limited resources: it is our perceptions and ways of living that rob us of natural abilities such as learning, healing, walking, conversing, and making sense of experiences.

THE PARROT BY TAGORE: IMPROVING THE CAGE AND IGNORING THE BIRD

A few years ago, Claude Alvarez handed me a publication that included a short story *The Parrot* written nine decades ago by Tagore (see Intermezzo, this volume). That short story clarified for me aspects in education more than tens of books critical of it. One aspect I want to raise is related to structures. There is a radical difference between cages and nests as structures. Building a cage requires one kind of mathematics; making a nest embodies a totally different kind. This is similar to the difference between structures I used in my mathematics and those used by my mother. The bodies of women she made dresses for were the structures she worked with. Unlike the geometrical shapes I studied and taught, my mother had to start with the geometrical figure of the woman she was working with. Every woman had her own figure/ geometry. This led to my conviction that the shallowest learning is the one acquired through teaching (which is usually related to cages). This reflection raises the question of how to gain instrumental knowledge without losing the rich and tremendous mathematics that we can learn without institutions, professionals, technology, and without instruction. We need to avoid what makes our bodies obsolete, unable to do what they can do on their own, such as learning, healing, walking, conversing, writing, and expressing.

Most projects and programs concerning improving the teaching of mathematics in Palestine today are related to improving cages: developing new curriculum, upgrading equipment, evaluation, training teachers, gaining technical skills, building schools, etc. Focusing on cages is covered up by claims that the purpose of improvements is to help and serve students. I want to stress again that I am not talking here about intentions but about perceptions. Claims of helping and serving usually rob people of dignity, natural abilities, and what they have. For example, the well-being of children, communities, and nature is not treated as a value in "quality education" espoused today by many organizations.

What can be measured belongs in general to the cage; what cannot belongs to what happens to people, communities, and nature. What can be measured can easily be transformed into commodities, such as what is happening to people and knowledge. We need to end the monopoly of using numbers to measure students' worth and look for alternatives. Luckily such an alternative has existed in the Arab culture for 1400 years – but totally ignored. I will present it shortly.

UNIVERSALS VS. DIVERSITY

Analysis and searching for universals are very strong within Western culture. This, no doubt, was very beneficial at many levels. However, it has reached today a dangerous state that the world cannot ignore anymore. Searching for universals today is not treated as a search that adds to our understanding but it has become a list of facts that are imposed on peoples and countries as the only absolute truth. What I wrote earlier compels us to regain "pluralism" as a most distinctive characteristic in life, nature, human beings, and knowledge and to realize that universals are contrary to humanity and Nature. This means that we need to start demanding to retrieve part of education's budget and use it in various diverse settings. I already mentioned the centrality of talking about pluralism in relation to knowledge. One way I used in working with mathematics teachers to bring out pluralistic attitudes into teaching was stressing that every child is logical; there are many logics which are not compatible. This was very hard for teachers to accept. One positive aspect (still more potential than actual) of modern technology (especially the Internet) is it is increasingly making isolation impossible and making the pluralistic nature in life more obvious. We need to build paths that take us out of our provincialities, without dumping us into a single global culture. I suggest that we stress not the right to education but to educations – in the plural. The age of information does not add up to knowledge just like the age of knowledge, sciences, and technology did not add up to wisdom.

Western civilization is not the only one that believes in universals. However, it is the only one that succeeded in producing tools (such as schools and grades) that claimed to be universal, objective, and neutral. People around the world adopted them as such. This led to "evangelising" the world through education, which was done mainly through people with good intentions but who carried the virus of the one and only one path for learning.

THE IMPORTANCE OF CO-AUTHORING MEANINGS AND MEASURES

Another thing we need to do in mathematics classes is to encourage students to co-author meanings of words they use, read, or hear. I experimented with the idea with children in the 1970s (Fasheh, 1982). One question I asked was "What is a point?" I will never forget an answer by a 7-years old girl, "A point is a circle without a hole"; so creative and imaginative! Co-authoring meanings in the light of our actions, experiences, reflections, and conversations is most fundamental in learning. Without it, learning would be seriously deficient.[12]

A main struggle I always felt has been between what people can do by and for themselves and those who impose on them ways that rob them of all that in the name of helping, assisting, and serving. I lived such parting of ways several times in Palestine, especially in the 1970s and the first *intifada* (1987–91). In both periods, people asked what they could do and went ahead and did it. Dignity rather than rights, hope rather than expectations, mutual support rather than self-serving, self-rule rather than ruling or being ruled, giving rather than demanding, and being attentive to life rather than to distractions were what moved us. In both periods, cages were dismantled and life vibrations moved us.

Shifting the source of one's worth from a universal measure controlled by institutions back to the person and community has been a main struggle for me since the early 1970s when I was in charge of mathematics instruction in the schools of the West Bank. However, it was in 1997 (as part of trying to find guiding principles for the Arab Education Forum which I established in 1998 at Harvard University's Center for Middle Eastern Studies) that I came across a 1400-year old statement by Imam Ali that made much sense to me. I chose it as the title of the Forum's vision and publications. I find it relevant in the world today. In Arabic, the statement is: *qeematu kullimri'en ma yuhsenoh* قيمة كل امرئ ما يحسنه which means: the worth of a person is what s/he *yuhsen*. *Yuhsen,* in Arabic, has several meanings, which *together* constitute the worth of the person: the first meaning refers to how well the person does what s/he does, which requires technical knowledge and skills; the second refers to how beautiful/pleasing what one does is to the senses, the aesthetic dimension; the third refers to how good it is for the community; the fourth refers to how much one gives of self rather than what one transfers (as a commodity) from one place to another; and the fifth refers to how respectful of people and ideas one is. According to the statement, a person's worth is not judged by professional committees or "objective and universal" measures but by the five meanings embedded in the word *yuhsen.* Since I read it in 1997, I have been amazed by its simplicity, profundity, depth, insight, and by how it embodies diversity. It has been a main principle guiding my thinking and work. It is crucial, however, that we do not treat it as constituting another "super system" but as a statement about taking a stance in the effective presence of others. It reflects the belief that one is in relation, rather than in opposition, to others, or better than others.

When I read Tagore's story around 2005, I felt the same way as when I read Imam Ali's statement in 1997. Both were like a "magical pen" that drew a clear painting of my experiences. Both helped clean what was left over within me in terms of perceptions and assumptions that I acquired through education. Ali's statement exposes shallow and harmful aspects of current evaluations and replaces them by a principle that is diverse and has respect for people; its essence is dignity. Tagore's story provides a map that helps us see the way clearly through the boundaries created by education; it helps us see whether what we do is related to improving cages or to well-being of children and communities. It helps us heal from distractions and self-deceptions which may appear to be good but in fact destroy biological regenerative abilities such as learning, making sense of experiences, and being sources of understanding. By perceiving "learning from life" as insignificant, educational authorities help destroy people's biological ability to learn[13], transforming it into a need, a right, and a service – to be provided by professionals.

Healing from the dominant ideology and regaining sanity necessitate two things: first, stop using numbers to decide people and community's worth and, second, regain a pluralistic attitude towards knowledge. This necessarily treats mathematics as a means rather than as a master. As an artistic and scientific tool, mathematics can help us realize the harm we as humans do to nature, to ourselves, and to communities as a result of our ways of living and, thus, help us in how we can reverse this process. Where I currently work (Shufaat Refugee Camp), there are about 4,000 students in the four schools. If we assume (and this is a minimum) that every child urinates 1 litre a day, and uses the flush toilet only once (9 litres), that means 40,000 litres are wasted every day. It is pure madness to go on using the flush toilet in such a place where water is very scarce. Obviously, we need to find alternatives. This is part of what we are talking about and trying to replace at the camp.

CURRENT SITUATION IN PALESTINE

In schools, we become conditioned to care only for ourselves, against others. This ideology of "each person for oneself" is the worst that can happen to people-in-community. This was the worst outcome of the Oslo agreement that took place in 1993 between the Israeli government and the PLO. The spirit of the 1970s and the first *intifada* was replaced by how much a person can get for himself. Before 1993, competition was confined to schools; after 1993, it permeated all aspects of life. It became a main tool in creating and spreading the ideology of "each man for himself". Competition (in the sense of comparing people along a vertical line) can never be in harmony with the well-being of people, communities, and nature. It turns things around where people and communities become agents in their own destruction. A challenge we face, thus, is how to teach without competition; this is harmonious with wisdom.

This fragmentation, where every individual cares only for self regardless of what happens to others, to community, and to nature, is accompanied by fragmentation due to restrictions on movement, which was actually started by the British and French occupations of the region after the First World War. Over the years, I have been experiencing how my world was shrinking and getting more fragmented. Borders have been increasing and tightening over the years – not only at the physical level but also at the perceptual, social, and cultural levels. It seems that fragmentation of knowledge into subjects at school, where every class/field is not related to any other, paved the way for other aspects of life to be fragmented. This logic led eventually to the age of information where every piece of information stands alone. One thing I am increasingly noticing is that when young people interact through language, their sentences hardly intersect – they are like parallel sentences that never meet. The inability to connect aspects and phenomena is reaching the level of a plague. This is ironic because a main role of mathematics is to see the logic that underlies reality and that brings the various parts into a system; a main role is to help us see the whole clearer and be able to make sense of our experience. Fragmentation and borders are always connected to occupation and conquest, not only of land but also of worthiness, knowledge, perception, language, and culture. To give just a small example, with my doctorate degree I only know two languages, whereas my father left school in fourth grade and started working. He learned four languages. The concept of identity (which we are now experiencing in the form of Palestinian identity and which is loaded with mathematical connotations/aspects) is a very good example that combines borders, fragmentation, occupation, and conquest. The concept of identity was the prelude to the apartheid wall built by Israel. But while the wall is a very visible aspect, identity is embraced as a positive aspect. It seems it is much easier to fool the mind than fooling the eye. It always puzzled me that we as human beings brag about – the mind – is the easiest to deceive. I believe that this is so whenever the mind works without wisdom.

When I was growing up, my sense of belonging was to the Arab World. Now, I can only move within a very tiny part of Palestine. I am not even allowed to visit Jerusalem where I was born and which is less than 10 minutes away! Despite all the talk about "we are now living in a global village" and despite the tremendous technological advances in communication and movement, people everywhere seem to be confined by one form or another of border, fragmentation, occupation, and conquest. Those that are less visible are more dangerous.

Although all current indications point to Israel's unwillingness to allow for a Palestinian state to emerge, still it would be very beneficial exercise, for us as well as for others, to be involved in reflections and discussions concerning the teaching of mathematics in anticipation of a state. The notion of foreground that Ole Skovsmose talks about is relevant in this regard. I believe that the Palestinian situation – because of its dynamic nature – can provide a rich case in exploring issues such as the relation between mathematics and the state (especially before its formation, even though it is not likely to happen). One question, in particular, is how we can avoid the pitfalls that others got into in relation to teaching mathematics in a way where it can help us live in harmony with values that protect life and enhance understanding.

USING ARABIC LANGUAGE AND CULTURE TO ENRICH THE TEACHING OF MATHEMATICS IN ARAB SCHOOLS

Whenever possible (and this requires teachers who appreciate both), it would be very enriching to teach mathematics and Arabic together. The Arabic language has many characteristics that make it close to mathematics. First, the language is built on patterns, which means once you know the three-letter root of any meaning, one can form many other words according to patterns (for example, the words write, book, office, library, desk, etc. are all different in English, while in Arabic they are all formed from a 3-letter root. In addition, the same roots of different meanings show connections in the minds of Arabs in ancient times that can be inspiring today. For example, "discussion" and "chiselling" have the same root, which reflects the fact that ancient Arabs saw that the purpose of discussions – just like in chiselling a rock – is to beautify what you are dealing with; i.e. the purpose is for the two persons to come out more beautiful as a result of the discussion. Second, *al-muthanna* (which has no synonym in any European language, with the exception of ancient Greek, and loosely means "dual") embodies logic different from Aristotle's and Hegel's. Whereas anything is either A or not A and cannot be both in Aristotle's logic, and whereas A and not A can form a synthesis which is higher than both, *al-muthanna* can be thought of as a triad where A remains A and B remains B but a relation between the two develops that is very important to both. This is extremely relevant to the concept of the "other" which is so popular in the world today. Third, Arabic poetry falls into 16 basic patterns. Fourth, Arabic script has harmony and beauty that can bring out the aesthetic dimension in mathematics; where art, Arabic, and mathematics can meet. Fifth, the sounds of reciting the Koran reflect enchanting patterns (but for one to feel it, s/he has to hear it from a real person, not from a recording). In addition, which is the sixth point, the movement of the moon is central in Arab culture, which makes it excellent material for teaching mathematics.

NOTES

1 The role of "elite" universities in creating current crises, in the US and UK as well as around the world, cannot be ignored. Most decisions that were taken in the various fields – politics, economy, finance, food, agriculture, education, military, raising children, health – were taken by graduates of elite universities.

2 The plural for knowledge is not used in English. When Arabs first encountered Europeans and heard for the first time the phrase "Department of Education" the way they translated it was "House of Knowledges".

3 What happened to my mother's mathematics happened to her Christianity. Missionaries from the US came to our home to "convert" my Christian family into their Christianity! She was one of the last Christians that carried the spirit of Jesus as it was lived through generations over 20 centuries. It is hard to find Palestinians today who carry that spirit in their hearts and daily lives. Religious missions conquered my mother's Christianity the same way educational missions conquered her knowledge. Both her mathematics and Christianity were wiped out in the name of progress! Just like I could not see my mother's mathematics, missionaries could not see her Christianity and educators could not see her wonderful ability for upbringing children. The same thing happened to the knowledge of Palestinian peasants in farming.

4 In the sixth grade, the British textbook which was used included shares. I got full grades in all tests but up till now I never owned a share and never experienced its meaning.

5 Those who hold high degrees (consciously or not) hide this fact and blame religions and politicians for the mess in the world. It wasn't Roosevelt who convinced Einstein to build the atomic bomb but the opposite: it was Einstein who sent letters to Roosevelt convincing him to support making (and using) the bomb!

6 There is a word in Arabic *al Ahaali* which does not have a synonym in English. Briefly, it means people who are connected to a place, history, collective memory, and to one another through a social-cultural-economic fabric.

7 A Palestinian anthropologist, Sharif Kanaaneh, uses the analogy to grafting fruit trees on bitter almond trees. Bitter almond trees provide strong basis against any diseases that can hit the trees.

8 I use the phrase self-rule in the way Gandhi used it to mean *swaraj* in Hindi. I worked with 7th grade girls in Shufaat refugee camp at two levels: by every girl keeping a record of what she eats and how

she spends her day (such as watching TV, talking over mobiles, using computers); and by every 4 or 5 girls producing a "magazine" (not for the market) without needing permit, money, or editors (to discover what they can do without institutions and professionals but with what they have).

9 It is worth comparing the destruction done to Ramallah by trucks and banks with the destruction done to Gaza through tanks and warplanes. Although the role of sciences and mathematics is crucial in both destructions, Gaza can recover while Ramallah is highly improbable.

10 An old Palestinian woman once said, "Anything that can be bought with money is cheap". This is so true about education: anything we can measure is not significant. Aspects such as happiness, well-being, dignity, compassion, responsibility, and wisdom cannot be measured.

11 It is worth mentioning that the World Bank's first priority (after it was allowed in Palestine after Oslo) was education. Once the mind accepts numbers as measure of people's worth, other myths fall in place.

12 We need to ask why this fundamental right and biological ability is absent from the Universal Declaration of Human Rights, the UN Convention on the Rights of the Child, and institutions in general.

13 This brings up the question as to how to acquire the "rational" form without losing the natural biological ability to learn; how can we gain instrumental knowledge without losing what we can gain without institutions, professionals, and technology. It is important not to lose the rich and tremendous mathematics that we can learn without instruction such as sewing, hand-writing, planting, playing, building, hiking, climbing…

REFERENCES

Berry, W. (1990). *What are people for?* San Francisco: North Point Press.

Fasheh, M. (1982). Mathematics, culture, and authority. *For the Learning of Mathematics*, *3*, 2–8.

Fasheh, M. (1990). Community education: To reclaim and transform what has been made invisible. *Harvard Educational Review*, *60* (1), 19–36.

Munir Jamil Fasheh
Director, Arab Education Forum, Center for Middle Eastern Studies (retired)
Harvard University

CHAPTER 4

THE USA MATHEMATICS ADVISORY PANEL: A CASE STUDY

BRIAN GREER

In the USA, the National Mathematics Advisory Panel (NMAP) was set up by President George Bush. Its stated aims centre on the importance of mathematics education to the perceived national priorities of the USA. Close analyses of the commendably extensive public documentation generated reveal further aspects of the workings of political processes, the nature of ideological alliances and conflicts among numerous interest groups, and philosophical/ideological positions about the nature of mathematics as a discipline, as a school subject, and in relation to society.

In this chapter, I begin with a brief description of how NMAP was carried out, followed by an overview of the educational/political context. I comment on what I consider to be key points raised, and how mathematics and mathematics education are portrayed. From all of this, I suggest lessons for political engagement for mathematics educators in the face of a blatant negation of their work.

BRIEF DESCRIPTION OF THE PROJECT

In April 2006, President Bush established NMAP, which issued its Final Report in March, 2008. The Final Report (a summary), plus reports of three subcommittees and five Task Groups, and a wealth of other information, can be found on the official website (http://www2.ed.gov/about/bdscomm/list/mathpanel/index.html).

Short official biographies of the panellists can be found there. The panel was chaired by a chemist who is also an emeritus university president. The rest of the panel (allowing for some fuzziness in categorization) comprised six psychologists (from various subdisciplines), four mathematics educators (including the president of the National Council of Teachers of Mathematics), four mathematicians, one middle school mathematics teacher, one special educator, one policy researcher, and one reading researcher. A number of consultants provided support.

In the Executive Summary of the Final Report (United States Department of Education, 2008, p. xi), by way of background, it is stated that:

> During most of the 20th century, the United States possessed peerless mathematical prowess … But without substantial and sustained changes to its educational system, the United States will relinquish its leadership in the 21st century.

O. Skovsmose and B. Greer (eds.), Opening the Cage: Critique and Politics of Mathematics Education, 107–124.

> Much of the commentary on mathematics and science in the United States focuses on national economic competitiveness and the economic well-being of citizens and enterprises. There is reason enough for concern about these matters, but it is yet more fundamental to recognize that *the safety of the nation and the quality of life* – not just the prosperity of the nation – are at issue. (Emphasis added)

The Presidential Executive Order specified that the report should contain recommendations "based on the best available scientific evidence" on many aspects, of which the first was "the critical skills and skill progressions for students to acquire competence in algebra and readiness for higher levels of mathematics".

The Panel held meetings at twelve locations around the USA, at which interested individuals made presentations. Submitted written presentations, together with transcripts of the meetings, are available on the website, as are statements submitted by e-mail.

Subcommittees addressed and wrote reports on *Standards of Evidence*, *Instructional Materials*, and a *National Survey of Algebra Teachers*, and the task groups dealt with *Conceptual Knowledge and Skills*, *Learning Processes*, *Teachers and Teacher Education*, *Instructional Practices*, and *Assessment*. Henceforth "the reports" refers to these documents collectively, and "Final Report" to the summary document.

POLITICAL CONTEXT

Recent educational politics in the USA

The dominant power bloc within educational politics in the USA (and beyond) has been characterized by Apple (2000, p. 244) as "conservative modernization":

> This power bloc combines multiple fractions of capital who are committed to neoliberal marketized solutions to educational problems, neoconservative intellectuals who want a "return" to higher standards and a "common culture", authoritarian populist religious fundamentalists who are deeply worried about secularity and the preservation of their own traditions, and particular fractions of the professionally oriented new middle-class who are committed to the ideology and techniques of accountability, measurement, and "management". Although there are clear tensions and conflicts within this alliance, in general its overall aims are in providing the educational conditions believed necessary both for increasing international competitiveness, profit, and discipline and for returning us to a romanticized past of the "ideal" home, family, and school.

The election of Barack Obama in 2008, and his subsequent choice of education team and policies has not significantly changed the governmental approach to education (Spring, 2010), reflecting the relative lack of disagreement on educational issues between the Republican and Democratic parties most clearly manifest in bipartisan support for the *No Child Left Behind* legislation of 2002.[1] Despite devastating criticism (e.g. Hess & Petrelli, 2006; Nichols & Berliner, 2007), which includes a notable conversion by a former believer (Ravitch,

2010), the basic fantasy of market forces driving change through the levers of test scores, accountability, choice, and financial sticks and carrots remains intact. Indeed, in several respects it has intensified under Obama's Secretary for Education, Arne Duncan, whose experience was in business, educational politics, and their intersection (Brown, Gutstein, & Lipman, 2009).

In this section, I focus on two central themes within this complex story. First is the hegemonic struggle of which the "Math Wars" (e.g., Schoenfeld, 2004) represent a particularly important front. Second is the degree to which education has become even more linked with national economic welfare and, indeed, military power (Giroux, 2007; Gutstein, 2008, 2009).

Hegemonic struggles

Much of educational politics in the USA can be framed in terms of hegemonic struggles to gain the dominant voice in media, education, and other mass institutions. The expression "culture wars" was introduced in that context by Hunter (1991) in his book *Culture Wars: The Struggle to Define America*. Hunter argued that USA politics and society in general polarized into two groups with ideological world-views holding highly correlated opinions in relation to a gamut of defining issues, these groups being "conservatives" and "liberals" (Lakoff, 1996).

Although, in accordance with established usage, the phrases "culture wars" and "math wars" have been used here, using war metaphorically in this context is problematic on several grounds. It diminishes the suffering of people in actual wars. It frames our thinking in harmful ways, such as justifying symbolic violence and invoking nationalism and xenophobia. It encourages and enables the media to promote the debate theatrically as a gladiatorial contest and by reference primarily to polarized positions. Multiple examples of the extended military metaphor can be found, an example being the book by Rochester (2002), beginning with the title *Class warfare: Besieged schools, bewildered parents, betrayed kids and the attack on excellence*, and extending from the first chapter, *How I became a soldier in the great American education war*, to the (marginally more constructive) final chapter *A modest proposal for a cease-fire.*

Education, for obvious enough reasons, is a major arena in which the hegemonic struggle has been contested, in particular in relation to history (Nash, Crabtree, & Dunn, 1997), reading (Pearson, 2004), and mathematics (Schoenfeld, 2004). There are clear parallels across these disciplines (e.g. Schoenfeld & Pearson, 2009). Commonalities can be as specific as terminology (for example, "whole math" as a derogatory term by analogy with "whole reading"[2], and protagonists, such as Lynne Cheney, wife of Bush's vice-president (Cheney, 1997).

NMAP was explicitly framed as following the example of the National Reading Panel (National Institute of Child Health and Human Development, 2000). In particular, that report set a precedent of narrowly defining what counted as scientific research and could be taken into consideration (Pearson, 2004).

Within mathematics education, there is a long history of debate between, broadly speaking, those who emphasize skills and those who emphasize understanding (with parallel debates in other disciplines). The debate became sharpened considerably following the publication of *Principles and Standards for School Mathematics* (National Council of Teachers of Mathematics, 1989) and the subsequent development of aligned curricula. First in California, and then more generally, there was a sustained and ongoing backlash (see Schoenfeld (2004), and a set of painstaking analyses by Jerry Becker, Bill Jacob, and others (e.g., Becker & Jacob, 2000) as events unfolded). One manifestation of this backlash was the number of groups that were formed in various parts of the USA, such as *Mathematically Correct* in California (http://www.mathematicallycorrect.com), and *Honest Open Logical Decisions on Mathematics Education Reform* (HOLD) (http://www.nychold.com) in New York City. Perhaps the most representative expression of this movement is an impassioned attack on constructivism and its effects on mathematics education, with the title *Ten Myths About Math Education And Why You Shouldn't Believe Them* (http://nychold.com/myths-050504.html) by a number of authors including two members of NMAP, Sandra Stotsky and Vern Williams.[3]

The military-industrial-academic complex

The rhetoric of mathematics and science education serving the cause of continuing military and economic dominance contrasts starkly with the warnings presciently delivered by President Eisenhower on leaving office in 1961 when he warned against the rise of the military-industrial complex (the speech is available at http://avalon.law.yale.edu/20th_century/eisenhower001.asp). As pointed out by Giroux (2007, pp. 14–15), Eisenhower more precisely warned about the rise of the "military-industrial-academic complex", stating that (and remember this was more than half a century ago):

> … the free university, historically the fountainhead of free ideas and scientific discovery, has experienced a revolution in the conduct of research. Partly because of the huge costs involved, a government contract becomes virtually a substitute for intellectual curiosity… *The prospect of domination of the nation's scholars by Federal employment, project allocations, and the power of money is ever present and is gravely to be regarded.* (Emphasis added).

A recurrent theme in recent discussions of education in the USA is that of crisis, often phrased in terms of external enemies (Gutstein, 2008). Typical of the tone is the report *A Nation at Risk* (United States Department of Education, 1983), which opens thus:

> Our nation is at risk. Our once unchallenged pre-eminence in commerce, industry, science, and technological innovation is being overtaken by competitors throughout the world.

and further states that:

> If an unfriendly foreign power had attempted to impose on America the mediocre educational performance that exists today, we might well have viewed it as *an act of war*. (Emphasis added)

The elision of supposedly deteriorating education with an act of war by a foreign power is hardly accidental, and is of a piece with the metaphorical framing of "culture wars" and related rhetoric.

Further reports with doom-laden titles followed (Gutstein, 2008, 2009) forming the backdrop to the *American Competitiveness Initiative* of 2006 (http://www2.ed.gov/ about/inits/ed/competitiveness/index.html), of which NMAP forms a part. Gutstein argues that the common thread is the fear that the United States will lose its dominant position in the world (see above quotation). Mathematics and science education achieve importance as means necessary to avoid this catastrophe.

A further, not unrelated, issue that I will simply touch on is the call for high levels of mathematics and science education supposedly driven by the needs of a highly trained work force, claims that are patently exaggerated (Gutstein, 2009) Hacker (2009) pointed out that jobs requiring college-trained scientists and engineers will be outnumbered by those that require high school education and on-the-job training in the use of instruments. Actually, as pointed out by Noddings (1994) in response to an earlier report making exaggerated claims about the need for people to have mathematical qualifications "one should not need a statistical study to be convinced that most people use virtually no algebra or geometry in their personal or work lives" (I take her to mean algebra or geometry as typically taught in school).

I found it interesting to take a more detailed look at the twelve consultants appointed to help the project. One is an expert with a great deal of clearly relevant experience in social intervention, program evaluation research, and applied research methodology.

Otherwise, five consultants were from Widmeyer Communications, a PR company (http://www.widmeyer.com). At the risk of appearing naïve, why does such an exercise *need* a PR company?

Three others were from Abt Associates. According to its website (http://www.abtassociates.com/page.cfm?PageID=99), it is "a mission-driven, global leader in research and program implementation in the fields of health, social and environmental policy, and international development". This company was in the news a few years ago when it was found to have mishandled a $43.8 million contract to improve Iraq's health-care system after the 2003 invasion (Dilanian, 2005) (http://www.corpwatch.org/article.php?id=12379). The other reference I found to Abt Associates is in Herman (1995), who relates how a computer simulation called *Politica* was designed by Abt to be used for monitoring internal wars in Latin America. Herman reports that, according to one of its inventors, the game's results gave the green light to policy-makers to remove Chile's democratic leftist government in 1973.

The remaining three consultants were from the Science and Technology Policy Institute that "assists the Executive Branch of the US government as it formulates federal [Science and Technology] policy by providing objective, high-quality analytic support to inform policymakers". It is operated by the Institute for Defense Analyses (http://www.ida.org/stpi.php), which is "a non-profit corporation that operates three federally funded research and development centers to provide objective analyses of national security issues, particularly those requiring scientific and technical expertise, and conduct related research on other national challenges." This organization, together with Abt, carried out reviews of literature and other relevant materials for the Task Groups (Final Report, pp. 82–83). It is easy to speculate why this task was not given to a consortium of academic experts from the field of mathematics education.

Thus, the outlines of the military-industrial-academic complex can be clearly discerned in both the framing of discourse around maintaining dominance in the world and in the choice of organizations recruited to help in the project.

KEY POINTS RAISED BY NMAP

What is mathematics education for?

> Consider the way in which the National Council of Teachers of Mathematics (NCTM) begins its draft of Standards 2000. No Socrates-like character asks "And shall we teach mathematics?" Even if the answer is a preordained "Of course, Socrates," asking the question raises a host of others: "To whom shall we teach mathematics? For what ends? Mathematics of what sort? In what relation to students' expressed needs? In what relation to our primary aims? And what are these aims?" (Noddings, 2003, p. 87)

In the NMAP Final Report, the main aims are clearly stated. Mathematics and science education are seen as key to economic competitiveness, with implications for national security (in the final meeting of the panel, one of its members commended his colleagues for their patriotism). Nationalistic motivations of this kind are by no means confined to the USA, but are most strongly expressed in this country. Yet, there is an alternative worldview in which mathematics and science are seen as having a central role in solving the problems of humankind in general. D'Ambrosio has written passionately about the ethical responsibilities of mathematicians and mathematics educators and declared that:

> It is widely recognized that mathematics is the most universal mode of thought and that survival with dignity is the most universal problem facing humanity. To investigate if or the extent of any correlation between these two world-views is a task for mathematicians and mathematics educators. (D'Ambrosio, 2010, p. 52)

The disappearance of an academic field

Mathematics education has emerged as a scholarly field in its own right (De Corte, Greer, & Verschaffel, 1996). It now has a well-developed, increasingly global, identity, expressed through the usual networks of scholars, organizations, and mechanisms such as conferences and publications. In the United States, K-12 teachers of mathematics, and mathematics educators and researchers are represented by the National Council of Teachers of Mathematics, with close to 100,000 members. Yet, in NMAP, mathematics education as a field was, to a bizarre extent, negated.

First, consider the make-up of the panel. Confrey, Maloney, and Nguyen (2008, p. 631) characterize it as "unusual, considering the charge". Having listed the backgrounds of the members), Confrey et al., continue (p. 631):

> This means that, of 19 members, only 5 … regularly had sustained interactions with mathematics instruction at the K-12 level; fewer than half the Panel members had documented academic preparation in mathematics. In the area of research, more of the experience on the Panel was based in psychology (cognitive and developmental) than in any other area. Mathematics education as a field includes experts in sociology, anthropology, and critical theory, but the Panel was deficient in expertise in those areas. Because of the documented and sizable achievement gaps in education related to race, socioeconomic status, and second-language learners, representation of those fields of expertise would ordinarily be expected for a panel of this kind.

More extreme omissions characterize the literature cited by the panel. I searched the reports for references to the work of the leading scholars in the field; in very many cases, there are none. This disappearing trick was, to a large extent, made possible by imposing methodological criteria for what counts as research (see below). As one consequence, Confrey et al., (2008, p. 633) report that only slightly more than 10% of the 466 journal articles referenced in the report of the Task Group on Learning Processes (Geary, Berch, Boykin, Embretson, Reyna, Siegler, & Graban, 2008) were from mathematics education journals, whereas articles published in journals on psychology, child development, and educational psychology constituted about 70% of the references.

The most blatant manifestation of exclusion of literature concerns the second *Handbook of Research on Learning and Teaching Mathematics* (Lester, 2009), an NCTM publication. While the panel was ongoing the handbook was being assembled and its editor, Frank Lester, offered to make the chapters available. His offer was ignored, save for a formal acknowledgment of receipt (Lester, personal communication). It is hard to interpret the negation of the significance of this handbook, as an important summary of research in the field, as other than an ideologically motivated piece of foul play.

What is accepted as research

The primary mechanism for the exclusion of the field just illustrated was an extremely narrow and positivistic definition of what can be called research in this field – as Boaler (2008, p. 588) put it "the methodological restrictions imposed by the Panel rendered the field of mathematics education virtually invisible." Cobb and Jackson (2008, p. 573) distinguish between experimental research as a methodology that can make valuable contributions to the improvement of mathematics education, and what they call "experimentalism as an ideology that holds that only studies conducted using this methodology constitute a trustworthy basis for making recommendations to policy makers and practitioners". To paraphrase the early Wittgenstein, whereof one cannot do randomised experiments, thereof one must be silent.

There were many harbingers of this position, to the extent that it looks very much like a planned campaign. The phrase "scientifically based research" appears more than 100 times in the *No Child Left Behind Act* of 2001 (Hess & Petrilli, 2006, p. 94). In February, 2002, the Assistant Secretary for Elementary and Secondary Education hosted a seminar in which leading experts in the fields of education and science discussed the meaning of scientifically based research and its status across various disciplines[4]. After a presentation by Valerie Reyna (subsequently a member of NMAP), entitled *What is scientifically based evidence? What is its logic?*, another future member, Russell Gersten, spoke on *Math education and achievement*, beginning by saying "This is actually an easy topic to be brief on because there isn't a lot of scientific research in math." He does cite, with approval, the work of Siegler and Geary, who along with Boykin, Embretson, and Reyna (all psychologists) were in the *Task Group on Learning Processes*.

Also in 2002, the November issue of *Educational Researcher* dealt with the theme *Scientific Culture and Educational Research* (Jacob & White, 2002), following hard on the heels of the requirement in the *No Child Left Behind Act* that federal grantees use their funds on evidence-based strategies. Criteria for what counts as scientific research, and therefore warranted consideration by the Panel were discussed early and formalized by the Subcommittee on Standards of Evidence, chaired by Reyna, which did not include any mathematics educators.

The issue of characterizing what is to be taken as legitimate research is the most recurrent criticism among the papers in the issue of *Educational Researcher* in December, 2008, devoted to reactions to NMAP. Mathematics education research now uses a very wide range of methodologies (Lesh and Kelly (2000) – not cited in NMAP – is a comprehensive and authoritative overview).

As Thompson (2008) and Confrey et al. (2008) pointed out, the panel did not adhere to its own, impossible, stated standards. Although an appeal to scientific objectivity was used to filter out many "mere" opinions, they did resort to their own opinions where necessary. More fundamentally, little consideration was given to the fact that many of the most important questions about (mathematics) education are not capable of being answered by research. On what grounds, for example, was it decided that algebra should enjoy such a privileged position?

Ideological alliances within and beyond the panel

It is not hard to discern commonalities in views about the nature of mathematics, education, and mathematics education in members of disciplines other than mathematics education that are represented in the panel – mathematics, cognitive psychology, neuroscience, statistical methodology, psychometrics, special education. It is also possible to discern more or less implicit alignments with external economic and political interests, including the testing industry.

On the other hand, consider disciplines that are increasingly part of the discourse in mathematics education as a field – history of mathematics (including social history), sociology, cultural psychology, anthropology, linguistics and semiotics, (modern) philosophy of mathematics, that are typically found in contemporary writings on mathematics education. What is common about these disciplines is that they characterize mathematics and mathematics education as human activities and the problems of improving mathematics education as human rather than technical problems (Kilpatrick, 1981).

Groups such as Mathematically Correct and HOLD were well represented both on the panel and in public testimonies. To give the flavour of the rhetoric deployed, I quote at some length from the testimony by Martin Rochester (whose 2002 book has already been alluded to) at one of the public hearings held by the panel:

> I should note that I am a political scientist, not a mathematician, but nonetheless I am someone who has spent over 30 years as a professional educator, and also as a parent, observing one failure after another in K-12 education, as every so-called "progressive" fad presented as a magic bullet has only added to our shooting ourselves in the feet…
>
> Fuzzy math (or integrated math, or whatever you want to call … reform math curricula now dominant in K-12) has been driven by the same constructivist paradigm and same dumbing-down, populist impulses that gave us the now discredited "whole-language" pedagogy in English. That is, in place of the old maxim "no pain, no gain," we now have the new maxim in K-12, "if it ain't fun, it can't be done." Under the guise of "critical thinking" and "problem-solving," which are ubiquitous buzzwords in every discipline in today's schools, fuzzy math is trying to make math more "interesting," i.e., enjoyable and entertaining and accessible to the masses, to the bottom, to the lowest common denominator. The new math de-emphasises and devalues direct instruction, drill and practice, basic computation skills, and getting it right – getting precise, correct answers. Forget rigor – the key concern here is to alleviate boredom and drudgery for mathphobes and those who suffer from math anxiety…

This fulmination contains much standard rhetoric, with many of the usual talking points – nostalgia for some vague Golden Age in the past, hyperbole (he goes on to say that universities are having to teach students what 2 plus 2 is!), the use of quotation marks around "critical thinking", "problem-solving" and "interesting" (I hope he is in favour of critical thinking, problem-solving, and learning that is interesting). As I interpret his argument, he starts from the premises that: (a) standards of performance in US education have declined greatly (let us say from the end of the 60s), and (b) that the predominant style of teaching is permissive. While both premises are extremely dubious, let us accept them for the sake of argument. He then implies that (b) is the cause of (a). Such a shallow argument is surprising from a political scientist who must be familiar with the concept of multiple interacting causes. During the forty or so years under consideration, very many other changes have fundamentally altered society and education within that.

Why would many diverse interest groups line up on one side of the argument about mathematics education? Various deep-level explanations have been offered. In political terms, there is Apple's analysis of "conservative modernization" quoted near the start of this chapter. Schoenfeld & Pearson (2004) refer to two very deeply contrasted views about the goals of education. A general, deep-level explanation has been offered by George Lakoff (1996) in terms of two basic metaphors – for Conservatives, the authoritarian father, for Liberals the nurturing family. Soros (2006), a student of Karl Popper who has developed Popper's concept of the Open Society, invokes as a unifying factor the ideology of certainty as opposed to modern philosophical conceptions of fallibility (see Skosmose & Greer, Chapter 17, this volume) that are prominent in modern views of the philosophy of mathematics.

HOW MATHEMATICS AND MATHEMATICS EDUCATION ARE PORTRAYED

Mathematics as "the A subject"

In this chapter, I am trying to make clear the political nature of the NMAP exercise. In critiquing its representation of mathematics and mathematics education, I have done so from an "enlightened mainstream" perspective, not from the critical perspective of this book. There is essentially nothing in NMAP that addresses mathematics from a historical, sociocultural, or political point of view. Mathematics is portrayed as "the A subject" – acritical, ahistorical, acultural, and apolitical. The portrayal of mathematics and mathematics education as apolitical is itself a political act.

In particular, Martin (2008), commenting on the omission of discussion in NMAP about race in mathematics education, stated that it is very apparent but not surprising since "the omission of race, on one hand, and its conceptual underdevelopment, on the other, are epidemic to mainstream mathematics education research and policy discussions" (p. 389).

The tenor of the report throughout reflects an unexamined belief in what Valero (2004, p. 13) called "the unquestioned intrinsic goodness of both mathematics and mathematics education [that represents] the core of its 'political' value". She continued:

> If students and citizens come to learn a considerable amount of mathematics properly, they will become per se better people and better citizens; that is, mathematics and its education empower or have the capacity of giving power to people… The problem with this kind of assumption is that there is no necessity for a further examination, neither of mathematics as a knowledge and of mathematics education as practices, nor of power.

and later (p. 15) asked:

> Is it possible to assume that mathematics is a knowledge associated exclusively with progress and the well being of humanity? Or do we need to consider the involvement of that knowledge in the creations of both wonders and horrors in our current technological society?

Algebra to the rescue

In 1957, the director of one of the major mathematic curriculum projects in the United Kingdom was quoted in a newspaper as saying: "Up went Sputnik and down came all the pure mathematicians saying we must do sets and be saved." Perhaps the corresponding message from NMAP is "Up went the economic performance of China and India and down came all the pure mathematicians saying we must do algebra and be saved". President Obama, in calling for more investment in education and science recently framed the situation as "another Sputnik moment" (Calmes, 2010).

Consider the question posed by Kilpatrick (2009): "Did George W. Bush … wake up one morning and say, 'Competence in algebra ought to be the first order of business'?" Let me make a conjecture. Bush[5] was influenced by an alliance of interest groups, prominent among whom are certain mathematicians, for whom the route through algebra to calculus is the royal road into mathematics. In any case, the domination of the report by algebra is such that it would more fittingly have been called the National Algebra Advisory Panel. Admittedly, the assignment given to the panel specifically mentioned algebra as a basis for readiness for higher levels of mathematics.

Other areas of mathematics are accorded importance only insofar as they serve algebra. In geometry, for example, teaching about similar triangles is singled out because they enlighten the graphical representation of linear functions, specifically the invariance of the slope of a linear graph. The subject of combinatorics, apparently, owes its importance to its relationship to the binomial theorem, not its role in developing probability theory.

The treatment of algebra is idiosyncratic, incoherent, and contains bizarre statements such as "Let X be a symbol". As Thompson (2008, p. 585) pointed out:

> … the vision of algebra reflected in the Panel's content recommendations is a skills-based foundation for advanced symbolic manipulation and abstract algebra (especially the algebra of polynomial forms). It completely ignores algebra as a preparation for calculus, which would entail strong emphases on variable as varying magnitude … covariation and function … rate of change and accumulation … and modeling.

By contrast, Kaput (1999, p. 134) proposed the following broad outlines for a more productive approach to the teaching of school algebra:

- begin early (in part, by building on students' informal knowledge);
- integrate the learning of algebra with the learning of other subject matter (by extending and applying mathematical knowledge);

- include the several different forms of algebraic thinking (by applying mathematical knowledge);
- build on students' naturally occurring linguistic and cognitive powers (encouraging them at the same time to reflect on what they learn and to articulate what they know);
- encourage active learning (and the construction of relationships) that puts a premium on sense making and understanding.

The omission of reference to Kaput's work on algebra and other aspects of mathematics education in NMAP is one of the most egregious examples of negation of the field.

Missing mathematics

The extreme concentration on algebra is complemented by the absence, or minimal treatment, of many aspects of mathematics, including probability and data handling, applications and mathematical modelling, even geometry, and problem solving (in the sense of Polya, another great mathematics educator purged). There are many prominent mathematicians who disagree with this imbalance, such as Steen (2003) who declared that: "One of the many ills afflicting mathematics education is its excessively narrow focus on algebraic symbol manipulation to the detriment of more widely useful aspects of the mathematical sciences" (p. 53). The role of computers in mathematics and their uses in mathematics education are unacknowledged (add Papert to the list of the disappeared).

It surprises me that the mathematicians on the panel did not object to the impoverished representation of their subject, and that mathematicians not on the panel did not raise more noise about it. The word "proof" does not appear in the Final Report, and the treatment of problem solving is trivial. Moreover, the portrayal of mathematics affords no hint that it could be intellectually, even aesthetically, exciting.

There are several respects in which the tone and content of the report should also disturb scientists – in the disproportionate emphasis on the mathematics of numbers as opposed to the mathematics of quantities, and in the inadequate treatment of probability, data handling, and mathematical modelling as core parts of mathematics. This failure to serve science is ironic, given the emphasis throughout on "the best science". Even more ironic is that, given the emphasis throughout on statistical methodology, the panel apparently did not consider it important to seek to lay down within school mathematics a foundation in probabilistic and statistical thinking (Shaughnessy, 2010).

As I study the NMAP documents, one overwhelming reaction I have is that it projects mathematics as a subject lacking in intellectual excitement. In the rhetoric of the various constituencies that make up the conservative alliance in contemporary educational politics in the USA, there are many statements to the effect that mathematics should be hard and uninteresting (see, for example, the quotation from Rochester cited earlier) and castigating progressive mathematics educators for being concerned with making mathematics more accessible, more engaging, more relevant to students' lived experience.

LESSONS FOR POLITICAL ENGAGEMENT

Aftermath

While several colleagues have expressed to me sanguine opinions that the report will gather dust and be ignored, I am not so optimistic. The struggle over mathematics education continues both within the United States and beyond – for example, in the Netherlands (Van den Heuvel-Panhuizen, 2010). Kelly (2008, pp. 561–562) mentions a number of follow-up activities, including the distribution of 160,000 pamphlets to parents and schools[6]. This 2-page brochure is entitled (with the obligatory pun) *Counting on Excellence*. Parents are encouraged to promote a positive attitude towards mathematics, and to do things to help their children learn mathematics. Even more that the report itself, this distillation of its findings privileges algebra. And this is what it says about technology: "Using technology can make a difference. Technology-based drill and practice activities can improve student performance in specific areas of mathematics."

Against the general political background of the educational policies of the Obama administration that extend those of Bush, the conclusions and recommendations of NMAP will be used as ammunition in the ongoing ideological struggles, and in national initiatives[7]. Confrey et al., (2008, p. 636) provide a clear example. Reacting to proposed K-12 mathematics curriculum standards for Missouri, a group of 39 university faculty, mostly mathematicians, wrote a critical letter in which NMAP is cited several times as authoritative, including the statement that:

> The proposed Missouri K-12 document is based narrowly and almost exclusively on the (National Council of Teachers of Mathematics) standards that were the motivation of much of the mathematics curriculum work of the LAST decade rather that the work of the (National Mathematics Advisory Panel) for the NEXT decade. (Catchings, 2008)

(See Confrey et al., (2008) for a refutation of the letter, and further details).

Resisting negation of the field

As illustrated above, the NMAP exercise represents a negation of mathematics education as a scholarly field in many respects, including the narrow scope of the exercise (with exclusion of a historical perspective, cultural and social issues), the composition of the panel, and the selection of literature by narrow criteria, enabling omission of the work of very many of the most important scholars and researchers in mathematics education.

Responses from the field came first in the form of representations to the panel in open meetings (see e.g. O'Brien and Smith, 2008 for a number of examples) and in e-mailed submissions. In July, 2008, six critical papers (Greer, 2008a, 2008b; Gutstein, 2008; Martin, 2008; O'Brien & Smith, 2008; Roth, 2008) were published in *The Montana Mathematics Enthusiast*. A rejoinder by Stotsky (2009) was published (see Greer, in press), not in an academic journal, but in *City*

Journal, a publication of the Manhattan Institute, a conservative think tank (Spring, 2010, pp. 136–141). In December, 2008, many important scholars from the field contributed to a special issue of *Educational Researcher* on NMAP. Among the most important criticisms raised in that issue were the political nature of the enterprise, the limitations imposed by criteria for what counts as valid research, the narrow focus on algebra, and the inadequate conceptualisation of assessment.

Speaking for myself, and recognizing that many colleagues work tirelessly, both academically and politically, for a more progressive mathematics education, I hoped for a stronger reaction. I am not aware of any significant public dissent on the part of the mathematics educators on the panel, yet various courses were open to them – in their comments during meetings or closing comments in the final meeting, or even by issuing a minority report (as happened in the case of its predecessor, the Reading Panel). Confrey et al., (2008, p. 635) are crystal clear:

> … the report as a whole … lacks intellectual integrity…
>
> … it is the responsibility of each panel member to refuse to sign off and permit such a report to proceed to publication if it contains breaches in the conditions for success.

I agree. Quite simply, I believe that the mathematics educators on the panel let their colleagues in the field down; it would not be appropriate to comment on their motivations. A similar comment applies to the mathematicians on the panel.

The engagement of NCTM with the exercise, and their subsequent reaction to its products, was timid (of course, I am not privy to the context of *realpolitik* in which NCTM operates). The organisation was represented on the panel by its then president, "Skip" Fennell. I was surprised that there was not a strong reaction from NCTM to the rejection of their research handbook (see above). Interestingly, in a follow-up radio discussion in 2009[8], Fennell reacted to a question about the research base of the panel by saying that "Were the panelists given more time to work and were the 'guidelines' different, a more complete range of research – qualitative through all versions of quasi-experimental and experimental design – could have (and I would hope would have) been explored and examined". Good comment, bad timing.

FINAL COMMENTS

Why bother? A number of colleagues whom I respect have told me, with varying degrees of forcefulness, that NMAP is history, and I should "get over it". I respectfully disagree, and not just because I take it personally – by which I don't mean that I have a problem with my work not being cited or anything of that nature, but that the field in which I have worked throughout my career, and some of the best people within it, many of whom I have been privileged to know, have been very badly served. The establishment and operation of NMAP need to be set in the context of a continuing ideological struggle. We are supposed to learn from history.

Given the commendably very complete and publicly accessible documentation, illustrated throughout this chapter, it affords an opportunity for a sociopolitical analysis of a complex political process involving many actors, an opportunity that I hope will be taken up.

In Greer (2008b, p. 366) I commented that the members of the Panel, instead of ploughing through masses of empirical work selected in accordance with narrow criteria, could have spent their time better in reading rich reflections of some of the greats of our field, such as Hans Freudenthal. Had they read Freudenthal, how would they have reacted, I wonder, to his conclusion that the advice that comes from an exercise such as this will would not be used "for anything else than rationalizing a politically based decision" (1991, p. 150)?

NOTES

1 At the time of writing, NCLB is still alive, though seriously wounded, unable to survive its own hysterical demands, the most extreme of which is the expectation written into the law that *all* students would be proficient in reading and mathematics by 2014. In 2011, the Secretary of Education announced that an estimated 82% of schools were failing by the criteria incorporated in the legislation.

2 The derogatory terms used to refer to progressive mainstream approaches to mathematics education deserve serious sociological analysis. Apart from "whole mathematics", they include "rainforest mathematics" (referring to mathematics applied to environmental and other sociopolitical issues), and "fuzzy mathematics" (in which it is not assumed that there is always one exact numerical answer – see Cheney (1997)).

3 For a point-by-point rebuttal, see Matthews (2005).

4 http://www2.ed.gov/nclb/methods/whatworks/research/index.html

5 I strongly suspect that President Bush could not solve a quadratic equation. Possibly President Obama could, *but when would he ever need to?*

6 http://www2.ed.gov/about/bdscomm/list/mathpanel/parent_brochure.pdf

7 At the time of writing, there is a contested ongoing effort to create "Common Core State Standards" to align curricula in mathematics and English language arts across all 50 states of the USA.

8 www.edweek.org/ew/events/chats/2009/05/05/index.html

REFERENCES

Apple, M. W. (2000). Mathematics reform through conservative modernization? Standards, markets, and inequality in education. In J. Boaler (Ed.), *Multiple perspectives on mathematics teaching and learning* (pp. 243–259). Westport, CT: Ablex.

Becker, J., & Jacob, B. (2000). The politics of California school mathematics: The anti-reform of 1997–99. *Phi Delta Kappan, 81*, 527–539.

Boaler, J. (2008). When politics took the place of inquiry: A response to the National Mathematics Advisory Panel's review of instructional practices. *Educational Researcher, 37*(9), 588–594.

Brown, J., Gutstein, E., & Lipman, P. (2009). Arne Duncan and the Chicago success story: Myth or reality? *Rethinking Schools, 23*(3), 10–14.

Calmes, J. (2010). Obama calls for new "Sputnik Moment". Retrieved from thecaucus. blogs.nytimes.com/2010/12/06/obama-calls-for-new-sputnik-moment/

Catchings, E. (2008). Math professors seek change in state's K-12 math curriculum. Retrieved from missourimath.org/documents/ Math%20professors%20seek%20change.pdf

Cheney, L. (1997). President Clinton's mandate for fuzzy math. *Wall Street Journal* (June 11).

Cobb, P., & Jackson, K. (2008). The consequences of experimentalism in formulating recommendations for policy and practice in mathematics education. *Educational Researcher, 37*(9), 573–581.

Confrey, J., Maloney, A. P., & Nguyen, K. H. (2008). Breaching the conditions for success for a National Advisory Panel. *Educational Researcher, 37*(9), 631–637.

De Corte, E., Greer, B., & Verschaffel, L. (1996). Learning and teaching mathematics. In D. Berliner & R. Calfee (Eds.), *Handbook of educational psychology* (pp. 491–549). New York: Macmillan.

D'Ambrosio, U. (2010). Mathematics education and survival with dignity. In H. Alro, O. Ravn & P. Valero (Eds.), *Critical mathematics education: Past, present, and future* (pp. 51–63). Rotterdam: Sense Publishers.

Dilanian, K. (2005). Audit critical, but firm gets U. S. deals. *Philadelphia Inquirer* (June 11).

Freudenthal, H. (1991). *Revisiting mathematics education.* Dordrecht, The Netherlands: Kluwer.

Geary, D. C., Boykin, A. W., Embretson, S., Reyna, V., Siegler, R., Berch, D. B., et al., (2008). *Report of the Task Group on Learning Processes.* Retrieved from http://www.ed.gov/about/bdscomm/list/mathpanel/report/learning-processes.pdf

Gutstein, E. (2009). The political context of the National Mathematics Advisory Panel. *The Montana Mathematics Enthusiast, 5*(2&3), 415–422.

Giroux, H. A. (2007) *The university in chains: Confronting the military-industrial-academic complex.* Boulder, CO: Paradigm.

Greer, B. (2008a). Algebra for all? *The Montana Mathematics Enthusiast, 5*(2&3), 423–428.

Greer, B. (Ed.). (2008b). Reaction to the Final Report of the National Mathematics Advisory Panel. *The Montana Mathematics Enthusiast, 5*(2&3), 365–370.

Greer, B. (in press). Who needs mathematics educators for mathematics education, anyway? In C. Bergsten & B. Sriraman (Eds). *The "non-contributions" of Jablonka: Festschrift in celebration of the 51st birthday.* Charlotte, NC: Information Age Publishing.

Gutstein, E. (2008). The political context of the National Mathematics Advisory Panel. The Montana agendas. In B. Greer, S. Mukhopadhyay, S. Nelson-Barber & A. B. Powell (Eds.), *Culturally responsive mathematics education* (pp. 137–164). New York: Routledge.

Hacker, A. (2009, April). Can we make America smarter? *The New York Review of Books*, 37–40.

Herman, E. (1998). Project Camelot and the career of Cold War psychology. In C. Simpson (Ed.), *Universities and empire: Money and politics in the social sciences During the Cold War* (pp. 97–133). New York: The New Press.

Hess, F. M., & Petrelli, M. J. (2006). *No Child Left Behind primer.* New York: Peter Lang.

Hunter, J. D. (1992). *Culture wars: The Struggle to define America* New York: Basic Books.

Jacob, E., & White, C. S. (Eds.). (2002). Theme issue on scientific research in education. *Educational Researcher*, November.

Kaput, J. J. (1999). Teaching and learning a new algebra. In E. Fennema & T. A. Romberg (Eds.), *Mathematics classrooms that promote understanding* (pp. 133–155). Mahwah, NJ: Lawrence Erlbaum.

Kelly, A. E., (2008). Reflections on the National Mathematics Advisory Panel Final Report, *Educational Researcher, 37*(9), 561–564.

Kilpatrick, J. (1981). The reasonable ineffectiveness of research in mathematics education. *For the Learning of Mathematics, 2*(2), 22–29.

Kilpatrick, J. (2009). Algebra is symbolic. In S. L. Swars, D. W. Stinson & S. Lemons-Smith (Eds.), *Proceedings of the 31st annual meeting of the North American Chapter of the International Group for the Psychology of Mathematics Education* (pp. 11–21). Atlanta, GA: Georgia State University.

Lakoff, G. (1996). *Moral politics.* Chicago: University of Chicago Press.

Lesh, R. A. & Kelly, E. A. (Eds.). (2000). *Handbook of research design in mathematics and science education*. Mahwah, NJ: Lawrence Erlbaum.

Lester, F. K. (Ed.), *Second handbook of research on mathematics teaching and learning*. Greenwich, CT: information Age Publishing.

Martin, D. B. (2008). E(race)ing race from a national conversation on mathematics teaching and learning: The National Mathematics Advisory Panel as white institutional space. *Montana Mathematics Enthusiast, 5(2&3),* 387–398.

Mathews, J. (2005). 10 myths (maybe) about learning math. *Washington Post* (May 31). Retrieved from http://www.math.rochester.edu/people/faculty/rarm/ Mathews_myths.htm

Nash, G. B. Crabtree, C., & Dunn, R. E. (2000). *History on trial: Culture wars and teaching of the past*. New York: Vintage.

National Council of Teachers of Mathematics (1989). *Curriculum and evaluation standards for school mathematics*. Reston, VA: Author.

National Institute of Child Health and Human Development (2000). *Report of the National Reading Panel, Teaching children to read: An evidence-based assessment of the scientific research literature on reading and its implications for reading instruction*. Washington, DC: Author.

Nichols, S. L., & Berliner, D. C. (2007). *Collateral damage: How high-stakes testing corrupts America's schools*. Cambridge, MA: Harvard University Press.

Noddings, N. (1994). Does everybody count? Reflections on reforms in mathematics education. *Journal of Mathematical Behavior, 13*, 324–336.

Noddings, N. (2003). *Happiness and education* [sic]. New York: Cambridge University Press.

O'Brien, T., & Smith, M. (2008). Three strikes. *The Montana Mathematics Enthusiast*, *5*(2&3), 399–414.

Pearson, P. D. (2004). The reading wars: The politics of reading research and policy, 1988 through 2003. *Educational Policy, 18*, 216–252.

Ravitch, D. (2010). *The death and life of the great American school system: How testing and choice are undermining education*. New York: Basic Books.

Rochester, J. M. (2002). *Class warfare: Besieged schools, bewildered parents, betrayed kids and the attack on excellence*. San Francisco, CA: Encounter Books.

Roth, W-M. (2008). Mathematical cognition and the Final Report of the National Mathematics Advisory Panel: A critical, Cultural-historical Activity Theoretic analysis. *The Montana Mathematics Enthusiast*, *5*(2&3), 371–386.

Schoenfeld, A. H. (2004). Math wars. *Educational Policy, 18*, 253–286.

Schoenfeld, A. H., & Pearson, P. D. (2009). The reading and math wars. In G. Sykes, B. Schneider, & D. N. Plank (Eds.), *Handbook of education policy research* (pp. 560–580). New York: Routledge.

Shaughnessy, J. M. (2010). Statistics for all – the flip side of quantitative reasoning. *Summing Up* (August). Retrieved from http://www.nctm.org/about/ content.aspx?id=26327

Soros, G. (2006). *The age of fallibility*. New York: Public Affairs.

Spring, J. (2010). *Political agendas for education (4th edition)*. New York: Routledge.

Steen, L. A. (2003). Data, shapes, symbols: Achieving balance in school mathematics. In B. Madison & L. A. Steen (Eds.), *Quantitative literacy: Why numeracy matters for schools and colleges*. Princeton, NJ: National Council on Education and the Disciplines.

Stotsky, S. (2009). Who needs mathematicians for math, anyway? *City Journal* (13 November).

Thompson, P. W. (2008). On professional judgment and the National Mathematics Advisory Panel Report: Curricular content. *Educational Researcher, 37*(9), 582–587.

United States Department of Education (1983). *A nation at risk*. Washington, DC: Author.

United States Department of Education (2008). *Foundations for success. The final report of the National Mathematics Advisory Panel*. Washington, DC: Author.

Valero, P. (2004). Socio-political perspectives on mathematics education. In P. Valero & R. Zevenbergen (Eds.), *Researching the socio-political dimensions of mathematics education* (pp. 5–23). Dordrecht, The Netherlands: Kluwer.

Van den Heuvel-Panhuizen, M. (2010). Reform under attack – forty years of working on better mathematics education thrown on the scrapheap? No way! In L. Sparrow, B. Kissane & C. Hurst (Eds.), *Shaping the future of mathematics education: Proceedings of the 33rd annual conference of the Mathematics Education Research Group of Australasia* (pp. 1–25). Fremantle, Australia: Mathematics Education Research Group of Australasia.

Brian Greer
Portland State University
Oregon, USA

PART II

BORDERLAND POSITIONS

CHAPTER 5

MATHEMATICS TEACHING AND LEARNING OF IMMIGRANT STUDENTS: AN OVERVIEW OF THE RESEARCH FIELD ACROSS MULTIPLE SETTINGS

MARTA CIVIL

This chapter is based on a paper I presented at the 11th International Congress on Mathematical Education as part of *Survey Team 5: Mathematics education in multicultural and multilingual environments*, chaired by Alan Bishop. Our team ended up dividing the survey team theme into four topics and I was in charge of surveying research related to the mathematics teaching and learning of immigrant students. The other topics were: Multicultural teacher education, particularly with indigenous teachers (Maria do Carmo Domite); A review of research on multilingualism in mathematics education in Africa, 2000–2007 (Mamokgethi Setati); Cultural conflicts, ethnomathematical developments, and marginalized learners (Alan Bishop). To address my topic, I drew on proceedings of recent international conferences that I knew had presentations related to mathematics education and immigrant students. Much of this research takes place in Europe. I also received information from various researchers who responded to the Survey Team's call for contributions. Some of these researchers sent a summary of their most relevant work; others addressed the following two questions (suggested by Alan Bishop):

> A. How do you think the situation has changed/improved/deteriorated etc. in your research/development work in the last few years?
>
> B. What problems/challenges do you see in the next few years?

Finally, I looked at other publications with a particular emphasis on research in the mathematics education of Latino/a students in the United States, and more specifically students of Mexican origin, since that is the largest group of immigrant students in the USA.

In this chapter I focus on the main themes that emerged from going though all these sources. Although I have organized this chapter in themes, I want to point out that there is considerable overlap across the different themes.

The premise guiding this chapter is that the mathematics teaching and learning of immigrant students is of utmost importance. As Gates (2006) writes:

> In many parts of the world, teachers – mathematics teachers – are facing the challenges of teaching in multiethnic and multilingual classrooms containing

O. Skovsmose and B. Greer (eds.), Opening the Cage: Critique and Politics of Mathematics Education, 127–142.

> immigrant, indigenous, migrant, and refugee children, and if research is to be useful it has to address and help us understand such challenges. (p. 391)

A detailed description of the patterns of immigration is beyond the scope of this chapter. King, Black, Collyer, Fielding, and Skeldon (2010) provide an insightful account of the many faces of migration in the world, paying particular attention to contemporary migration. In their introduction they write:

> The rhetoric of migration exaggerates its scale. We are told, in newspapers, by politicians, and even by academics who should know better, that "massive" numbers of people are on the move in the world today, that there is a "global migration crisis", and that migration is leading to a "clash of civilizations." This hype does enormous disservice to migrants, who, already poor in many cases, are further vilified for their poverty and for trying to improve their lives through hard work in an unwelcoming environment, where, furthermore, they are often scapegoated for the ills of the society they have joined. (p. 13)

In this excerpt, the authors point to a view of immigration as a problem rather than a resource, a view that is quite prominent in schooling contexts, as the studies I reviewed will show. In 2010, the total number of international migrants was 214 million; this number represents 3% of the total population of the world. Each year from 2005 to 2010, 2.7 million people emigrated from poor to rich countries (King et al., 2010). Of interest for this chapter are the changes in patterns of migration in countries such as Greece, Italy, Portugal, and Spain. Between the 50s and 70s these countries experienced emigration to Northwest Europe for economic reasons. Yet, since the 80s, these same countries are experiencing immigration from countries in Africa, Latin America, South and East Asia, and, more recently East Europe (King et al., 2010). Several of the studies I found on my topic for this Survey Team were actually located in these Southern European countries, which have changed over the years from sending to receiving communities. Also of interest to the work I present in this chapter are these authors' observations on the concepts of assimilation and multiculturalism as two models of integration. While some countries in Europe may have had multiculturalism as a model, in recent years "a swing back to assimilation has occurred, with greater demands on immigrants to learn the host-country language and subscribe to core national values" (p. 92). The educational policy implications of views of immigration as a problem, and of the need for immigrants to assimilate, directly impact the mathematics teaching and learning of immigrant students, as this chapter shows.

DIFFERENT FORMS OF MATHEMATICS

Several studies address issues related to everyday mathematics, critical mathematics, community mathematics, school mathematics, and so on. For example, researchers in Greece have been looking at Gypsy / Romany students' use of mathematics in everyday contexts, in particular computation grounded in children's experiences with their involvement in their families' businesses (Chronaki, 2005, 2009; Stathopoulou & Kalabasis, 2007). (Moreira (2007) reports on a study along the same lines but with Portuguese Gypsy children). Some aspects of this research remind us of the work with street vendors in Brazil described in Nunes, Schliemann, and Carraher (1993). However, particularly relevant to our theme is the observation by the researchers in Greece that schools and teachers seem to show little interest in what knowledge minority students (in this case Gypsy) bring with them and, thus, in how to build on this knowledge for classroom teaching. It may be due to little interest on the part of the teachers, or it may be due to a lack of awareness on how to build on this knowledge. Stathopoulou and Kalabasis (2007) argue for the need for schools to recognize and build on the oral tradition and experiences with mental arithmetic that Romany children have from their participation in their community's everyday activities. Chronaki's (2005, 2009) research with Greek Roma children uses the concepts of learning identities and Roma funds of knowledge as resources for instruction to gain a better understanding of Gypsy children as mathematical learners.

The topic of bridging in-school mathematics and out-of-school mathematics has received quite a deal of attention in research in recent years (e.g., Abreu, Bishop, & Presmeg, 2002; Civil, 2007; Nasir, Hand, & Taylor, 2008). My work, and that of my colleagues, building on the concept of Funds of Knowledge, provide important insights into the development of modules and teaching approaches in mathematics that are based on students' and families' knowledge and experiences. The pedagogical transformation of the findings from the household visits into mathematical learning modules for the classroom is quite challenging (Civil, 2007; Civil & Andrade, 2002; González, Andrade, Civil, & Moll, 2001). Some of these challenges have to do with time, support, and, most relevant to this discussion, the notion of what counts as mathematics. As we consider different forms of mathematics and whose mathematics to bring to the foreground, issues of power and valorisation of knowledge become prominent. Abreu has written extensively on the concept of valorisation of knowledge (Abreu, 1995; Abreu & Cline, 2007). Another body of research that is relevant here is that dealing with mathematics education and indigenous students. It seems that there is much that we could learn for the mathematics teaching and learning of immigrant students from the ethnomathematics projects that develop curriculum and teaching approaches with indigenous communities. Meaney (2004) writes about issues of power and whose knowledge gets recognized, in the context of her work within a Maori community. Zevenbergen (2008) brings up several dilemmas in relation to the mathematics education of indigenous students that seem relevant to the education of immigrant students:

> How relevant are particular forms of knowing mathematics, what forms or aspects of the mathematics curriculum are needed or should be included in curriculum for the students, or should the expectation be one where they are exposed to the same curriculum as their urban counterparts. (p. 5)

TEACHER EDUCATION

Much of the research I reviewed for this topic addressed teachers' attitudes towards, and knowledge of, immigrant students. This body of research presents a rather grim picture and thus opens the door to several possibilities for further research. Reports on a European project that is looking at the teaching of mathematics in multicultural contexts in three countries, Italy, Portugal, and Spain, point out that teachers feel unprepared to work with immigrant students (Favilli & Tintori, 2002). César and Favilli (2005) report that teachers in this study underscore the issue of language as being a problem and do not seem to recognize the potential for richer learning grounded in different problem-solving approaches and experiences that immigrant students may bring with them. They also note that teachers seem to have different perceptions of immigrant students based on the students' countries of origin. Overall, my reading of these reports points to a deficit view by teachers of their immigrant students.

Abreu (2005) reports that most teachers in the studies she examined tended to "play down cultural differences", arguing for general notions of ability and equity, as in "treating everybody the same". Accordingly, she points out the need for teacher preparation programs to pay more attention to the cultural nature of learning. Gorgorió and Planas (2005) discuss the role of social representations in teachers' images and expectations towards different students. In particular, they write, "unfortunately, too often, 'students' individual possibilities' do not refer to a cognitive reality but to a social construction. Teachers construct each student's possibilities on the basis of certain social representations established by the macro-context" (p. 1180). The influence of the macro-context, and, more specifically, the public discourse around immigration as being a source of problems rather than a resource for learning, is a common theme in this review. Researchers are critical of this discourse (e.g., Alrø, Skovsmose, & Valero, 2005), as it is counterproductive to the education of immigrant children. Unfortunately, as Gorgorió and Planas (2005) point out, some teachers use this public perception as their orientation to assess immigrant students in their classrooms, rather than a direct knowledge and understanding of their individual students and their families.

My colleagues and I have addressed the need for teachers to gain a better understanding of their students and their families. Our work points to the potential of teachers engaging in action research in multicultural contexts in terms of teachers developing curriculum and approaches that build on students' knowledge and experiences (Civil, 2002, 2007; Civil & Andrade, 2002; Kahn & Civil, 2001). In Civil and Andrade (2002), we describe the impact of ethnographic household visits on teachers' views of their students. Teachers gain an understanding of their students' context; they also learn about the different activities, networks, and resources that their students and families draw on. This may be a promising direction for teachers to learn first-hand about their students' contexts, instead of relying on public discourse about them. In Civil and Bernier (2006), we highlight some of the challenges in having teachers and parents (Latino/a parents in this case) work together to present mathematics workshops to the school community. Issues of power, and of whose knowledge is valued, are very present. In Quintos, Bratton, and Civil (2005) we bring up the need for parents and teachers to value the different approaches to doing mathematics that are likely to be present in multicultural settings. We point out that often teachers do not recognize parents'/ home approaches as a valuable mathematical contribution.

There is a clear need for teachers to understand other ways of doing and representing mathematics (Abreu & Gorgorió, 2007; Civil & Planas, 2010; Moreira, 2007). As Gorgorió and Abreu (2009) write, in relation to a teacher's reaction to differences between representations of division in Ecuador and in Spain, "the important issue ... is not whether there are or are not differences in the way the division algorithms look, but the reaction of the teacher to this difference" (p. 72). Related to the need for teachers to know about others' ways of doing mathematics is a need for an expanded view of what mathematics is. Teachers tend to view mathematics knowledge as culture-free and universal (César & Favilli, 2005; Gorgorió & Abreu, 2009). This point relates directly to the previous section where I discuss different forms of mathematics. It seems that teacher preparation programs and professional development experiences should address this view of mathematics as being culture-free. Moreira (2007) brings up the need for teacher education programs to prepare teachers to research "local" forms of mathematics (e.g. everyday uses of mathematics). In referring to the mental computation strategies used by Portuguese Gypsy children, Moreira writes:

> If teachers are not aware of children's mental calculation processes and do not use them to know more about the role of local mathematics in mathematical knowledge in contemporary society, a good opportunity to educate the citizens of the world is lost. (p. 1594)

Furthermore, by seeing mathematics in a universal and culture-free way, teachers tend to believe that the only issue that immigrant students have when learning mathematics is learning the language of instruction (Gorgorió & Abreu, 2009). For a contrasting example of a teacher's effective use of the home language as a resource for the teaching and learning of mathematics see Khisty and Chval (2002).

ISSUES RELATED TO EDUCATIONAL POLICY

Researchers from different countries are critical of educational policies that push towards assimilation of immigrant students. These policies convey a deficit view on immigrants' language and culture, instead of promoting diversity as a resource for learning.

Anastasiadou (2008) writes:

> The de facto multiculturalism (…) which now describes the Greek society, … [which] continues to function with the logic of assimilation (…). In the field of education the adoption of the policy of assimilation means that it continues to have a monolingual and monocultural approach in order that every pupil is helped to acquire competence in the dominant language and the dominant culture. (p. 2)

The work of Alrø et al. (2005) is particularly relevant here as these authors take a socio-political approach to the discussion of the teaching and learning of mathematics with immigrant students. The influence of public discourse and, in particular, of the view of immigration as a problem rather than a resource is well captured in what these authors write:

> In Denmark, the sameness discourse has spread into a variety of discourses, which highlight that diversity causes problems – it is not seen as a resource for learning. And this idea brings about a well-defined strategy: Diversity has to be eliminated. (p. 1147)

Then, as researchers in other parts of the world have also noted, these authors point to the emphasis in educational policy on students' acquisition of the Danish language as the priority. Alrø, Skovsmose, and Valero (2007) argue for the need to look at the complexity of the situation, rather than at just one aspect (e.g., language). In their study with 8th graders at a school with a considerable number of immigrant students, the authors note that students did not seem to pay much attention to the topic of multicultural diversity: "diversity is not used as any resource for teaching and learning" (p. 1571). They report that:

> The students are well aware of cultural differences, but they seem to agree that to take specific notice is not important.... This indicates that they represent the "sameness" approach in the Danish public discourse about integration, which implies making "them" just like "us". (p. 1571)

The idea that mathematics education is political is particularly true when studying the mathematics education of immigrant students. An implication from the research that looks at the macro-context, and in particular the influence of public discourse and educational policy on the education of immigrant students, is the need for interdisciplinary research teams, where in addition to the expected expertise in mathematics education, there is expertise on the political and policy scene (social, educational, language, in particular with respect to immigrant students) in the context (country, region) of research.

LANGUAGE, MATHEMATICS, AND IMMIGRANT STUDENTS

In the eyes of educational policy-makers and many teachers, not knowing the language of instruction is seen as a major (and, in most cases, the main) obstacle to the teaching and learning of mathematics of immigrant students. Hence, the push is for these students to learn the language(s) of instruction as quickly as possible. As Alrø et al. (2005) point out, the emphasis on learning the language of the receiving country may occur at the expense of these students' learning of mathematics. Gorgorió and Planas (2001) have documented a similar situation in Catalonia. In 2008, there was ongoing debate in Catalonia around the proposed education policy for immigrant students that would keep them in separate school buildings apart from the local students (or from those who already know the language of instruction) with the aim that they learn the language of instruction. As Planas writes:

> There are still "reception classrooms" though we are in a transition time and the length of stay in these classrooms varies from half a year to one year. Right now these reception classrooms are in regular schools, but the talk is about creating separate spaces [different buildings], thus increasing the segregation of immigrant students. It is very controversial and it is not clear that they will be able to do it. (N. Planas, personal communication, May 22, 2008)

In my local context (the U.S.) there is a long history of changes in language policy for education, with some states now having banned or severely limited bilingual education, including the one I was in until recently, Arizona. In 2000, Proposition 203 was passed in Arizona. This proposition severely limits bilingual education. In 2006, further legislation in language policy was passed. The result was that, in 2008, schools in Arizona started implementing a state-mandated program towards the education of English Language Learners (ELLs) that essentially segregates ELLs for 4 hours a day to learn English (and in some cases, particularly in elementary schools, this segregation is for the whole day as students are placed in designated English Language Development classrooms). This 4-hour English block has serious consequences for the learning of other subjects and limits the opportunities for ELLs to be with English-speaking peers who could serve as supports for their learning of this language. In Civil (2011) I discuss the case of the implementation of a program very similar to the 4-hour block in a middle school by focusing on a mathematics classroom composed of ELLs, most of them recent immigrants from Mexico.

Language policy in many countries reflects the push for assimilation that often characterizes their policies towards immigration. Macedo, Dendrinos, and Gounari (2003) capture this when they write, "American monolingualism is part and parcel of an assimilationist ideology that decimated the American indigenous languages as well as the many languages brought to this shore by various waves of immigrants" (p. 23). These authors write about the inhumanity of language policies such as English-only that devalue other languages, and, in particular, the home languages of many immigrant students in the USA (and I will add, the same applies in other parts of the world that have repressive language policies). Although their writing is not about mathematics education, their insights into issues of power associated with the dominance of one language (English, in this case) have implications for the education of immigrant children, including their mathematics education. The parents of immigrant children are (or should be!) an important voice in the issue of language policy in schools. I turn my attention to them next.

What are immigrant parents' views on issues of language policy and mathematics education? An interesting theme emerging from our research with immigrant parents is that, for many of them, language also seems to be the main obstacle to their children's learning of mathematics (this parallels what teachers think, as we have illustrated earlier). This pattern is the case in our research with mostly Mexican parents (Civil, 2006, 2008) but is also the case with immigrant parents in Barcelona (Civil, Planas, & Quintos, 2005). The reason why I think this is an important topic to pursue is that as immigrant parents focus on the language as being the main obstacle, we wonder whether they are aware of the actual mathematics education that their children are receiving. In particular, I am referring to issues of placement. Are the students placed in the appropriate mathematics classroom (based on their knowledge and understanding) or are schools basing their placement on their level of proficiency in the language of instruction? In Civil (2006, 2008, 2011) I present the case of Emilia, a mother who seems satisfied with the fact that her son is mostly "learning" content that he already knew from Mexico because, as she says:

> That is, for them it's perfect what they are teaching them because in this way it's going to help them grasp it, to get to the level, because for them, with the lack in English that they have, and if to that we were to add, ... Right now, what he is learning, what I see is that it's things that he had already seen, but if he gets stuck, it's because of the language, but he doesn't get stuck because of lack of knowledge. (Emilia, interview #1, March 2006)

I wonder about the thinking behind these placement policies. Not only are parents not aware of the implications of this policy for their children's learning (or not) of mathematics, but also teachers often are not either, as Anhalt, Ondrus, and Horak (2007) show. In their article, they describe the reactions of a group of middle school mathematics teachers after they played the role of students in a mathematics class taught in Chinese. As the teachers reflected on this experience in relation to their own work with English Language Learners (ELLs), they realized that in some cases they had not paid any attention to the Chinese language and had focused on the mathematics that they already knew. Hence, they wondered about a placement policy that places ELLs in lower level mathematics with the idea that it will help them learn English. Teachers questioned whether, through this practice, students would learn either English or mathematics. Experiences such as this one can be quite powerful in addressing some of the beliefs that teachers seem to have about language and the teaching and learning of mathematics.

RESEARCH WITH IMMIGRANT PARENTS

In general terms, research on parents and mathematics education is rather limited. In the U.S., for example, there has been some research on parents' views of reform mathematics. For this review, however, our goal is to focus on research with immigrant parents and their views of mathematics education. Most of the research I found on this topic was done by Abreu and her colleagues in the United Kingdom (Abreu & Cline, 2005; O'Toole & Abreu, 2005) and by my colleagues and myself in the USA (Civil & Andrade, 2003; Civil & Bernier, 2006; Civil & Menéndez, 2011; Civil & Planas, 2010; Civil & Quintos, 2009; Quintos et al., 2005). In Civil et al. (2005) we look into immigrant parents' perceptions about the teaching and learning of mathematics in two different geographic contexts, Barcelona, Spain, and Tucson, USA.

Besides these studies in the UK, USA, and the study with immigrant parents in Barcelona and in Tucson, I found one study with immigrant parents in Germany. Hawighorst (2005) presented a study on parents' conceptions and attitudes towards mathematics. She focused on three groups of parents: German parents, resettler parents of German descent (from the former Soviet Union), and Turkish parents. Her interviews covered topics also addressed in the research by Abreu and her team, and by my team, namely, the importance and uses of mathematics in their everyday life, parents' experiences with their own learning of mathematics, as well as parents' views on their children's mathematics instruction.

There are three (related) themes that emerged and that cut across all immigrant parents in these studies. Overall, immigrants in the four geographic contexts shared a concern for a lack of emphasis on the "basics" (e.g., learning of the multiplication facts) in the receiving country, a perception that the level of mathematics teaching was higher in their country of origin, and a feeling that schools are less strict in their "new" country. There is something quite remarkable in reading some of the quotes from the parents in that while they are coming from very different countries (e.g., Pakistan, Mexico, Morocco, Turkey), what they say is almost identical, when sharing their perceptions about their children's mathematics education in their "new" country.

Abreu and colleagues, as well as my colleagues and I, have looked at these themes in some depth, thus providing an analysis related to issues of differences in approaches, issues of valorisation of knowledge, and potential conflict as children are caught between their parents' way and the school's way. As we write in Quintos et al., (2005), "the knowledge that working class and minoritized parents possess is not given the same value as that which middle-class parents possess" (p. 1184). We go on to address this topic of valorisation when it opposes home and school knowledge:

> Alternative approaches are often not treated equally…. In this context, the parents' or home method is not given the same value as the teacher's or textbook method. Historical relations of power at the schools can not only be reproduced but also exacerbated through mathematics education. (p. 1189)

The research with immigrant parents and their perceptions of the teaching and learning of mathematics underscores the need for schools to establish deeper and more meaningful communication with parents. Parents tend to bring with them different ways to do mathematics that are often not acknowledged by the schools, and conversely, parents do not always see the point in some of the school approaches to teaching mathematics. Although this may be the case with all parents (e.g., in the case of reform vs. traditional mathematics), the situation seems more complex when those involved are immigrant parents and their children. As our research shows, differences in schooling (different approaches to doing mathematics), and in language, influence parents' perceptions of, and reactions to, practices related to their children's mathematics education. In particular, there is evidence that when English is the only language of instruction (e.g., in the case of our local context in the U.S.), there are cognitive and affective implications.

LIMITATIONS AND IMPLICATIONS FOR FURTHER RESEARCH

Very likely, this survey of research has many gaps. I had to find a way to organize how I was going to conduct the survey, and in so doing I made choices that put some contributions in and left some out. To begin with, I focused primarily on articles and papers that explicitly talked about mathematics education and immigrant students. This means that many articles that present important research on issues of diversity, or papers addressing theoretical or methodological approaches that should be relevant to this topic of study, may have been left out. I did try to include some of those, but my emphasis was on empirical work done with immigrant students, immigrant parents, or teachers of immigrant students.

Most of my focus was on research with immigrant students in Europe, hence, many other parts of the world are not represented (indeed, many parts of Europe are not represented!). I relied on proceedings from recent conferences and on contributions from researchers on this topic. I know I missed many. For the work in the USA, I focused only on the research with Latino/a students, as this is the largest group of immigrants at the moment. Even there, I most likely missed research studies that should have been mentioned. There is certainly a need to address the mathematics education of less talked about immigrant groups in the USA, such as the Hmong students or the Somalis. Susan Staats (University of Minnesota) reflects on one of the trends in mathematics education that argues that using students' home language can help them in their learning of mathematics. She wonders what happens when students do not really know their home language, as is the case of the Somali students with whom she works. She writes:

> With the educational history of Somalis they do not know their math vocabulary. It is a point of sadness, in fact, for many young people that they feel they do not know any language well, they might know parts of Somali, Swahili, Arabic, Italian, or English but feel insecure speaking any of these. (S. Staats, personal communication, June 8, 2008)

She adds, "I think the Somali situation highlights the need for different kinds of interventions for different students." (S. Staats, personal communication, June 8, 2008). (See Staats (2009) for more on her work in mathematics education with Somali students). Her comment points to the need to not view all immigrant students as "the same“. In thinking of immigrant students, there is a risk of essentialising these students, as we often essentialise students within other groups, such as Latinos/as, when it is well known that there are vast differences among the various groups of Latinos/as. Or, as Swapna Mukhopadhyay (Portland State University) writes about Asian immigrants in the USA, we need to:

> Critically examine the myth that all Asians excel in school math. Is class (and social capital) an integral part of what makes the Asian kids do well? (Also,

> what role does complacent/highly adaptive behavior play in attributing their "success"?) Asian-Americans signify a very large group of people from highly diverse cultural and economic backgrounds. Is there a difference between the blue-collar/transient Asians versus the affluent middle and upper middle class professional? For example, how do the children of taxi drivers do versus that of the doctors? I have a feeling that the achievement/school performance is bounded by class, privilege and access. (S. Mukhopadhyay, personal communication, June 8, 2008).

My hope is that through this review, we will hear from other researchers who are working in mathematics education with immigrant students and that we will be able to elaborate further on this work, which I consider a work in progress. Having said that, however, there are several implications that this review points to and that I want to briefly mention here.

Abreu, César, Gorgorió, and Valero (2005) raise two important questions that should frame, I think, further research in this field. They ask "why research on teaching and learning in multiethnic classrooms is not a bigger priority" and "why issues of teaching in multicultural settings are not central in teacher training" (p. 1128).

Based on the research reviewed, there seems to be a clear need for action-research projects with teachers of immigrant students engaging as researchers of their own practice to counteract what appears to be a well-engrained deficit view of these students and their families. Through a deeper understanding of their students' communities and families (e.g., their funds of knowledge), maybe teachers can work towards using different forms of doing mathematics as resources for learning instead of the current trend that seems to view diversity as an obstacle to learning (there are of course exceptions to this view and I have addressed those in the review). Related to this idea of understanding immigrant students' communities, there is very little research looking at the sending communities. That is, what do we know about the teaching and learning of mathematics in the countries/communities that these immigrant students came from? (For a recent publication that addresses some aspects of this issue, see Kitchen & Civil, 2011).

There is also a need to analyse the learning conditions in schools with large numbers of immigrant students. What Nasir et al. (2008) write in reference to African American and Latino and poor students is likely to be the case with immigrant students in many countries:

> African American and Latino students, and poor students, consistently have less access to a wide range of resources for learning mathematics, including qualified teachers, advanced courses, safe and functional schools, textbooks and materials, and a curriculum that reflects their experiences and communities. (p. 205)

Issues of valorisation of knowledge and different forms of mathematics need to continue to be explored, as there are still many open questions. Related to this issue is the idea of non-immigrant students' views of immigrant students. This topic has received very little attention (a notable exception is the work of Planas (Planas, 2007; Planas & Civil, 2008)), yet it seems like it would be important to understand how all the students see and understand the experience of being in a multicultural classroom (Alrø et al. (2007) address this topic to a certain extent).

Another area that needs further research is that of immigrant parents' perceptions about the teaching and learning of mathematics. It is certainly interesting to note the similarities in these perceptions across very different contexts of immigration. "What now?" is my question. Furthermore, an important and under-researched area is that of interactions between immigrant parents and teachers, and perceptions of each other, in terms of the children's mathematics education. Civil and Bernier (2006) address this to a certain extent, but much more work is needed, especially given the need for a holistic approach to the education of immigrant students that really includes multiple voices, and the different participants in this education (parents, teachers, school administrators, community representatives, and the students themselves).

Language is a prominent theme in the research with immigrant students and mathematics education. More research is needed that focuses on multiple languages as resources for the teaching and learning of mathematics, once again to counteract the deficit perspective, particularly in the public discourse that sees the presence of other languages and not knowing the language of instruction as obstacles to the mathematics education of immigrant children. Issues of placement based on language proficiency, and the impact that these decisions have on students' learning of mathematics, also need to be studied further.

In reflecting on what needs to happen next, Philip Clarkson (Australian Catholic University) writes:

> We know quite a lot about multilingual kids' learning. But we have not been very good at looking at the varying multilingual contexts in which that learning takes place. Are we making too many assumptions there? In the last 10–15 years we are moving to look at teaching, but much more needs to be done with this. What models are there available that teachers can use as guides? Are they sorted according to the multilingual context of the classroom? What do we need to do now? What do we think would be good places to search that we think will move us forward to greater insight? (P. Clarkson, personal communication, May 27, 2008) (see Clarkson (2009) for more on teaching models for multilingual classrooms).

Finally, a clear implication from the research reviewed on this topic is the need for interdisciplinary teams with expertise in different areas, including mathematics education, immigration policy, linguistics, socio-cultural theories, anthropology, just to name a few. There is a clear need for this interdisciplinary expertise, as well as for the development (or refinement) of theoretical and methodological approaches.

ACKNOWLEDGMENTS

This chapter is based on a presentation given at the 11th International Congress of Mathematical Education, as part of Survey Team 5: Mathematics Education in Multicultural and Multilingual Environments, Monterrey, Mexico, July 2008.

Some of the work presented here is part of the research agenda of CEMELA (Center for the Mathematics Education of Latinos/as), which is funded by the National Science Foundation –ESI 0424983. The views expressed here are those of the author and do not necessarily reflect the views of the funding agency.

REFERENCES

Abreu, G. de (1995). Understanding how children experience the relationship between home and school mathematics. *Mind, Culture, and Activity, 2*, 119–142.

Abreu, G. de (2005). Cultural identities in the multiethnic mathematical classroom. In M. Bosch (Ed.), *Proceedings of the fourth Congress of the European Society for Research in Mathematics Education* (pp. 1131–1140). Sant Feliu de Guíxols, Spain: FUNDEMI IQS, Universitat Ramon Llull.

Abreu, G. de, Bishop, A. J., & and Presmeg, N. C. (Eds.). (2002). *Transitions between contexts of mathematical practices*. Boston, MA: Kluwer.

Abreu, G. de, César, M., Gorgorió, N., & Valero, P. (2005). Issues and challenges in researching mathematics education in multicultural settings. In M. Bosch (Ed.), *Proceedings of the fourth Congress of the European Society for Research in Mathematics Education* (pp. 1125–1130). Sant Feliu de Guíxols, Spain: FUNDEMI IQS, Universitat Ramon Llull.

Abreu, G. de, & Cline, T. (2005). Parents' representations of their children's mathematics learning in multiethnic primary schools. *British Educational Research Journal, 31*, 697–722.

Abreu, G. de, & Cline, T. (2007). Social valorization of mathematical practice: The implications for learners in multicultural schools. In N. Nasir & P. Cobb (Eds.), *Improving access to mathematics: Diversity and equity in the classroom* (pp. 118–131). New York, NY: Teachers College Press.

Abreu, G., de, & Gorgorió, N. (2007). Social representations and multicultural mathematics teaching and learning. In D. Pitta-Pantazi & G. Philippou (Eds.), *Proceedings of the Fifth Congress of the European Society for Research in Mathematics Education* (pp. 1559–1566). Larnaca, Cyprus: Department of Education, University of Cyprus.

Alrø, H., Skovsmose, O., & Valero, P. (2005). Culture, diversity and conflict in landscapes of mathematics learning. In M. Bosch (Ed.), *Proceedings of the fourth Congress of the European Society for Research in Mathematics Education* (pp. 1141–1152). Sant Feliu de Guíxols, Spain: FUNDEMI IQS, Universitat Ramon Llull.

Alrø, H., Skovsmose, O., & Valero, P. (2007). Landscapes of learning in a multicultural mathematics classroom. In D. Pitta-Pantazi & G. Philippou (Eds.), *Proceedings of the Fifth Congress of the European Society for Research in Mathematics Education* (pp. 1567–1576). Larnaca, Cyprus: Department of Education, University of Cyprus.

Anastasiadou, S. (2008). The effects of representational systems on the learning of statistics between Greek primary school students and immigrants. In J.F. Matos, P. Valero & K. Yasukawa (Eds.) *Proceedings of the Fifth International Mathematics Education and Society Conference* (pp. 167–176). Lisbon: Centro de Investigação em Educação, Universidade de Lisboa & Department of Education, Learning and Philosophy, Aalborg University.

Anhalt, C. O., Ondrus, M., & Horak, V. (2007). Insights from middle school teachers' participation in a mathematics lesson in Chinese. *Mathematics Teaching in the Middle School, 13*(1), 18–23.

César, M., & Favilli, F. (2005). Diversity seen through teachers' eyes: Discourse about multicultural classes. In M. Bosch (Ed.), *Proceedings of the fourth Congress of the European Society for Research in Mathematics Education* (pp. 1153–1164). Sant Feliu de Guíxols, Spain: FUNDEMI IQS, Universitat Ramon Llull.

Chronaki, A. (2005). Learning about "learning identities" in the school arithmetic practice: The experience of two young minority Gypsy girls in the Greek context of education. *European Journal of Psychology of Education, XX*, 61–74.

Chronaki, A. (2009). An entry to dialogicality in the maths classroom: Encouraging hybrid identities. In M. César & K. Kumpulainen, (Eds.), Social interactions in multilingual settings (pp. 117–143). Rotterdam: Sense Publishers.

Civil, M. (2002). Culture and mathematics: A community approach. *Journal of Intercultural Studies, 23*(2), 133–148.

Civil, M. (2006). Working towards equity in mathematics education: A focus on learners, teachers, and parents. In S. Alatorre, J.L. Cortina, M. Sáiz, & A. Méndez (Eds.), *Proceedings of the Twenty Eighth Annual Meeting of the North American Chapter of the International Group for the Psychology of Mathematics Education* (Vol. 1, pp. 30–50). Mérida, Mexico: Universidad Pedagógica Nacional.

Civil, M. (2007). Building on community knowledge: An avenue to equity in mathematics education. In N. Nasir & P. Cobb (Eds.), *Improving access to mathematics: Diversity and equity in the classroom* (pp. 105–117). New York, NY: Teachers College Press.

Civil, M. (2008). Language and mathematics: Immigrant parents' participation in school. In O. Figueras, J.L. Cortina, S. Alatorre, T. Rojano, & A. Sepúlveda (Eds.), *Proceedings of the Joint Meeting of PME 32 and PME-NA XXX* (Vol. 2, pp. 329–336). México: Cinvestav-UMSNH.

Civil, M. (2011). Mathematics education, language policy, and English language learners. In W. F. Tate, K. D. King & C. Rousseau Anderson (Eds.), *Disrupting tradition: Research and practice pathways in mathematics education* (pp. 77–91). Reston, VA: National Council of Teachers of Mathematics.

Civil, M., & Andrade, R. (2002). Transitions between home and school mathematics: Rays of hope amidst the passing clouds. In G. de Abreu, A. J. Bishop, & N. C. Presmeg (Eds.), *Transitions between contexts of mathematical practices* (pp. 149–169). Boston, MA: Kluwer.

Civil, M. & Andrade, R. (2003). Collaborative practice with parents: The role of the researcher as mediator. In A. Peter-Koop, V. Santos-Wagner, C. Breen, & A. Begg (Eds.), *Collaboration in teacher education: Examples from the context of mathematics education* (pp. 153–168). Boston, MA: Kluwer.

Civil, M., & Bernier, E. (2006). Exploring images of parental participation in mathematics education: Challenges and possibilities. *Mathematical Thinking and Learning, 8*(3), 309–330.

Civil, M., & Menéndez, J. M. (2011). Impressions of Mexican immigrant families on their early experiences with school mathematics in Arizona. In R. Kitchen & M. Civil (Eds.), *Transnational and borderland studies in mathematics education* (pp. 47–68). New York: Routledge.

Civil, M., & Planas, N. (2010). Latino/a immigrant parents' voices in mathematics education. In E. Grigorenko & R. Takanishi (Eds.), *Immigration, diversity, and education* (pp. 130–150). New York, NY: Routledge.

Civil, M., Planas, N., & Quintos, B. (2005). Immigrant parents' perspectives on their children's mathematics. *Zentralblatt für Didaktik der Mathematik, 37*(2), 81–89.

Civil, M., & Quintos, B. (2009). Latina mothers' perceptions about the teaching and learning of mathematics: Implications for parental participation. In B. Greer, S. Mukhopadhyay, S. Nelson-Barber, & A. Powell (Eds.), *Culturally responsive mathematics education* (pp. 321–343). New York, NY: Routledge.

Clarkson, P. C. (2009). Mathematics teaching in Australian multilingual classrooms: Developing an approach to the use of classroom languages. In R. Barwell (Ed.), *Multilingualism in mathematics classrooms: Global perspectives* (pp. 145–160). Tonawanda, NY: Multilingual Matters

Favilli, F., & Tintori, S. (2002). Teaching mathematics to foreign pupils in Italian compulsory schools: Findings from an European project. In *Proceedings of the Third International Mathematics Education and Society Conference*, Helsingør, Denmark. Retrieved from http://www.mes3.learning.aau.dk/All_Doc.htm

Gates, P. (2006). The place of equity and social justice in the history of PME. In A. Gutiérrez & P. Boero (Eds.), *Handbook of research on the Psychology of Mathematics Education: Past, present and future* (pp. 367–402). Rotterdam: Sense Publishers.

González, N., Andrade, R., Civil, M., & Moll, L.C. (2001). Bridging funds of distributed knowledge: Creating zones of practices in mathematics. *Journal of Education for Students Placed at Risk, 6,* 115–132.

Gorgorió, N., & Abreu, G. de (2009). Social representations as mediators of practice in mathematics classrooms with immigrant students. *Educational Studies in Mathematics, 7,* 61–76.

Gorgorió, N., & Planas, N. (2001). Teaching mathematics in multilingual classrooms. *Educational Studies in Mathematics, 47,* 7–33.

Gorgorió, N., & Planas, N. (2005). Norms, social representations and discourse. In In M. Bosch (Ed.), *Proceedings of the fourth Congress of the European Society for Research in Mathematics Education* (pp. 1176–1181). Sant Feliu de Guíxols, Spain: FUNDEMI IQS, Universitat Ramon Llull.

Hawighorst, B. (2005). How parents view mathematics and the learning of mathematics: An intercultural comparative study. In M. Bosch (Ed.), *Proceedings of the fourth Congress of the European Society for Research in Mathematics Education* (pp. 1165–1175). Sant Feliu de Guíxols, Spain: FUNDEMI IQS, Universitat Ramon Llull.

Kahn, L., & Civil, M. (2001). Unearthing the mathematics of a classroom garden. In E. McIntyre, A. Rosebery, & N. González (Eds.), *Classroom diversity: Connecting school to students' lives* (pp. 37–50). Portsmouth, NH: Heinemann.

Khisty, L. L. and Chval, K. (2002). Pedagogic discourse and equity in mathematics: When teachers' talk matters. *Mathematics Education Research Journal, 14,* 154–168.

King, R., Black, R., Collyer, M., Fielding, A., & Skeldon, R. (2010). *People on the move: An atlas of migration.* Los Angeles, CA: University of California Press.

Kitchen, R., & Civil, M. (Eds.). (2011). *Transnational and borderland studies in mathematics education.* New York: Routledge.

Macedo, D., Dendrinos, B., & Gounari, P. (2003). *The hegemony of English.* Boulder, CO: Paradigm Publishers.

Meaney, T. (2004). So what's power got to do with it? In M. Walshaw (Ed.), *Mathematics education within the postmodern* (pp. 181–199). Greenwich, CT: Information Age Publishing.

Moreira, D. (2007). Filling the gap between global and local mathematics. In D. Pitta-Pantazi & G. Philippou (Eds.), *Proceedings of the Fifth Congress of the European Society for Research in Mathematics Education* (pp. 1587–1596). Larnaca, Cyprus: Department of Education, University of Cyprus.

Nasir, N. S., Hand, V., & Taylor, E.V. (2008). Culture and mathematics in school: Boundaries between "cultural" and "domain" knowledge in the mathematics classroom and beyond. *Review of Research in Education, 32,* 187–240.

Nunes, T., Schliemann, A., & Carraher, D. (1993). *Street mathematics and school mathematics.* New York: Cambridge University Press.

O'Toole, S. & Abreu, G. de (2005). Parents' past experiences as a mediational tool for understanding their child's current mathematical learning. *European Journal of Psychology of Education,* XX, 75–89

Planas, N. (2007). The discursive construction of learning in a multiethnic school: perspectives from non-immigrant students. *Intercultural Education, 18,* 1–14.

Planas, N., & Civil, M. (2008). Voices of non-immigrant students in the multiethnic mathematics classroom. In O. Figueras, J.L. Cortina, S. Alatorre, T. Rojano, & A. Sepúlveda (Eds.), *Proceedings of the Joint Meeting of PME 32 and PME-NA XXX* (Vol. 4, pp. 121–127). México: Cinvestav-UMSNH.

Quintos, B., Bratton, J., & Civil, M., (2005). Engaging with parents on a critical dialogue about mathematics education. In M. Bosch (Ed.), *Proceedings of the fourth Congress of the European Society for Research in Mathematics Education* (pp. 1182–1192). Sant Feliu de Guíxols, Spain: FUNDEMI IQS, Universitat Ramon Lull.

Staats, S. (2009). Somali mathematics terminology: A community exploration of mathematics and culture. In R. Barwell (Ed.), *Multilingualism in mathematics classrooms: Global perspectives* (pp. 32–46). Tonawanda, NY: Multilingual Matters.

Stathopoulou, C., & Kalabasis, F. (2007). Language and culture in mathematics education: Reflections on observing a Romany class in a Greek school. *Educational Studies in Mathematics, 64,* 231–238.

Zevenbergen, R. (2008). The dilemmas of Indigenous education: The passion for ignorance. In J.F. Matos, P. Valero & K. Yasukawa (Eds.) *Proceedings of the Fifth International Mathematics Education and Society Conference* (pp. 505–514). Lisbon: Centro de Investigação em Educação, Universidade de Lisboa & Aalborg: Department of Education, Learning and Philosophy, Aalborg University.

Marta Civil
School of Education
University of North Carolina at Chapel Hill

CHAPTER 6

LEARNING OF MATHEMATICS AMONG PAKISTANI IMMIGRANT CHILDREN IN BARCELONA: A SOCIOCULTURAL PERSPECTIVE

SIKUNDER ALI BABER

Through a case study of a Pakistani girl currently living and studying in the multicultural context of Barcelona, Catalonia, I examine transitions on several levels – between cultures in the broadest sense, and between school and classroom cultures in particular – and their relationships with successful learning trajectories. The data show how she interprets and gives meaning to her own trajectory as a (mathematics) learner in a context that is culturally and socially distinct from the one she experienced in her home society. Drawing on concepts such as "foreground" (Skovsmose, 2005), and theories of transition processes and social representations, I show how, as a student in a Barcelona secondary school, she is making sense of her past experiences in the light of both the present and her expectations about the future, negotiating these meanings both with herself and with significant others. Moreover, I show how personal and cultural processes such as the negotiation of meaning, rethinking and social repositioning of the self, and identification of significant others may contribute to her successful educational transitions.

THE CONTEXT OF PAKISTANI IMMIGRANTS IN BARCELONA

Immigrants in Spanish urban society

Heterogeneity and hybridity of populations due to the processes of immigration and globalization have created multiple situations where people from different cultures are bound to interact with each other. However, the interaction is much more complex than it would be if it was limited to cultural interaction, since immigration is also regulated through many other different regimes and discourses. It is important to locate these discourses in order to unpack the complex threads that structure the life worlds of the immigrants.

Broadly speaking, two groups of "immigrant students" may be distinguished. One group consists of the children of highly-skilled professionals who come to work at the high end of the occupational hierarchy, people who, in Spain, are commonly named "foreigners". Normally their children do not attend public schools. The other group includes children of those who come to Spain for the purposes of seeking employment at the low end of the scale where native workers shun jobs that are difficult or demanding or carry little prestige. Given the

O. Skovsmose and B. Greer (eds.), Opening the Cage: Critique and Politics of Mathematics Education, 143–166.

complexity of the immigrant family of the student in my case study, one could place the children born within this Pakistani family intermediate to these two categories. They come from a middle-class family background in Pakistan and have sufficient resources to send their children to good schools according to their socio-economic positioning in Pakistan, and this socio-economic positioning is maintained in Catalonia. Nevertheless, the student does attend a public school.

Due to the influx of immigrants, Spanish society faces a variety of challenges relating to diversity. Moreover, it has an issue of minority nations within the nation-state of Spain itself. Catalonia (capital city Barcelona), located in the north-eastern part of the Iberian Peninsula, is one of the 17 autonomous communities into which the Spanish state is divided. This makes the understanding of the positioning of immigrants within Catalan society more complex, since it requires considering the languages the minorities use in asserting their identity within Spanish society. Catalonia has two official languages, Catalan and Spanish. Catalan is not a dialect of Spanish but a distinct language of the Romance family. Many immigrants to Catalonia tend to learn to speak Spanish rather than Catalan, because of social relationships and the neighbourhoods they live in. The Catalan language has played a vital role in consolidating Catalan identity, and Catalan language immersion is one of the main policies of the Catalan Government with regard to immigrants' integration (Zapata-Barrero & Zaragoza, 2009).

Large-scale immigration into Spain is a relatively new phenomenon. It started at the beginning of the 1970's because of the oil crisis that required increasing the workforce in the charcoal mining industry. Nevertheless, until the middle of the 1990's, the majority of foreign people who came to Spain were still retirees from Western Europe that came in search of sunnier landscapes and cheaper prices than in their countries of origin. Since then, and because of the boom of the building industry, the percentage of immigrant people has constantly increased (Pérez & Medina, 2004) until the recent economic crisis.

Thus, the scenario for immigration is one of a country where, for a long time, the only shared meaning for "foreigner" was "tourist". When foreigners are no longer only tourists but also immigrants, and local people start feeling that their rights and privileges are at risk, the interactions among the different cultures may create fears of others, both imagined and actual. Thus, for example, native-born people may feel that their jobs are in jeopardy. They may easily feel that they are losing their identification with the neighbourhoods where they have lived for so long. These feelings, in turn, could lead to a sense of insecurity among the local population. On the other hand, immigrant workers consider it extremely important to secure their economic independence, a consideration that, in certain cases, may lead them to establish relationships only with people from the same country of origin. What initially was a move to gain acceptance in the reception society may contrarily have a ghettoising effect.

The growing cultural and economic power of immigrants is building a new type of metropolis, hitherto unknown in Spanish cities. Although a relatively new phenomenon, immigration has brought different cultures, languages, and customs with which natives of Spain have to interact (Perez & Medina, 2004). In this connection, Perez and Medina have suggested that "the immigration process is a key element in the analysis of the latest socio-cultural changes and their relation to the transformation of urban areas in the more dynamic Spanish cities" (Perez & Medina, 2004, p. 158). Moreover, immigration and the processes of modernization have changed the nature of major cities such as Barcelona and Madrid:

> … where the population concentration process runs the risk of converting large urban sectors into differentiated ghettos. In both cities, this phenomenon occurs mainly in the degraded historical centres, in the marginal zones that were middle class in 50s and 60s, and in traditional nuclei of the urban outskirts. In all of them, the multicultural city is found on the streets in public plazas, in parks, in malls, and in the everyday life on the commute to work (Perez & Medina, 2004, p. 158)

This spatial segregation of cities into ghettos and into commercial centres has created major social, cultural, and economic discriminations sometimes having roots in ethnic and cultural characteristics. Moreover, the politics of redevelopment and revitalization of cities in Spain has created power differentials between haves and have-nots (see, e.g., Degen, 2001; Degen & Garcia, 2008). Sassen (1996) suggests that corporate buildings and immigrant communities are the two extreme modalities that appropriate and form urban spaces in the cities of the developed world. On the other hand, Delgado (1996) observes that:

> Cities have been a place of co-existence (peaceful and conflictive), social mobility, and crossbreeding all throughout history. Heterogeneity and the mix of social forms in the city are not only possible, but are healthy, even structurally strategic. (Quoted in Perez & Medina, 2004, p. 169)

These are only a few of the components that, together with legal regulations, add to the complexity of immigration in an already culturally diverse society. Moreover, the fact that native-born people have always got a power differential in their favour cannot be ignored, as they are in a position to define the status of minorities according to their privileged position of being in the majority.

Pakistani immigrants in Barcelona

The majority of Pakistani immigrants in Barcelona come from a district called Gujrat in the province of Punjab. Almost 90% are males living along (either single, or with wives in Pakistan); about 10% live with their families in Barcelona, having been in Spain more than ten years. Most Pakistani immigrant children are first-generation. Among Pakistani immigrant families, the orientation towards education for their children varies. For example, one group of parents are only interested in the education of their children to the extent that they can study languages such as Spanish or, to a lesser degree, Catalan so that they can take part in the family business. Another group is more deeply interested in the education of their children, in particular wanting their children to go on to further education. Many other immigrants fall outside these two groups.

The number of Pakistani immigrants in Barcelona is in the range 25,000–50,000, a precise figure being hard to get. Most live in a neighbourhood called El Raval, one of the oldest neighbourhoods in Barcelona and, historically, a red light district. It is located in the neighbourhoods of La Rambla and the Gothic quarter, which are mostly tourist areas. The 1996 census revealed the registered population of El Raval as about 35,000, with a high proportion of older people, and 28% immigrants.

The schools in the locality of El Raval are full of children of immigrants and different nationalities. Often it is heard that it is not easy for schools to handle the multifarious demands of these children. There is also recognition that people often face lots of psychological, social, and economic problems in handling the challenges of life in this area. Moreover, local politics is impacting the overall look of the area. For example, Degen (2001) has reported regeneration of El Raval after 1980, and a power play in its redevelopment. Through this re-shaping, segments of El Raval were re-designed in order to create "designer heritage aesthetics", the main purpose being to turn Barcelona into a "city of style". The 1992 Olympic Games in Barcelona also accelerated this effort. The business-oriented move to create Barcelona as a city of the future has also brought effects on the landscape of El Raval. According to Degen (2001), this move created a tension between the historical landscape of El Raval and a modern conception of Barcelona mostly designed to bring it under control.

Schools in multicultural contexts

In countries such as Spain, Ireland, and Italy, the percentage of students born in other countries has increased threefold since the year 2000 (De Heus & Dronkers, 2009; European Commission, 2008). Schools in multicultural cities like Barcelona are also getting affected by changes in political, socio-economic, and cultural contexts in Spain. Furthermore, one can notice that schools in urban cities are increasingly changing due to the presence of students from different countries, and changing politics of education.

Within this context of change, the influence takes place in both directions: cultures influence schools and schools influence cultures. The changing composition of schools has generated a need to understand how teaching and learning are organized within the schools and how students from different cultures are affected during the processes of teaching and learning. In addition, this recognition of complexity of the schools within multicultural contexts also creates a need to conduct studies that can specifically document the conditions under which immigrant students learn to cope not only with educational demands but also the demands of life in general. Here, knowing the perceptions of learners, with their agency, about the processes of learning is important in order to understand how they are handling the demands of learning in multicultural contexts. Other actors, such as teachers and parents, also face varied challenges to cope with the new educational demands due to changing realities of the school under the influences of transnationality and globalization. Within this scenario of education in the contexts of globalization and the changing conditions of immigrants, I next elaborate the notion of transitions in order to capture the complexity of the lives of immigrant children (in this case, Pakistani) and then situate their learning in that context.

Negotiating transitions: Young immigrants in multicultural settings

Young people often experience ruptures as they have to go through deep changes in their lives. These ruptures can be the result of changes in their cultural milieux, in particular when moving from one country to other country. Such a change can bring different types of demands on a person. It could encourage the young person to reflect on his or her current positioning and develop some strategies to make sense of the new location or new country. It could also mean establishing new relations with others and new friendships, learning new communication tools such as the languages of the new country, becoming aware of general norms and practices in schooling in the new environment. At the same time she or he is meeting the challenge of organizing life in such a way as to have some sense of stability and hope about how to create a future in the face of uncertainty.

> *Uncertainty* can designate a person's experience of blurred personal reality, *relatively* to a previous state of apprehension of things. Experiencing uncertainty might be paralyzing or stimulating, but in most of the cases, it questions previous understanding, and might call for exploring possibilities and elaborating new conduct. These processes, which aim at reducing uncertainty, can be called *processes of transition* (Zittoun, 2007, p. 196).

Accordingly, I use transition processes among young people as important constructs in elaborating their engagement within educational processes. Here I draw on the work of Tania Zittoun (Zittoun, 2007; Zittoun, Duveen, Gillespie, Ivinson, & Psaltis, 2003). She and her colleagues have developed a theory of symbolic resources among young people, especially when they are in the processes of going through various transitions. She has shown how symbolic resources such as cognitive resources, knowledge construction, and meaning-making processes support learners to take on the challenge of transitions under the rapidly changing conditions of modern cities, in which immigrants are often located in the intersection of different social conditions. It is essential to take modernity as a background that influences the choices and perceptions of the immigrant young people (Mørch, 2003). Zittoun describes the transition processes in these terms:

> In terms of dynamics, transitions in the lifetime can be said to involve *three interdependent streams of processes*. First, transitions involve changes in the social, material, or symbolic spheres of experience of the person. ... Second, people's relocation might need social, cognitive and expert *forms of knowledge and skills*. Third, through these relocations, encounters with others, and learning, the person might be brought to engage in *meaning-making*, that is, to confer sense to what happens to him/her... These three processes are necessarily linked; in youth, learning difficulties are often linked to the fact that the person feels his/her identity put at stake or cannot find a personal sense in the learning situation (Zittoun, 2007, pp. 195–196)

Transition processes are influenced by social representations. Here I use the notion of social representations proposed by Serge Moskovici (1973) who focuses on: self in the context, self in relation to others, and how the self handles the challenge of social representations in the context. In this way, we consider social representations as "ways of world making". Moscovici (1973) elaborated the notion of social representations as:

> Systems of values, ideas and practices with twofold functions; first, to establish an order which will enable individuals to orient themselves in their material and social world and to master it; secondly to enable communication to take place among the members of a community by providing them with a code of social exchange and code for naming and classifying unambiguously the various aspects of their world and their individual and group history (Moscovici, 1973, p. xiii).

Moreover, social representations are responsible for the creation of intersubjective reality, which implies both human agency and social influence.

> On the one hand, social representations are created by human beings in order to conventionalize objects, persons, and events by placing them in a familiar social context... On the other hand, once established, these representations influence human behavior and social interaction by often subtly imposing themselves upon us and so limiting our socio-cognitive activities. Social representations are therefore not only the product of human agents acting upon their society but are equally prescriptive and coercive in nature. They become part of the collective consciousness, especially once they are "fossilized" in tradition and taken for granted in social practice. (Moscovici, 1984, p. 13)

Social representation brings our attention to dialectics between agency and structure, tradition and change (Jovechelovitch, 2007). Also, it encourages us to explore relationships between social representations, language, and communication.

Social representations are inseparable from the dynamics of everyday life, where the mobile interactions of the present can potentially challenge the taken-for-granted, imposing pockets of novelty on tradition coming from the past (Jovechelovitch, 1996, p. 124)

In addition to taking into consideration the elements of transition processes and the social representations that shape these processes, I also take into account the perceptions of learners of the processes of learning as important in understanding how learners are handling the demands of learning in modern societies. Here, I use the theory of foreground and background in order to shed light on the future aspirations of the young immigrant so that one can see how learners are interpreting their future possibilities and in what ways these possibilities are shaping their present living conditions (Baber, 2006; Skovsmose, 2005). Ole Skovsmose (2005) describes "foreground" in terms of:

> … the opportunities that the social, political and cultural situation provides for this person. However, not the opportunities as they might exist in any socially well-defined or "objective" form, but the opportunities as perceived by a person. Nor does the background of a person exist in any 'objective' way. Although the background refers to what a person has done and experienced (such as the situations the person has been involved in, the cultural context, the socio-political context and the family traditions), then background is still interpreted by the person. (Skovsmose, 2005, pp. 6–7)

Here I recognize the fact that the opportunities as perceived by a person are shaped or limited or reinforced by social representations. In consequence, special attention will be paid to the processes that open up possibilities for actions for the immigrants. I also pay particular attention to the points where immigrant children face the tensions and ruptures in the course of their transition phases. Thus, my specific questions are: *How do immigrants in general, and young Pakistani immigrant in particular, negotiate transition processes within the multicultural context of Barcelona? How do they perceive their learning opportunities through engagement in their educational processes in general, and in the learning of mathematics in particular, within the educational landscape of Barcelona?*

I examine these questions through the case study of Hina, a successful Pakistani immigrant girl.

A CASE STUDY FROM A SOCIOCULTURAL PERSPECTIVE

Hina's immigration story

I now present the background and the context that were responsible for the relocation of the family of Hina, a Pakistani immigrant girl living in Barcelona and studying at the high school there. Hina's family experienced living in various countries and cultures before they came to Barcelona, thus they have moved and continue to move through different types of transition in order to shape their present and future.

Hina has four sisters and a brother. Her brother is the oldest and she is the second oldest child in the family. She was born in 1987 in Gujrat, a city in Pakistan from which many people come to settle in Europe. Her father married from a family in the province of Sindh and brought his wife to Gujrat, where he joined the Engineering Corps of the Pakistan Navy. He got an opportunity to work in Saudi Arabia with a German factory producing heavy machinery. After several years of work, he brought his family there, including Hina. Some of her siblings were born in Saudi Arabia. She attended Kindergarten and primary school there. Hina's father now owns two grocery stores in Barcelona, which he runs with his son. He works extremely hard, typically from 8.00 am till midnight, often not having time to spend with his children. But through this hard work, he earns enough to support his children very well.

Travelling to Saudi Arabia, then back to Pakistan at the age of 11, exposed Hina to different types of transition, including getting engaged in educational processes in the two countries, participating in cultural practices, and also working out her positioning in response to the resulting changes. At the same time, Hina was, and is, developing awareness of the symbolic tools that she needs to choose in order to successfully navigate through such transitions.

In Pakistani society, it is normal practice that parents with sufficient means try to send their children to schools where their children have possibilities to learn English, schools that are usually private and expensive, and create different types of social inequality. Hina instead chose to go an Urdu-medium school. The main reason for her was to improve her Urdu, as well as having the possibility of speaking Punjabi, as there was social pressure within her family to speak Punjabi to ensure acceptance and recognition within the extended family. Such social exclusion based on language can get further reinforced, especially for girls, who normally have to stay within the family most of the time, getting no opportunity to interact with wider sections of society. Thus, she and her brother had unequal resources at their disposal to develop their competence in Punjabi. Her brother could easily go out with Punjabi-speaking friends but Hina did not have this possibility. In this way, Hina already knew the importance of knowing languages that could give her access to competence in communicating effectively with others. Besides this, she experienced different educational contexts within the country as well. For example, her father is from the province of Punjab and her mother from Sindh, where the language of instruction is Sindhi and the schooling is organized around this language. On the other hand, in the Punjab, the language of school instruction is Urdu and the daily life in school is organized around the Punjabi language.

Hina has also experienced diversity in the ways in which teachers interact with students. For example, when she went to school in Punjab she had to memorise things properly otherwise she would be punished, including beatings, whereas in Saudi Arabia, the children did not get any such punishment in school. Although such treatment could have had a negative impact, she reacted to it positively, as a reflection of, and encouragement for, taking studies seriously.

Hina's engagement with the educational landscape in Barcelona

In Spain, primary (age 6–12) and secondary (12–16) education is compulsory and free. Students who complete their secondary education successfully receive a graduation diploma and may continue their secondary education for another two years. Those whose grades are insufficient to obtain the diploma receive a school-leaving certificate, and may go on to the first level of vocational training. Post-compulsory secondary education consists of two tracks: mid-level vocational and technical training, and the baccalaureate, which prepares students for university or for high-level technical training. The legal age for starting work in Spain is 16 years, but the chances of getting a job, or a good one, are strongly connected with your school certificates.

Here I present how Hina is interacting with the educational landscape in Barcelona, describing her school, the situatedness of the school within the policy framework of Catalonia, the challenge of diversity and the way the school deals with it. I also describe Hina's response to constraints that schooling imposes on her, her engagement in the schooling processes, her assertive agency in shaping her learning conditions and her role, and how she draws on her teachers as role models for success in the educational context, and, above all, her engagement with the processes of learning mathematics.

Hina came to Barcelona when she was in grade 10. In accordance with school policy, she was placed in an "open classroom" (La Aula de Oberta) designed specifically for the children of immigrants. The main priority in such classes is to teach immigrant children the Catalan language. Even in these classes there are two categories of children, one comprising those children who come from countries where Romance languages are prevalent and the other those from a non-Romance linguistic background. According to Hina's teacher, the students in the latter group face lots of difficulties in adjusting to the Catalan educational system. However, there are always exceptions to the rule. Hina has been a student highly motivated to succeed in the system. Here she describes her entry into the Catalan educational system:

> I was in tenth grade in Pakistan. When I came here (to Spain), I had to study according to age. Initially, we faced problem in getting admission. With great efforts of my father I got admission to this school. There was a teacher in this school who also helped .... The first year I spent in learning the language. I also got good results. With this progress, teachers decided to place me in the B section, the best section in the school.

Every school has a particular culture in organizing learning and teaching practices. These cultural practices also create several mechanisms of exclusion and inclusion for learners in the school. For example, the placement of learners in different sections can open possibilities for the future for some and, at the same time, close possibilities for others to move to the next stages. In Hina's school, access to sections D and B means enhanced possibility to go to high school and then having a possibility to sit for the university entrance examination (*Batxillerat*)[1]. Here one can also identify two trends salient in the school, namely the discourse of academic competence and the importance of mastering the language of instruction, i.e. gaining an access to language that supports learning and subsequent competent performance.

The placing of students into different sections points to the fact that school divides students according to performance. Such a system in the school reflects the competitive trends that are prevalent in society. Specifically, it reinforces the legitimacy of unequal access to resources – that is, it legitimises the idea that those who follow the norms and do well within the norms are granted access to resources such as the higher educational possibilities that society can make available. Within this competency-oriented discourse, it is not easy for immigrants to live up to these expectations. Those who are in possession of good social resources, like parents who are educated, have more possibilities to be successful in the school system. Further, if a sibling is educated or well aware of the processes of education then he or she can provide guidance, a circumstance very visible in the case of Hina. For example, she is well aware about the sections where one should be in order to be successful. Because of this awareness, she advised her siblings how to work hard to get to the "right" section in the school. Conversation with Hina also points to the fact that mastery of the language of schooling is very important to be successful. For example, Hina's sister was facing difficulties when she had to learn many languages at the same time – Catalan, Spanish, and English. The focus of the school on learning these languages made Hina's sister very confused, and she could not get a grip on a language needed to communicate her learning for evaluation. Hina's school provided an opportunity to pay greater attention to the learning of Catalan, and subsequently Hina's sister was able to transfer to the desirable section where she has more possibilities for realizing her goal of getting higher education.

How Hina's school meets the challenges of diversity

Hina's school is located in El Raval and gets children from families who have migrated to Barcelona from many different countries. It has gone through a big transformation due to this major influx, which brings lots of challenges. One of the challenges is how to craft pedagogies that can support teachers to relate with this new body of immigrant students. According to an interview with one of the teachers at the school, he has learned how to treat students with respect. In this way, he has an opportunity to learn about particular needs of his students, as well as helping them to develop competencies and motivation for study. Also, he has been involved in the process of teaching students from different nationalities for several years, so he knows what can support students to stay motivated. According to Hina, this teacher helped her a lot in supporting her to learn mathematics while, at the same time, he always encouraged her to continue to pursue studies more seriously. Whenever she had any difficulties in learning certain topics in mathematics, he was always available to clarify her confusion.

I came as a postdoctoral research fellow to Barcelona to work with the interdisciplinary research group *Educació Matemàtica i Context Sociocultural* (EMiCs, Mathematics Education and its Sociocultural Context) in the Department of Mathematics and Science Education at the Autonomous University of Barcelona. Given my background of being from Pakistan, with knowledge of Pakistani society and Urdu, the national language of Pakistan, and my PhD work with Pakistani immigrant families in Denmark, I set out to do research with young Pakistani immigrants. With the help of Professor Nuria Gorgorio, leader of the EMiCs, we contacted the head of the school and thereby the coordinator of the Mathematics Section of the school, and through her we got connected with Hina and other young Pakistani immigrants. In our initial conversation, we explained the focus of the research and negotiated a schedule for interviews, which I conducted in Urdu. During my conversations with these young Pakistani immigrants, I then made requests to meet with their parents. In this way, I got an opportunity to meet with Hina's father. Below, I present some of the interview data that were generated through my conversations with Hina.

Hina shares with us her ways of looking at the school. Here she compares her schooling experience in Barcelona with her previous schooling experience in other countries. She has experienced different roles of teachers in Catalonia and Pakistan. She found that teachers in Catalonia could be friendly, so that a teacher could assume a position of a significant other for her, whose advice she would take seriously. Asked about her experience in the school, Hina explained that in the first month she had some difficulties, due to having to cope with the language, and with a different way of scheduling classes.

> Hina: When one goes to a new class, it takes time to make friendships with teachers. Normally, people take a longer time to make friendship. In my case, I make friendship very quickly. … [After the first month] I developed a friendship with a teacher. Later on I realized that she was the principal of this school. This was a surprise for me!! I became friends with this teacher. We often met with her. She often asked me about my studies … You know that, being the principal, she had access to the report of each child. In this way she knew my results and performance at the school. She also knew about my behaviour. I developed a good friendship with her. She often asked me questions as what would I do in my future and what subject would I choose etc. And her discussions with me were very frank and friendly. … When I came to the first year of the *Batxillerat* she advised me that if I would like to take law for my future then I should study Latin and Greek. I took this advice very seriously and I thought that she must have thought this through before she gave me her advice. Therefore, I decided to opt for Greek and Latin.

Hina reports that her teachers in the school take the learning of students very seriously. Teachers are always there to help students to clarify their concepts if they have difficulties in understanding. Students can approach teachers in the classroom settings or outside the classroom. According to her, teachers also use different kinds of resources to make their learning meaningful. This engagement of teachers in her learning is different from what she experienced in Pakistan.

> Sikunder: So one thing that you like about the school is that principal is friendly with you. What else do you like about the school?

> Hina: Teachers are also very good at the school. They explain things very well to students. For example, if a student has not understood a thing well then teachers will explain this thing to you again. Even if you have not understood it then they would try to explain you this thing again. And every time they explain you in such a way that they are explaining this thing to you in a new way; they will try their best to explain things in different ways so that student can fully understand this thing. Even to the extent that if you are on the staircase and if you ask a question to the teacher that you had not understood, then she/he will try to explain you this thing on the staircase. I

really like this behaviour of the teachers at this school. This is really different here. In Pakistan it was not the case. There, if you had asked a teacher more than two times they could become annoyed and could punish you.

Moreover, Hina is also active in creative activities at the school:

Sikunder: What are your favourite activities in the school?

Hina: The activities in which I have participated up till now: to participate in the Saint Jordi Day. On this day, we write in English, Catalan, Spanish and French. One can write poems or stories in any one of these languages. I wrote a poem in English and a story in Catalan, and won prizes in these activities.

Hina is appreciative of what is being offered at the school, but she is also aware that her school is not doing well in comparison to other schools. In this regard, she considers others in the school as responsible for the overall positioning of her school. She termed them lazy and said that they are not taking their studies seriously. That is, she is aware that laziness in study could prevent success in life.

Hina gives particular importance to the process of education in general, and schooling in particular. For example, she considers schooling as a part of the processes of knowing and opening up the mind. She is of the opinion that education supports one to live a successful life and that it facilitates progress in the work that one chooses to do. Here, she also recognizes the importance of pursuing education beyond finishing secondary school. That is, her thinking about education goes far beyond "instruction" and is much more than about getting useful practical knowledge that you get at school, or knowledge to get jobs:

Sikunder: In your opinion, why is it important to go to school?

Hina: The way the world in which we live is, it has become increasingly important for us to be educated in any subject. Because, until you get educated your mind would not get opened and it is difficult for you to understand things around you. Without having clear understanding you cannot make progress in any type of work. In my opinion, when you study you come to know the reasons of anything that you do. Also, education can help us to differentiate between right or wrong. Here I do not say that uneducated people are not intelligent. They are also intelligent.

Here, Hina gives the example of her grandfather as someone who did not get the opportunity to go to school but is still an intelligent man. Despite not going to school, he was very good in mathematics. If you give him information about the business, about expenditure and income, then he could quickly tell you how much profit you have earned. He is very fast in doing these calculations, which he does in his head. Even though her grandfather is intelligent despite being uneducated, she feels that education through schooling is also very important and it offers comparative advantage.

Hina: I feel that education is extremely important. In my opinion, if you finished grade 10 then it does not count at all in today′s world. In particular, grade 10 is nothing in Pakistan. In Pakistan, people would think that you had

> not studied at all. So the grade does not count at all in Pakistan. Here in Barcelona, finishing grade 10 has significance. For example, if you pass the grade 10 exam, then you can easily join the police force.

For her, just passing the secondary school exam is not enough. Instead she aspires to be well educated and wants to pursue higher education. Here she is more influenced by the image of being educated in the context of Pakistan as compared with the importance of finishing secondary school in the context of Barcelona. Besides attaching value to higher education, which is rooted in the cultural context of Pakistan, she is also critical about the quality of the provision of education in Pakistan. On one question about her expectations from the school, she made the following comments:

> Hina: In my opinion, yes, school is matching my expectations. … If someone asks me the question whether I would prefer to study here or in Pakistan, I would say that I prefer studying here. The reasons are that the teachers here are very good and they teach us very well. They explain to you in such a way that you would hardly forget what you have learned. … By comparison, in Pakistan it is difficult to understand even if the teacher explains several times.

Hina is actively involved in shaping her foreground. As one can see, her engagement in educational processes came about in the course of her relocation in different cultural contexts and consequent re-positionings. She is aware of the advantage of learning Catalan as compared with Spanish, and that her success at school depends on mastery of Catalan. This awareness shows that her access to educational opportunities is very much shaped by the political policies of the Catalan government – consolidation of Catalan identity through the promotion of Catalan language is perceived to be essential within the socio-political landscape of Spain. These influences of larger discourses are reflected clearly in the schools, which are not located in a vacuum but very much an integral part of the socio-political landscape of Catalan society.

Hina's engagement with the processes of learning mathematics at school

Just as, for Hina, mastery of Catalan gives her access to getting good grades and hence access to making her career, at the same time she considers learning of mathematics very important to read the world. Here, I present my conversations with Hina about the role of mathematics in her life, why one learns mathematics, how she is engaged in learning mathematics, and her experiences of learning mathematics in different schools. The conversations show that she has brought with her different social representations around mathematics, some of rooted in her religious ideas (being part of a Muslim family), and others derived from the need to understand the world in which she lives and numbers are important.

Mathematics is important for accountability to God (religious influence) Hina believes that mathematics is associated with the concept of accountability that one can feel when one faces God after death. This comes with thinking informed by religious beliefs.

> Sikunder: How do you think that this subject is important for life?
>
> Hina: Whatever you do in life, you need calculations. I can go even further – if you die then you have to be accountable to God as well. That is, calculation or accountability goes from birth to death and beyond, it is something that goes on continuously. I think mathematics is important. You take any aspect of life in the world, you will find there the importance of mathematics.

Learning mathematics is strategically important and has advantages over other subjects Hina also thinks that mathematics has a strategic importance. For example, one can get good grades if one is good at doing mathematics. She also thinks that study of mathematics is much easier that learning languages. According to her, learning of languages requires extra effort.

> Sikunder: In your opinion, do you like mathematics? If yes, then what do you like most in mathematics?
>
> Hina: Mathematics is easy for me. If I compare mathematics with my current subjects (which are in social sciences and humanities), I often share with my teachers that if I had mathematics or science subjects then I would have got better grades than the grades that I have now. I often give advice to my sisters that they should not leave mathematics or science and choose languages because languages can make one mad [laughs]! We have studied mathematics in Saudi Arabia, we have also studied mathematics in Pakistan and also we are studying here as well. Thus, we have been very well in touch with mathematics. On the other hand, languages are different. As you go to a new country, the language is different so you have to learn it afresh, and this requires more effort than learning mathematics.

Mathematics as numbers and procedures Hina has specific social representations of mathematics. For example, when she thinks of mathematics, she thinks of it as numbers and then she relates mathematics with algorithms and solving procedures. She finds algorithms interesting.

> Hina: Mathematics is a kind of game. … That is, whatever you do in the process, the result will follow automatically in front of you. Whether you do the operations of plus, or minus, or multiplication, or division, you will get the result after doing certain operations. … Numbers is another type of world. This is a different world and I find it interesting.

Her conception of mathematics is to work with algorithms. She gets upset if she has not solved a mathematics problem immediately.

> Hina: When I do not get the expected answer while solving the problem then I really get angry and I really feel that I should throw my notebook away and

> I think why is this answer not coming out yet? I do not think there is anything else that I dislike in mathematics.

Learning Mathematics can provide a competitive advantage within multicultural settings Hina talks about wanting to compete with others, in particular a Chinese student.

> Sikunder: Now I need your comment on this: some students are better than others in mathematics. Can you describe to me why these students are better than others?
>
> Hina: I want to share with you an example. In my class, there was a student from China. I often become angry with myself. For example, when we were solving a mathematical problem, we normally take 30 minutes to solve the problem while this Chinese boy solved the same problem in 15 minutes. We used to say about this kind of students that these children are born for mathematics. He does mathematics so quickly it was very difficult for us to understand how he could do so. He could not understand Catalan or Spanish, but he was excellent in mathematics. … Maybe the reason is that in his life mathematics has been more than anything, or he comes with a background where he is used to doing mathematics.

Influence of conditions of learning of mathematics in different cultural contexts (role of significant others) Hina gets inspiration or support for learning mathematics from the teachers she had in different schools in different countries. For example, she gets her inspiration for doing mathematics from her teacher called Mr. Zulfiquar in a Pakistani school. Mr. Zulfiquar prepared her well in mathematics – in fact, she came first in the Grade 8 exams in Pakistan. He not only taught her mathematics, but also Urdu, the national language of Pakistan. He was kind of a role model for her. Mr. Zulfiquar never gave his students full marks as he thought there was always room for improvement and giving full marks might give an impression to the student that he or she is the best student in the class and there is a chance that the student could misconstrue these marks and become overproud of being an excellent student. According to Mr. Zulfiquar, there is always a possibility that someone is better than you. Hina takes this philosophy of Mr. Zulfiquar very seriously in her life. For example, when she saw that her sister got 10 out of 10 marks in the mathematics test she was surprised. On enquiring, Mr. Zulfiquar shared with her the fact that he had never given 10 out 10 marks in mathematics to any student. According to him, one must not feel superior to others. Mr. Zulfiquar's advice is present in Hina's mind about the highest marks and she gave the same advice to her siblings as well.

> Hina: That is the reason when sometime my sisters get 10 marks in mathematics; I always wonder how come that their teachers have given them full marks. Then I used to tell the sisters that there would some students who might be better than them in doing mathematics who deserve a mark of 10 rather than them. I used to tell them that no one should get 10 marks rather

> you should get 9 marks because I believe that no one can claim intellectual superiority over others. … I have always taken this advice of our teacher Zulfiquar very seriously. I always follow this philosophy in my life.

The influence of Mr. Zulfiquar travelled with her to Barcelona. He used to tell the students that mathematics is the kind of subject that requires daily practice; then you will be able to make progress in it. If you study, then you will be ready for tomorrow. If you do not prepare it well today, then you will forget even the mathematical problems that you did yesterday. Also, Mr. Zulfiquar encouraged her not to leave the solution of any question incomplete. This learning encouraged her to develop the principle that one must understand every idea as fully as possible.

> Hina: I have developed thanks to my mathematics teacher Mr. Zulfiquar from Pakistan. He guided us very well. To us, he was like a father figure. Now I have heard that he died but I am not sure about this, but I cannot confirm this. My heart is not satisfied that a teacher like him could die. But he was an excellent teacher for me.

When, during a two-hour lesson, they asked for a break, Mr. Zulfiquar used to advise them that they should understand and complete the mathematics right away, then they would get a break. According to him, if he gave them a break without completing the question then whatever they had done so far would slip from their mind and they would not be able to recall it. Through this method, Mr. Zulfiquar developed a habit among his students that if they had started solving a question, then they had to complete it.

One can see that Mr. Zulfiquar's advice has been crucial for Hina, not only to engage in the processes of learning mathematics but also her learning at school in general. Despite this, Hina also faces difficulties in learning certain topics in mathematics. She mentioned difficulties in understanding some algebraic concepts. But she has an approach for addressing these difficulties. In this regard, she considers the role of the teacher very important. Here she also recognizes the fact that different teachers could have different approaches to support students to learn mathematics and address their difficulties. For example, Hina knows that the way Miguel and Luis teach mathematics in different ways. She likes the way Miguel teaches mathematics. He explains every step of the problem very carefully and supports students to clarify all the steps involved in solving a problem.

> Hina: I remember that I faced difficulty when we started learning algebra, especially when one moves from one equation to another equation, there I had difficulty. My teacher Miguel teaches this to us very well. There is another teacher whose name is Luis. I do not follow his way of explaining things. It could be the case that other children may follow him well. … The difference is that our teacher Miguel would explain you the whole procedure of solving the problem from point A to point B. … Our teacher Luis does not do that. … He leaves many steps unexplained and it is our responsibility to construct them. In this way, you are left in doubt…. In short, our teacher Luis

> explains things to us as if we are university students. In my opinion, a student at this stage is a student, who cannot take the place of the teacher. Therefore, it is an obligation of the teacher to explain things to students at the level where he or she is studying. The main thing is that student must not left in any doubt. … For this reason, I like my teacher Miguel very much in comparison to my teacher Luis.

When Hina still has problems with mathematics, she gets help from different people. For example, in Barcelona there is an educational support system called Casal. Casal is organized in such a way that it provides an opportunity to teachers to come in the evening time to volunteer themselves to support students who want either to further strengthen the learning that they do in school or resolve difficulties in subjects like mathematics. If she does not get help from Casal, then she could also seek help from her friends or from her father who is good at mathematics. When one gets support from different sources to resolve difficulties in mathematics, it is possible that one might learn mathematics in a different manner. Using different methods could create conflict in the minds of students. The teacher might also encourage or discourage solving of mathematical problem in a particular way, which could create conflict in the minds of the students. But Hina feels that her mathematics teachers do not discourage learning mathematical problems in a different way. Instead the teachers discuss the relative merits of different methods.

Hina's parents are also actively involved in supporting her to stay engaged in her educational processes, so they take keen interest in her learning of subjects like mathematics. She mentioned that her father used to bring mathematical games for them so that they could hone their mathematical skills. Also, her father is an active listener as well. He always asks how Hina and her siblings are doing in their studies. For example, when Hina's father is at home, he engages his children deliberately in discussing certain topics concerning their education or about the society in which they live. Or he reads a story to them wherein his children could find some kind of problem. In this way, his children normally spend time together and they try to find out what could be possible solutions to that problem. He always gives them plenty of time to solve these problems, which encourages them to think deeply about the problem, and from multiple perspectives. The children normally come up with different solutions than their father, but his habit is never to tell them the right answer. In this regard, Hina cites one example:

> When we were children our father used to bring different toys for us. These were not just toys. They always stimulated us to think more deeply while playing with them.… For example, he brought blocks for us to create some models and encouraged us to put these blocks together thoughtfully. I recall once my father gave a gift to all of us. In this toy, there were different things made up of wires. My father did not give us the wrapper as it included instructions for how to make different things. In whatever way we put together the things in it we would get a new object. We were all working on it and getting different things out of it. We then shared these discoveries with

> our father and also what were the problems we were facing to complete a design or particular shape. Then our father helped us to think how we could resolve the puzzle that we were facing.

Hina's father created positive competition among his children. They learned a lot from this competition. The main lesson that Hina's father wished to develop among his children was that they had to work and it is not possible to be good at what you are doing without putting hard work into it. This attitude would give them an edge over others. Hina shares a story about her participation in a running race game when she was at a school in Saudi Arabia. Here her father deliberately gave her advice when she was running against other children that if someone came close to her she could hinder them. But Hina reacted against that advice and stated that it was not fair in the game. Her father responded that he was very happy that she had recognized the unfairness of the act and objected. From then on, her father encouraged her not to accept any bad act and to be ready to communicate her objection to the person concerned.

Here one can recognise that students' social representations about learning of mathematics are formed in a variety of schooling contexts and in the home and society in which they live. For Hina, mathematics is important not only to be successful in this world but also because one will be accountable to God. This suggests that the importance of mathematics and the learning of mathematics is continuously shaped and changed with the experiences of learners and the significant others who inspire them to learn. At the same time, collective social representations also affect the formation of the perceptions of learners about mathematics. Social representations are again shaped and reshaped in the interaction of a person with others and also with the multiplicity of discourses that are available in society. That is, the social representations that students develop about mathematics are constructed from when they are born, when they start to establish their first social contacts with others, and they go on being developed under the influence of school, family, friends, and media (Abreu & Gorgorió, 2007; Piscarreta, 2002; Piscarreta & César, 2004).

Hina's future expectations

Hina wishes to become a lawyer. This aspiration is something she has been thinking about since she entered school. In order to enter law school, Hina has to score a certain average in her school grades. For public universities in Barcelona, one has to have an average of 5 in order to be accepted as a law student. Getting this average at the school is not easy. She has to maintain good grades in all the subjects she is studying at the school and besides this she has to sit the entrance examination for the public universities in Barcelona.

Her parents' expectations also play an important part in the shaping of her aspirations. Her father and mother are worried that law is a professional field that is full of danger. For example, as a lawyer one has resolve many conflicts and parties who are affected by the results can bring harm to lawyers. This is how Hina's parents perceive the field of law and those practicing in it. Perhaps this perception is based on their experiences of seeing lawyers practicing in the context of Pakistan where lawyers could even be killed by the affected parties in the conflict, as had happened close to their home. Despite this opposition of her parents, she is persistent in her aim to become a lawyer and, given her determination, her father now supports her.

Hina is aware of the fact that it is not easy to become a successful lawyer. She has to work hard in order to achieve this professional goal. Besides this, she is aware that she has to do well in both Catalan and Spanish in order to enter into the University. She is also aware of the options for higher education options available in Barcelona. One is going to university where the focus of study is more on theoretical knowledge, while the focus at the *Graus* is more on practice-oriented knowledge.

DISCUSSION AND CONCLUSIONS

Hina`s success story demonstrates many elements that are responsible for her successful transitions in different cultural contexts. For example, Hina is in a constant dialogue with her different selves and also she is active in comparing her views with those of significant others. When she experiences conflict in her different selves, she tries to use her memories of success in other contexts as symbolic resources.

For example, she used her success in mathematics in a Pakistani school as a tool to be successful in the context of Barcelona, drawing on the advice of her teacher Mr. Zulfiquar as an internal voice guiding her to be persistent in the face of challenges that schooling in Barcelona brought to her. This observation is very much consistent with the idea of dialogical selves developed by Hubert Hermans (2001), who conceptualises dialogical self:

> … in terms of dynamic multiplicity of relatively autonomous *I*-positions. In this position, *I* has the possibility to move from one spatial position to another in accordance with changes in situation and time. The *I* fluctuates among different and opposed positions, and has the capacity imaginatively to endow each position with a voice so that the dialogical relations between the positions can be established. (p. 248)

At the same time Hina is clearly cognizant that it is not easy to become successful in new cultural settings. For example, she has been moving in three different national settings and each setting offered her different opportunities and resources for success. When she was in Saudi Arabia, she had to pay attention to become proficient in English and Urdu, as these were the cultural tools that gave her access to educational resources there. But when she came back to Pakistan, these two languages became a barrier for her as they created some kind of superior position for her in Punjab province where the dominant language is Punjabi. Her cousins who were well conversant in Punjabi were not ready to accept her positioning in Punjabi society due to her having access to English and Urdu. So, understanding the importance of Punjabi and the relative importance of Urdu, Hina went to a school that teaches in Urdu. For Hina, this was a strategic decision as, in this way, she could join the circle of her cousins, negotiate her positioning with them, and become assertive in influencing the social circle around her.

In different cultural settings, Hina always looks for possibilities affording opportunities to sharpen both her knowledge and skills, but also to construct interpretations of reality that can promote her educational success. For example, she appreciates the teaching methods of her teachers in her school in Barcelona. These methods allow her to do things more practically and she also gets an opportunity to get deeper into the subjects she is studying. Furthermore, Hina also likes her mathematics teacher who respects her and acknowledges her previous learning. If she solves a mathematical problem using a different method, the teacher accepts this method as legitimate. In other words, Hina is in a fortunate situation in which her mathematics teacher is aware of some of the needs of the immigrant students and also genuinely interested in the growth of his students.

If one looks at Hina's agency, one can see that she is a very confident person who knows very well what she is doing and how she can get to the stage where she can emerge as a successful professional. She is also constantly negotiating her positioning with other actors who are influential in her development. For example, she is actively engaged with her teachers; if she has any confusion relating to any topic, she always goes to a teacher and seeks clarification. She is also actively involved in the process of shaping the future possibilities of her siblings. For example, she proactively worked with her sister and convinced teachers at the school that her sister should go to the section where she gets more quality teaching and the opportunity to proceed through the Batxillerat in order to get to university. Here it is very important to recognise that Hina has very supportive parents who are keen that she should succeed educationally while, at the same time, being open to negotiation. One can see that this space of dialogue between Hina and her parents has opened considerable possibilities for them to negotiate the demands of their changing needs. Hina`s story demonstrates that an immigrant student's successful engagement in educational processes is a very complex process, embedded in several socio-cultural threads, not only in the school but also in the orientation of the family members, especially parents. The success of an immigrant student also depends on how actively this particular student is involved in shaping his or her future possibilities. Here we have seen that assertive individual agency could create possibilities of success despite facing a challenging educational situation framed by the politics of education in the situation. One of the advantages that a successful student may use in succeeding in the face of the educational/political context is his or her facility of moving through different cultures and using the cognitive or symbolic tools that this movement across cultures affords. One can recognize that the social, economic, and legal frameworks under which immigrants operate, local and global politics around immigrants, and educational policies and their interpretations at the school level, also affect the possibilities for young immigrants to gain access to symbolic resources. If the immigrants are not perceived well in the receiving country then this could highly constrain the life possibilities of immigrant children and greatly polarise the foregrounds of young immigrants (Baber, 2007). Moreover, the schools are normatively organized. There are value systems that schools develop to regulate the conduct of students and teachers within the school. If the teachers and administration of the schools are not ready to face the challenge of diversity at the classroom level then it is highly likely that the students from different cultural backgrounds will be misperceived and misinterpreted, and subsequently the formations of their lifeworlds would become highly restricted. In other words, young people who have strong potential to succeed within schooling processes could lose their energy and potential for productive engagement in the shaping of their learning trajectories. It is evident from the case of Hina that, without having supportive teachers, supportive parents, and supportive peers, her success story could have easily become a struggling story or even story of failure. Through Hina's story, we also learn that the lifeworld of any young immigrant is very complex. Accordingly, we need research tools and perspectives that could support us in understanding and unpacking the complex threads of the young immigrant's engagement with the educational processes. Here, one can detect the importance of their transnational experiences in formulating their approaches to handle the challenges of life, including school. In this connection, looking at learning in general, and learning of mathematics in particular, from a socio-cultural perspective could bring greater insight into the interplay of the conditions of learners and agency of learners in shaping their foregrounds. One can keep an eye on the dynamic nature of the educational processes impacting the learner and, at the same time, see how the learner is perceiving and creating his or her world through active engagement.

Implications for educational policy

The educational landscape in multicultural societies like Barcelona is changing very fast. These changes affect schools, teachers, students, and parents. Of course, the educational policy-makers are engaged in the processes of not only understanding changes in the educational landscape but also finding ways to initiate processes of changes that can enable all actors within the educational scenario to better cope with the changes. Within this context, this research study could offer some insights for informing the educational policy of multicultural cities like Barcelona. There follow some specific policy recommendations.

First, the educational policy should recognise that learners and other actors associated with the schooling processes, while engaged in the processes of education, are experiencing and making sense of changes in their own ways. It is important that policy-makers should give enough freedom to schools and teachers to understand the intentions and foregrounds of learners. This focus on intention, agency, and foregrounds of learners would allow teachers to make learning more meaningful for them. It would also help schools to find ways to provide both symbolic resources and material resources to cope with the specific demands of the learners in relation to their future goals. The notion of foreground allows us to examine how learners go through the processes of creating their futures. Often this future-making process is very complicated and creates lots of tensions and stress in students' minds. Young immigrants experience many challenges, as they have to grow through several transitions at the same time to make sense of their living in a new society. In other words, these young people are at the intersections of different cultures and they might not necessarily have tools to deal with the demands of the new society and get ready to face the challenges of the future. Therefore, it is an important consideration to focus on learners' engagement in a context where different cultures meet.

Second, the schools are not immune from the socio-political conditions within which they are located, and they are heavily affected by the general discourse on education. Furthermore, there are different social representations around immigrants in the society, through which schools often look at immigrants, and which could constrain or open the possibilities of dealing with the challenge of handling diversity. For example, a common political tendency in political discourse is to look at immigrants from a deficit perspective. That is, immigrants can be viewed as burdens. Or immigrant youths are perceived as trouble-makers or not interested in their studies. These general perspectives can block possibilities for the future-making of young immigrants. In this context, it is important that the teachers and educational authorities should be aware that the young immigrants are passing through not only different cultural planes but also are located in social and economic milieux. Here the cultural and religious influences of immigrant families could also bring influence on meaning-creation for the young people. Teachers and schools can form partnership with the learners and their families in order to provide more meaningful educational experiences. In this way, the educational processes could be closely aligned with future-making processes. In other words, interpretations of social representations about young immigrants have major implications for educational practice at the school level.

Third, during this study I discovered that the schools are not aware of the challenges of education in multicultural settings. They do not have a clear strategy for coping with the challenge of handling students from different cultures. This lack of strategy brings a huge challenge for teachers to meet with the needs of immigrant students in an appropriate way. The lack of understanding on the part of the schools and teachers could create situations where young immigrants might find school a boring and alienating place instead of considering it as a place that can prepare them for the future.

Finally, the case of Hina demonstrates that the collaboration between schools and families in creating educational learning possibilities is absolutely important. Understanding diversity in a multicultural society is very complex, requiring tools and resources, both symbolic and material, and the continuous engagement of all the actors to make the schooling experience successful and meaningful. Theoretical constructs such as transition, background/foreground, and social representations may help us understand immigrant students' perspectives, and provide insights for informing pedagogical actions in situations where different cultures interact.

ACKNOWLEDGEMENTS

This chapter emerged from my post-doc stay (January 2009 – January 2011) at the Department of Mathematics Education and Science Education of the Faculty of Education at the Autonomous University of Barcelona. Here I was part of the research group EMiCs led by Professor Núria Gorgorio. The EMiCS group – Educació Matemàtica i Context Sociocultural (Mathematics Education and its Sociocultural Context) is recognized by the Direcció General de Recerca of the Generalitat de Catalunya (the research office of the Catalan autonomous government) as a Consolidated Research Group (2005SGR-00211 and 2009SGR-00590) which has partially supported the research presented here. Initially, Nuria Gorgorio and I agreed to produce this chapter together but because of her unavoidable engagements and pressing responsibilities at the University and my particular personal and professional conditions, Nuria withdrew from the authorship. Nevertheless, I got very valuable and generous support from her and other members of the EMiCs in writing this chapter and I thank them all.

NOTE

[1] Batxillerat: This is the Catalan word. This indicates the students who are in high school and preparing to sit for the university entrance examination.

REFERENCES

Abreu, G. de, & Gorgorió, N. (2007). Social representations and multicultural mathematics teaching and learning. In D. Pitta-Pantazi & G. Philippou (Eds.) *Proceedings of the Fifth Congress of the European Society for Research in Mathematics Education* (pp. 1556–1559). Larnaca, Cyprus: University of Cyprus.

Baber, S. A. (2006) Investigating interactions of foregrounds and backgrounds of migrants from Pakistan. In *Tanker til tiden: Om studier af laering* (pp. 181–201). Aalborg, Denmark: Aalborg Universitietsforlag.

Baber, S. A. (2007) *Interplay of citizenship, education and mathematics: Formations of foregrounds of Pakistani immigrants in Denmark.* Doctoral thesis, Aalborg University.

Degen, M. (2001). Regenerating public life? A sensory analysis of regenerated public spaces in el Raval, Barcelona. In J. Rugg & D. Hinchcliffe (Eds.), *Advances in art and urban futures Vol. 2: Recoveries and reclamations* (pp. 19–35). Bristol, England: Intellect Books.

Degen, M., & Garcia, M. (Eds.) (2008). *La Metaciudad: Barcelona – tranformación de una metrópolis.* Barcelona, Spain: Anthropos.

De Heus, M., & Dronkers, J. (2009). *Immigrants' children scientific performance in a double comparative design: The influence of origin, destination, and community.* Meeting of the International Sociological Association Committee on Social Stratification and Mobility. Yale University, New Haven. Retrieved from http://www.eui.eu/Personal/Dronkers/English/Heus2.pdf

Delgado, M. (1996). *Qui pot ser "inmigrant" a la ciutat?. In Ciutat i inmigració (Il debat de Barcelona).* Barcelona: Centre de Cultura Contemporània de Barcelona.

European Commission (2008). *Migration and mobility: challenges and opportunities for EU education systems (Green Paper).* Brussels: Author.

Hermans, H. (2001). The dialogical self: Towards a theory of personal and cultural positioning. *Culture and Psychology*, 7, 243–281.

Jovchelovitch, S. (1996). In defense of representation. *Journal of Theory of Social Behavior, 25*(1), 81–102.

Jovchelovitch, S. (2007). *Knowledge in context: Representations, community and culture*. New York: Routledge.

Mørch, S. (2003). Youth and education. *Young, 11*(1), 49–73.

Moscovici, S. (1973). Foreword. In C. Herzlich (Ed.), *Health and illness: A social psychological analysis* (pp. ix–xiv). London/New York: Academic Press.

Moscovici, S. (1984). The phenomenon of social representations. In R. M. Farr & S. Moscovici (Eds.), *Social representations* (pp. 3–69). Cambridge: Cambridge University Press.

Perez, J., & Medina, J. (2004) The (re)creation of postmodern Spanish cities: The role of immigrants, *Dela, 21*, 157–170.

Piscarreta, S. (2002). *Malmequer, bem-me-quer, muito, pouco ou nada: Representações sociais da matemática em alunos do 9º ano de escolaridade*. Lisbon: APM.

Piscarreta, S., & César, M. (2004). Desafinado... ou o meu primeiro amor: A construção das representações sociais da matemática. *Vetor Neteclem, 2(s/n.º)*, 31–51.

Sassen, S. (1996). *Losing control: Sovereignty in an age of globalisation*. New York: Columbia University Press.

Skovsmose, O. (2005). Foregrounds and politics of learning obstacles. *For the Learning of Mathematics, 25*(1), 4–10.

Zapata-Brarrero, R. & Zaragoza, J. (2009). *Elements of Spanish philosophy of diversity management*. Barcelona: GRITIM (Grup de Recerca Interdisciplinari sobre Immigració), Universitat Pompeu Fabra.

Zittoun, T. (2007). Symbolic resources and responsibility in transitions. *Young, 15*, 193–211.

Zittoun, T., Duveen, G., Gillespie, A., Ivinson, G., & Psaltis, C. (2003). The uses of symbolic resources in transitions. *Culture and Psychology, 9*(4), 415–448.

Sikunder Ali Baber
Educational Consultant

CHAPTER 7

MATHEMATICS EDUCATION ACROSS TWO LANGUAGE CONTEXTS: A POLITICAL PERSPECTIVE

MAMOKGETHI SETATI AND NÚRIA PLANAS

There is a continuing debate in mathematics education research and practice regarding the use of languages in multilingual mathematics classrooms in which children are not yet fully fluent in the language of learning and teaching (LoLT). Research in mathematics education supports the use of the students' home languages for teaching and learning (e.g. Adler, 2001; Khisty, 1995; Moschkovich, 1999, 2002; Setati, Molefe, & Langa, 2008). The use of the students' home languages has been argued as a support needed while students continue to develop proficiency in the LoLT at the same time as learning mathematics. However, there is not much take-up on the ground. The question is why?

During a conversation that took place between two Latin American students that were working in the same small group in a mathematics classroom in Barcelona, one of the students insisted to the other, in Spanish, *"You'd better say it in Catalan."* The second student responded, *"You listen to the mathematics."* In our view, this exchange illustrates the complex relationship between language choice, participation, and mathematical learning in multilingual classrooms. Why would two students who share a first language argue about the use of a second language in a mathematical interaction? These kinds of conversations are not unique to students in Barcelona. Below are two extracts from Setati (2008) in which a teacher and a student, in data collected in South Africa, express similar views.

> If we changed our [mathematics] textbooks into Setswana[1] and set our exams in Setswana, then my school will be empty because our parents now believe in English. (Lindi, a Grade 4 mathematics teacher)
>
> English is an international language, just imagine a class doing maths with Setswana for example, I don't think it's good. (Tumelo, a Grade 11 mathematics student)

The introductory discussion above captures the essence and complexity of the debate regarding the use of languages in multilingual mathematics classrooms in which children are not yet fully fluent in the LoLT. In our view, the debate is not just about language and/or mathematics; it is also about social relations, the political role of language, and the context in which the mathematics is taught and

O. Skovsmose and B. Greer (eds.), Opening the Cage: Critique and Politics of Mathematics Education, 167–186.

learned. It is important to ask what political issues are at play in these interactions and how different they are for different political and multilingual contexts.

In this chapter, we draw on our individual journeys in this area of study to explore some political aspects of mathematics education and language diversity across two contrasting political and multilingual contexts. Through these narratives we provide a window into the political tensions and questions that illustrate the complexity of this work and illuminate/illustrate some of the research findings that seem to hold across our differing multilingual contexts. To put this debate in perspective we begin with a brief overview on the political role and use of language. We then comment on concrete data from students and teachers in our two research contexts, Catalonia and South Africa, and move to a more general discussion on issues of language, race, class, and power.

LANGUAGE, POWER, AND MATHEMATICS TEACHING AND LEARNING

An important connection exists between the languages of learning and teaching and the knowledge that is produced in the schools. Nevertheless, in our work the emphasis is on social access to school mathematics (Planas & Setati, 2009), and not just the mathematical knowledge that is produced through epistemological access. We argue that an important connection also exists between the LoLT and the nature and extent of the students' participation. Participation is highly influenced by questions related to the politics of language and the political roles of languages which include, among others, issues concerning who decides what language should be used for learning and teaching, what informs this decision, as well as whose participation in the classroom is supported or constrained as a result of the chosen LoLT.

The political role of language

Language, like multilingualism, is always political (Gee, 1999). It is a characteristic that is used in society to determine power (Gutiérrez, 2002). All over the world the issue of language has always been interwoven with the politics of domination and separation, resistance and affirmation. During apartheid in South Africa, the language of learning issue became a dominating factor in opposition to the system of Bantu Education.

The Bantu Education Act of 1953 stipulated that "mother tongue instruction" be phased in across all primary school grades in African primary schools, with English and Afrikaans as compulsory subjects from the first year of schooling. English and Afrikaans were the only two official languages, the latter having developed out of Dutch settlement. None of the African languages was recognised. In addition, both English and Afrikaans were also to be used as languages of learning and teaching on a 50/50 basis when transfer from main language learning took place in the first year of secondary school (Hartshorne, 1987, p. 70). The educational interests of the pupils became subordinate to ideological and political factors. The government's greatest concern at the time was that the constitution of South Africa required equality in treatment of the two official languages. These policies were centred on fears that the Afrikaans language, culture, and tradition might be overwhelmed by the older, more internationally established, English language, culture, and tradition (Reagan & Ntshoe, 1992, p. 249).

Though not unmindful or ashamed of African traditions per se, the mainstream African nationalists have viewed cultural assimilation as a means by which Africans could be released from a subordinate position in a common, unified society (Reagan & Ntshoe, 1987). They therefore fought against the use of African languages as languages of learning and teaching because they saw it as a device to ensure that Africans remain oppressed. Many analysts trace the 1976 uprising, which began in Soweto and spread all over South Africa, to belated attempts by the Nationalist government to enforce the controversial and highly contested 50/50 language policy for African learners. African teachers were given five years to become competent in Afrikaans.

In 1979, in the wake of the 1976 revolt, the government introduced a new language policy. This new policy emphasised initial main language learning with an eventual shift in LoLT to English or Afrikaans. As a general rule, the African child began his or her schooling in the main language, which remained the LoLT through the fourth year of schooling (Grade 4). During these first four years both English and Afrikaans were studied as subjects. Beginning in the fifth year of schooling (Grade 5) there was a shift in LoLT to either English or Afrikaans, which were by then the official languages of the country. This is the system under which Mamokgethi, who is fluent in nine languages, including English and her home language, Setswana, went to school. In her writings she describes her educational experience as follows:

> I learned mathematics in Setswana at primary school up to Grade 4. The switch to English as language of learning and teaching (LoLT) happened in Grade 5. Even though I passed my mathematics Grade 12 examinations in English I was not fluent in it. I proceeded to university where I took mathematics (in English) as my first major and passed it with a good grade in my final year. Reflecting back on my own learning of mathematics in English the greatest difficulty was learning in a language in which I was not fluent. As I look back, I am aware that much of my learning was based on memorisation, a function, in my view, of my limited fluency in English. (Setati, 2002a, p. 1)

As she explains, her fluency in English improved markedly during her postgraduate studies. As many South Africans who were in primary school in the eighties and nineties could testify, Mamokgethi's story is not unique. The new era for South Africa began in 1990 with the unbanning of liberation movements and the release of Nelson Mandela.

In 1997, three years after liberation, South Africa introduced a new language-in-education policy (LiEP) that recognises 11 official languages. According to this policy, not only can South African schools and students now choose their preferred LoLT, but there is a policy environment supportive of the use of languages other than one favoured LoLT in school, and so, too, of multilingual language practices like code-switching. While this new LiEP is widely acknowledged as "good", it is meeting significant on-the-ground constraints. Several South African researchers have argued that while this policy is intended to address the undervaluing of African languages, in practice English, the language of former colonisers, still dominates (Setati, 2002b; Setati & Adler, 2000; Setati, Adler, Reed, & Bapoo, 2002). Research shows that most schools are not opting to use students' home languages as LoLT, in both policy and practice (Taylor & Vinjevold, 1999, p. 216). This situation is not unexpected – home language as LoLT policy or "mother tongue instruction" has a bad image among speakers of African languages. It is associated with Apartheid and hence inferior education.

While the new language policy in South Africa is intended to address the overvaluing of English and Afrikaans and the undervaluing of African languages, various institutional arrangements and government policies have resulted in the dominance of English in the linguistic market. First, there has been the policy of English and Afrikaans-medium higher-education policy in South Africa for many years. The LoLT in most of the universities in South Africa is English and it seems that this policy will continue for many more years since it has not yet been challenged in higher education circles. Second, there is an English/Afrikaans language pre-requisite for anyone aspiring to become a professional in South Africa. Students need to pass a school-leaving examination in English as a first or second language, in addition to mathematics, to enter and succeed in the English-medium higher education and training programs. Third, there have been policies upholding English and Afrikaans as official, legal, and government languages. The nine African languages spoken by the majority of South Africans did not enjoy any official status until 1994. However, these languages are still in many ways secondary to English in reality; for example, most of the policy documents are written in English only. Fourth, there has been the imposition of an English/Afrikaans-language requirement for individuals aspiring to join the civil service. This is mainly because English and Afrikaans were the languages used before liberation in 1994. For instance, while Afrikaans has lost popularity, the ability to communicate in English or Afrikaans is one of the requirements for anyone willing to train as a policeman or policewoman. The fact remains that English is the most important criterion for selection for high-ranking officials, and knowledge of an African language is seen as an additional asset, but not an essential one. With these institutions and policies well entrenched in the various administrative, educational, and professional arenas of South Africa, a symbolic market has been formed where English constitutes the dominant, if not exclusive, symbolic resource. It is a prerequisite for individuals aspiring to gain a share of the socio-economic, material resources enjoyed by an elite group.

In Catalonia, an autonomous region in North-Eastern Spain, Catalan and Spanish are used as markers of social class and nativeness (Mar-Molinero, 2000). Catalan was a forbidden language during Franco's dictatorship. This same language is now being politically affirmed as a consequence of processes of Catalan nation-building that focus on the differences between Catalonia and Spain. The choice of Catalan as the LoLT was organized after 1983 as a controversial way to integrate a large portion of the population that had arrived from other parts of

Spain in successive immigration waves (Strubell, 2006). In the school system the tensions between the two official languages in the country have mostly been represented by the symbolic distance between the Catalan "native" people and those Catalans whose parents are Spanish and were born outside Catalonia. The arrival of people from Latin America in the nineties has introduced new power relations, as their accents are socially considered as being of a lower status in comparison to those of the Spanish speakers regarded as nearer to the so called standard Catalan language and culture.

The current situation in the Catalonian context is totally different in comparison to South Africa and what Mamokgethi experienced in her primary school. In the seventies, the only official language in all parts of Spain was Spanish, although during later stages of the Franco regime, certain uses of Catalan were "tolerated". Núria's home language is Catalan. She grew up being told at school that her language was a "variety" of Spanish and French, mostly spoken by working-class groups and peasants of villages from North-Eastern Spain. In the nineties, before getting her university position and having finished her studies in mathematics, Núria became a secondary school mathematics teacher in Barcelona. The long tradition of monolingual policies in Spain had already been changed by making the use of Catalan obligatory in many public domains, and by institutionally recognizing the importance of the Catalan academic grammar and literature. The reconsideration of status for the Catalan language was forced by the solid organization of Catalan nationalism in reaction to forty years of the Franco regime. The fact also helped that Catalan became the national and the only official language of Andorra, a small prosperous country in South-Western Europe. Since then, the issue of whether Catalan is a language or a dialect, closely related to the discussion of Catalonia being a nation or a region, has been the subject of political agitation several times. On July 10, 2010, for instance, more than a million people marched in Barcelona in support of the Catalan language and protesting a verdict by the Spanish Supreme Court imposing the co-officiality of Catalan and Spanish in the school system, and explicitly talking about the need to preserve the unity of Spain. Although the court was clear about its verdict, more than one year later Catalan still remains as the LoLT as a consequence of complex and strategic political alliances.

As described for our two cases, South Africa and Catalonia, the political nature of language is not only at the macro-level of structures but also at the micro-level of classroom interactions. Language can be used to exclude or include people in conversations and decision-making processes. Zentella (1997), through her work with Puerto Rican children in El Barrio, New York, shows how language can bring people together or separate them. Language is one way in which one can define one's adherence to group values. Therefore, decisions about which language to use in multilingual mathematics classrooms, how, and for what purposes, are not only pedagogic but also political (Setati, 2005). While research in general education on language and minority students is strongly rooted in the socio-political context of learning (Cummins, 2000), most research on mathematics education and language diversity has been framed by a limited conception of language as a tool for thinking and communication. As Setati (2005) argued, to ignore the political role of language in mathematics education research and practice would assume that power relationships do not exist in society.

The political use of language

The historical exclusion of speakers of minority languages in mainstream schools and classroom practices has been extensively documented (Barwell, 2005, 2009; Cummins, 2000; Gutstein, 2006; Khisty, 1995; Planas & Civil, 2010a). In particular, the language policies and ideologies in Catalonia and South Africa mark in concrete ways the use of languages in mathematics classrooms and the broader school context.

In Catalonia, for instance, the terms "Catalan students", in opposition to "non-Catalan students", are common expressions that tend to be accepted as neutral. However, the term "non-Catalan" is applied to students who do not have Catalan as their first language, although they may have been born in Catalonia and they may not be children of immigrant families. This fact indicates exclusion based on language issues. Moreover, different official documents refer to students who are not predominantly Catalan speakers as "students with low language proficiency", which suggests a deficiency on the part of students without clarifying that this proficiency is considered in relation to Catalan. Hence, exclusion is orchestrated through a use of language that points to what some groups do not "have". These, and other common expressions in the Catalonian context, make language minorities live as a "perpetual underclass" based on language, a situation that predominantly Catalan-speaking students and teachers have never experienced.

Since "Catalan native speakers" are unlikely to disadvantage themselves for reasons of language, one should ask what is hidden behind classroom practices and why the words themselves (e.g., "Catalan students", "native speakers") become the issue. In Planas (in press), the analysis of interactions in a mathematics secondary classroom indicates different practices of exclusion based on language. Gutstein (2003, 2006) and Khisty (2006) have also documented practices of exclusion based on language in the United States with Latino and Chicano students in their learning of mathematics. Setati's work (e.g. 2005, 2008) on exclusion based on language use is particularly illuminating due to the multilingual complexity of South Africa.

Setati (2008) argues that, given the dominance of English, the choice that the South African LiEP offers is a chimera. In her view, the assumption embedded in this policy is that mathematics teachers and students in multilingual classrooms, together with their parents, are somehow free of economic, political, and ideological constraints and pressures when they apparently freely opt for English as LoLT. The LiEP seems to be taking a structuralist and positivist view of language, one that suggests that all languages can be free of cultural and political influences.

In this chapter, the hegemony of English is mainly related to the discussion of the South African context, while the hegemony of Catalan is related to the discussion of the Catalonian context. Nevertheless, in Catalonia the number of mathematics classrooms with Content and Language Integrated Learning (CLIL) methodologies, with English as the LoLT, is an increasing phenomenon. Although the official positioning is that CLIL methodologies seek to support multilingualism and additional language(s) learning, in practice the political tensions between Catalan and Spanish point to the progressive promotion of fluent Catalan and English bilinguals as a way to reduce the representation of the Spanish language. We argue that the idea of a hypothetical and future balanced situation between Catalan and English is also a chimera. Phillipson's (1992) theory of English language imperialism, and his more recent discussion on the well-organized attempts to establish the hegemony of English in Europe (Phillipson, 2003), indicate that the value of any sort of multilingualism is code for appreciation of the importance of English. We need to wait for at least one decade to examine how the language and political situation in Catalonia evolves. On one hand, the verdict by the Spanish Supreme Court in 2010 and, on the other, the local initiatives to progressively substitute Spanish by English through CLIL methodologies, suggest significant changes.

ANALYSING EMPIRICAL DATA TO UNDERSTAND THE PROBLEM

In our two contexts of research we work with classroom data, teachers and students to examine the political use of language in mathematics teaching and learning. In this section, we put together data from the teachers and the students' perspectives in South Africa and Catalonia with the aim of identifying practical orientations to common challenges. There are other significant works in mathematics education and language diversity that also draw on the idea of searching for differences and commonalities across contexts (see, for instance, Setati & Moschkovich, 2010; Planas & Civil, 2010b; Setati & Barwell, 2006). This is a very useful approach because, when researchers work only within their countries, the context can be taken for granted to some extent and important elements can be unintentionally dismissed.

Teachers' perspectives

In each of our studies we have explored the perspectives of teachers in multilingual mathematics classrooms through semi-structured interviews and conversations. In Catalonia, these are teachers from secondary schools, while in South Africa they were selected from primary schools. Our analysis of part of the knowledge gained from these interactions is presented below.

What can be seen is that, while different preferences are differently argued, teachers have a clear preference for one language. From a poststructuralist perspective (Gutiérrez, 2010), students and teachers are not "agentive" on their own; rather the social and political structures in which they participate impose constraints on, and enable their agency. In particular, the teachers' perspectives are influenced by these structures. These structures may explain shared discourses on "choosing the right language of teaching", which in turn convey an ideological position against other languages that are thought of as "not appropriate" for instruction. The interpretation of these discourses on language is, however, problematic. On one hand, it can be said that exclusion appears by the fact of reproducing the language status quo. But, on the other hand, it also can be argued that inclusion is sought through the efforts toward "democratising" the language privileges that the dominant groups have enjoyed for a long time.

Perspectives of teachers in South Africa Teachers were asked the following question in English during individual interviews: "Which language do you prefer to teach mathematics in? Why?" Over and above all else, "English is international" emerged as a dominant discourse that shaped the teachers' language choices. All the six teachers interviewed stated ideological and pragmatic reasons for their preference to teach mathematics in English. Like many teachers in South Africa, they are aware of the linguistic capital of English and the symbolic power it bestows on those who can communicate in it. One of the teachers for example said: "I prefer to teach in English because it is a universal language." All of the teachers used similar language referring to English as an international or universal language. Awarding such a status to English suggests that they have accepted the dominance and power of English. They do not have any control over the international nature of English. All they can do is to prepare their students for participation in the international world, and teaching mathematics in English is an important part of this preparation. One of the teachers expressed the reasons for her preference for English as follows: "It is an international language… The textbooks are written in English, the question papers are in English…". Another one argued that: "If they [the students] do not learn the language how will they be able to cope in higher classes?" All of the reasons that the teachers gave for their preference for English were unrelated to mathematics learning but, rather, were about the need to ensure that students can gain access to social goods that fluency in English makes available.

The analysis presented above highlights the teachers' preference for English as the LoLT. A glaring absence in the teachers' discourses is any reference to how learning and teaching in English, as they prefer, would promote their students' access to mathematics knowledge and success. The teachers interviewed regarded teaching mathematics in English in these multilingual classrooms as another opportunity for students to gain fluency in English. Explanations for preferred language(s) for mathematics teaching focussed on English and not mathematics. These teachers positioned themselves in relation to English (and so socio-economic access) and not mathematics (and so epistemological access).

Perspectives of teachers in Catalonia The perception that languages spoken by immigrant students are directly connected with underachievement in school mathematics appears in many teachers' discourses. When asked about how the teaching of mathematics can be improved in multilingual classrooms, teachers first suggested increasing the teaching of the LoLT, that is, to have a stronger focus on the learning of Catalan. This is expressed with sentences like: "When they [immigrant students in mainstream Catalonian schools] speak Catalan, their problems will be solved." Other teachers put the emphasis on the students' transition processes by pointing to discourses on "normality" and "difference": "They are good at learning the way things are here quickly."

Below is part of the conversation in one of the meetings with the teachers of the Critical Mathematics Education Group (for more details on this group see Planas & Civil, 2009). Since 2005, Núria has been co-ordinating this group of secondary mathematics teachers with experience in multilingual classrooms. In one of the sessions, two of the Catalan teachers, Cesca and Anna, talked about the idea of having only one language that "helps", and that language being Catalan.

– Núria: What could you tell other teachers from your experience in multilingual mathematics classrooms? You said it's hard but it's worth it…
– Cesca: My experience tells me that Catalan helps with the mathematics. I'd like very much knowing their languages but I'm not good at that. And I really could not lead a mathematical discussion with so many languages all together. Catalan helps, though they can summarize the main ideas to each other by using Spanish, Arabic, or whatever.
– Anna: Yes, Catalan… it really helps.
– Núria: In your case, Anna, most students are Latin American, right?
– Anna: Yes. It's not clear why it's difficult for them to speak Catalan. Maybe because we also make efforts to speak a sort of mixed language that is neither Catalan nor Spanish. I talk to them in Catalan because I want them to know the language of the place they are living in. How long would you expect their learning to be if we, the teachers, use Spanish?

These teachers and others in the group referred to achievement and underachievement questions in terms of advances in the learning of Catalan. The challenge was identified as being the fact that some students are dominant in their home languages while they learn mathematics in an additional language (Catalan), which they are still learning. As shown in the extract above, the students' home languages were not valued by the teachers, who could not imagine a mathematical discussion with "so many languages all together." Somehow this means that the teachers did not recognise the power of thinking mathematically in a language different to Catalan, except during the time devoted to small group work.

The teachers' focus on the social dimension of "access" When examining the teachers' perspectives a shared finding has to do with the teachers putting the emphasis on the importance of gaining fluency in the LoLT (English in South Africa and Catalan in Catalonia). Despite the fact of being teachers of mathematics and responding to questions during the interviews in which the mathematics was made explicit, the teachers argued in favour of using either English or Catalan to ensure access to social goods, and rarely referred to issues concerning epistemological access. In Barcelona, Cesca talked about "Catalan helping with the mathematics" but she did not explain why it should be easier to learn "fractions" or "isometries" in Catalan in the cases in which this is not the students' home language. This differs from Anna's argument when saying, "I want them [the Latin American students] to know the language of the place they are living in." In the interviews with the South African teachers, a similar positioning is expressed by exemplifying concrete social goods such as "coping in higher classes" where English is the LoLT.

Gaining epistemological access in a multilingual mathematics classroom includes how the mathematics curriculum is mediated as well as the teaching and language practices used. The fact that these mathematics teachers considered "access" only from a social perspective and with very weak connections to the mathematics and the curriculum is itself problematic. For instance, one of the South African teachers says that the textbooks and the question papers are written in English, but she does not reflect on the textbooks and question papers themselves. In this way, she positions herself as a helpless teacher who just implements what is prescribed rather than one who can shape what should be prescribed. A productive discourse would be one that, instead, engages with the possibilities of adapting the use of an English mathematics textbook with students who are still developing fluency in English. We argue that by showing a strong awareness of the social dimension of "access", the teachers in our studies have a "sensitivity" that is absolutely necessary. Nevertheless, at the same time they lose the opportunity to stress the complementary epistemological dimension of "access" and do not go into a more practical curricular discussion.

Together with the tension between gaining access to mathematics and gaining access to social goods, other dialectical tensions are represented by discourses of the teachers in the two settings. The dilemma between participation in the "normalized" international world and participation in the "diverse" local communities is real for many multilinguals. Most teachers position themselves in clear ways when arguing in favour of English as the international language and Catalan as the "normalized" LoLT in classrooms with students whose home languages are not, respectively, English or Catalan. Nevertheless, the adoption of these practices does not imply a disregard for practices that strive to retain the students' "minority" worldviews. Many teachers do not consider these sets of practices as incompatible. Vithal and Skovsmose (1997) indicate the invisibility of this particular tension between the promotion of dominant and minority worldviews.

> Cultural approaches to education in multicultural societies often assume that cultures are compatible and in harmony within themselves and with each other. In so doing they render invisible any conflicts that can and do exist and hence preclude the development of strategies for coping with them. The consequence is that teachers who employ cultural approaches appear not to notice conflicts which do exist or in the face of conflicts simply stop using such approaches. (Vithal & Skovsmose, 1997, p. 146)

Teachers may acknowledge the linguistic, cultural, and social capital of their students (e.g. "They are good at learning the way things are here"); and at the same time, they may perceive complications for the students' academic and social promotion linked to certain school practices (e.g., "How long would you expect their learning to be if we, the teachers, use Spanish?").

Students' perspectives

In our view, the voices of marginalised students are not adequately represented in mathematics education research, hence we consider students' perspectives. The work of authorizing students' perspectives is essential in our research because it helps introduce the voices of those who daily experience the effects of existing educational policies-in-practice. We want to help address and redress the fact that students tend to be silenced in the comprehension of the school context, and in the analysis of what elements of schooling need to be changed.

In our analysis below, what can be seen is that students have a clear preference for one language, as we have documented with the teachers in our two contexts of research. Again the social and political structures ground official discourses that favour the students' "enthusiasm" for learning in a language that is not their own, even in the cases of students who have difficulties understanding the LoLT, and this reality gives them a hard time. Accordingly, language diversity is not considered as an appropriate reality for the school context, and in particular for the mathematics classroom. These students' ideal picture of language use in their school is precisely what they are experiencing with having only one LoLT.

Perspectives of students in South Africa All of the interviewed students are multilingual as they have fluency in at least four languages. They were all at Grade 11 level and studying mathematics in English, which is not their home language. During the individual interviews, students were asked the following questions in English, "Which language or languages do you prefer to be taught mathematics in? Why?" Students were also given an opportunity to choose their preferred language for the interview. All of the students chose to be interviewed in either their home language or a mixture of English and their home language.

While there were conflicting discourses in the students' views, what was clear was that the majority of students expressed their preference to be taught mathematics in English. For these students, learning mathematics in English is not so much about choice; it is just how things should be. Examples of how the students expressed this sentiment are, "English is an international language; just imagine a class doing maths in Setswana" and "It is the way it is supposed to be because English is the standardized and international language." For these students it is unimaginable for mathematics to be taught in any other language. The use of English as a LoLT for mathematics is common sense to them; they simply cannot imagine mathematics without English. Among the reasons why they want to be taught in English is the fact that mathematics textbooks and examinations are in English, university lectures and job interviews are only in English, and communication with "white people" is in English. All this contributed to the discourse that without fluency in English a student would not have access to social goods such as higher education and employment, which suggests that, like teachers, these students saw mathematics learning as another opportunity for improving their fluency in English. This was the case even for the two students who indicated that for them it does not really matter what language is used for teaching and learning because mathematics is a language on its own. Below is an extract from the interview with one of the two students, whose home language is isiZulu.

> Mamokgethi: So if you had a group of students who want to do maths in isiZulu, what would you say to them?
>
> Lehlohonolo: That's their own problem because if they get out of high school, they cannot expect to find an Indian lecturer teaching mathematics in isiZulu. English is the simplest language that everyone can speak so they will have to get used to English whilst they are still here.

Throughout the interview, Lehlohonolo never connected success in mathematics or lack of it to fluency in English. However, in the above extract he argues for the importance of gaining fluency in English before completing high school. The sentiment that English is bigger than us and thus cannot be avoided or ignored because in higher education no lecturer will be able to teach mathematics in the students' home languages is evident in Lehlohonolo's discourse. He even draws in the issue of race by referring to a hypothetical Indian mathematics lecturer. Hidden in what Lehlohonolo is saying above is a suggestion that there are no isiZulu speaking mathematics lecturers at university. This indicates the strong connection between the politics of language and race and mathematics teaching and learning.

Despite the overwhelming discourse that foregrounds the hegemony of English and the need to gain access to social goods that English makes possible, there are differences in the manner in which different students positioned themselves. The students who explicitly indicated that it does not really matter what language mathematics is taught in positioned themselves in relation to mathematics. Their language preferences were connected to gaining proficiency in mathematics rather than gaining fluency in English. The rest of the students positioned themselves in relation to English in the sense that they were more concerned with gaining fluency in English so that they can access employment and higher education. Their desire to gain fluency in English was not connected in any way, at least explicitly, to improving their mathematics learning but to access to social goods. As a result, they saw mathematics teaching and learning in multilingual classrooms as an opportunity to gain fluency in English.

Perspectives of students in Catalonia In Catalonia, there are important similarities in the students' answers concerning their "preferred" language while learning mathematics in the class. In individual interviews with ten Latin American immigrant students in mainstream Catalonian schools, the students themselves do not seem to expect to learn mathematics through their home language.

They express their concern that using Spanish "too much" could have negative effects on their mathematical achievement by stating, for example: "I speak Spanish with my peers in the small group work but try to speak Catalan the rest of the time, to go on with the mathematics." The conversation below with Julio, a student aged 12 from Venezuela, indicates how he sees his home language, Spanish, as a weak resource in the teaching and learning of mathematics in Barcelona. At the time of the interview, he was still in the parallel system of special classes[2] with other late arrival immigrant students who had been classified like him as not having a sufficient knowledge of the LoLT.

- Núria: Why don't you like the idea of getting into the regular classes?
- Julio: It's not that I don't like it, but that I'm fine here.
- Núria: You mean in the special class, do you?
- Julio: Yes.
- Núria: Why?
- Julio: I'm learning Catalan.
- Núria: And what about the mathematics?
- Julio: That comes later.
- Núria: When?
- Julio: When I need fewer efforts with the understanding.
- Núria: The understanding of Catalan?
- Julio: Yes.
- Núria: But now we are speaking Catalan and you do it great.
- Julio: I can do it even better.

A conversation with Julio's father also reinforced this idea of Catalan being the "right" and "preferred" LoLT. Julio's father indicated that he does not want his children to be taught through their first language because it was not the language that was going to help them if living in Catalonia. This is consistent with what Julio and other Latin American students said about "wanting the opportunity" to learn basic Catalan as quickly as possible in the parallel system of special classes for late arrival immigrant students, and mathematics classes helping them to improve their knowledge of Catalan. The issue of focusing on the "right" to be exposed to the LoLT shows that these students are positioning themselves in relation to Catalan and not mathematics. As in the interviews with the teachers in the Catalonian context, these students did not connect the use of home languages with the learning of mathematics.

The students' choice to learn mathematics in English/Catalan Like the teachers, the students in our work put the emphasis on the language rather than on the mathematics. They also refer to the social dimension of "access" and more explicitly to the idea of having the right to choose the language for their learning of mathematics. They enact this belief in the power of English by indicating preference for English or Catalan as LoLT. This view is clearly represented by Julio, who tries to exclude himself from the system of regular classes in which students have more lessons of mathematics per week, because "he can do it even better" with the Catalan. In neither Catalonia nor South Africa have the students expressed a concern with the possibility of losing mathematics learning opportunities as a result of their limited fluency in the LoLT. For them, what is important is utilising their primary and secondary school years to gain fluency in Catalan/English.

As indicated earlier, most of these students do not directly connect success, or lack of it, in mathematics to fluency in Catalan/English. They actually position the learning of mathematics as secondary to gaining fluency in Catalan/English. It is not that they do not value the importance of mathematics in their schooling and the role of this subject in opening up their opportunities for social and academic promotion. However, they interpret the learning of Catalan/English as more important in their respective "perceived" school, social, and political contexts. Lehlohonolo says, "English is the simplest language that everyone can speak"; therefore it makes sense to argue for the right to learn in this language and not in isiZulu, for instance. They do not seem to appreciate the fact that the complex relationship between language proficiency and mathematical achievement still remains even after having acquired a good knowledge of the LoLT. In our view, while successful learning of mathematics is enabled in contexts where the students are fluent in the LoLT, we agree with Setati, Chitera, and Essien (2009) that student performance (and by implication, mathematical achievement) is determined by a complex set of interrelated factors.

> Poor performance by multilingual learners thus cannot be solely attributed to the learners' limited proficiency in English (suggesting that fluency in English will solve all problems) in isolation from the pedagogic issues specific to mathematics as well as the wider social, cultural and political factors that infuse schooling. (Setati et al., 2009, p. 73)

In general, what we have learned during the interviews with teachers and students is that within school mathematics teaching and learning there is no recognition of the political role of language and how it plays itself out in multilingual classrooms. There is a need to raise the level of these students' "empowerment" so that they ask for the same right to access to mathematics.

GENERAL DISCUSSION

So far, through our interpretation of the cases of Catalonia and South Africa, we have highlighted the complex relationship between language choice, participation, and mathematical learning in multilingual classrooms. We have paid special attention to the political dimension of this relationship. To understand the problem of unequal learning opportunities, we claim the need to explore different "marks[3]" that make it difficult for certain groups of students to participate in the mathematics classroom. We have focused on groups of language minority students in mainstream classes who do not have the LoLT as a "native" or home language. For these groups, we assume that things such as language choice, language accents, and language ideologies are at the origin of social marks and academic behaviours that have a strong influence on the students' achievement. We have examined political "common sense" realities in which languages other than English or Catalan are associated with lack of opportunities and discourses of difference.

Most of the students and teachers in our work in Catalonia and South Africa attach high value to being fluent in the language of power. Moreover, all of these students and teachers are multilingual, which means that they are able to communicate in both their home languages (e.g. Spanish, Setswana, isiZulu) and in Catalan/English. In the case of the students, the fact that their home languages are not their "preferred" languages of learning and teaching speaks of a "pragmatic" choice, and provides insight into the relative value that they attach to their home languages in general. There is therefore evidence of different valorisations of different languages.

Our data, however, tell a more complex story than the issue of "choice" itself. It is a story of languages than are not supported enough within the society, and of other languages that are marked with the ideals of social promotion and academic success. We have the status of English in South Africa as an international language, and that of Catalan as the language of the "native" middle social class in Catalonia. Hence it is not only a question of having different languages, but overall a question of these languages and their speakers having different valorisations and voices. The problematic issue is not the language per se, but the values given to the language and those who speak it. The extent to which students will be enabled to participate in the mathematics classroom has to do with the knowledge and acceptance of certain values attached to the official LoLT.

Discourses on the "inappropriateness" of minority languages in the school context provide arguments for the imperative to learn "standard" Catalan/English. In Catalonia this has led to two parallel systems of classes, one with an intensive Catalan as a Second Language program in the so-called "special classes" to facilitate fluency in Catalan as quickly as possible. The role given to language in the LiEP and in the students and teachers' perceptions, however, contrasts with the creative use of language diversity in some of the mathematics classrooms from our case studies in Catalonia and South Africa. While teachers may be in favour of monolingual settings, these settings are not easy to translate into practice. In our view, this is due to the challenges that are imposed on classroom settings where students learn mathematics in a language that is not their own.

In Catalonia, despite the absence of language guidelines for (mathematics) classroom practices, and the segregation in "special classes" for late arrival immigrant students who do not speak (sufficient) Catalan, some teachers express a need for understanding how practices of code switching would help the teaching and learning of mathematics. There is a pedagogic debate started on how a "controlled" incorporation of the students' home languages could be achieved. Hence an increasing tension appears between discourses on monolingualism in institutional contexts and more flexible and sometimes hidden discourses on language diversity in practice contexts. These tensions are also expressed through the experiences of teachers, who sustained in the interviews the monolingualism power discourse and at the same time promote situations of home language use in their teaching by drawing on practices such as code switching (see Planas, Iranzo, & Setati, 2009).

The situation in South Africa is different from Catalonia since there is a formal policy that prescribes eleven official languages and encourages multilingualism. The challenge, however, is in the valuing of these languages. While the South African language policy in South Africa is intended to address the undervaluing of African languages, in practice English continues to dominate. Although it is the main language of a minority, English remains both the language of power and the language of educational and socio-economic advancement; that is, it is a dominant symbolic resource in the linguistic market (Bourdieu, 1991, 1998) in South Africa.

> Cultural and linguistic unification is accompanied by the imposition of the dominant language and culture as legitimate and by the rejection of all other languages into indignity. (Bourdieu, 1998, p. 46)

The linguistic market is embodied by and enacted in the many key situations (e.g. educational settings, job situations) in which symbolic resources, like certain types of linguistic skills, are demanded of social actors if they want to gain access to valuable social, educational, and eventually material resources. It is now more than ten years since this language policy was instituted and English remains the language of politics, media, commerce, and higher education. It is not surprising, therefore that South African teachers and students think it unimaginable that mathematics can be taught in any other language than English. While the political landscape is now different and the language policy has changed, what remains is the hegemony of English, which is fuelled by the desire of the formerly oppressed masses to gain access to a language that they were denied access to, as well as the social goods that accompany it. It is this symbolic power of English that makes families, teachers, and students want to strive for proficiency in English, even when it is at the expense of what Morrow (1994) refers to as epistemological access, namely access to mathematical knowledge and information. Analysis of data from South Africa shows that the quest for access to social goods predominates over that for epistemological access. This makes the progressive LiEP and research hard to translate into practice in South African classrooms.

FUTURE RESEARCH

In our studies we work with students who not only learn mathematics in a language that is not their home language but who also come from low socio-economic backgrounds and are of a different race – for example, in South Africa it is black South African township students and in Catalonia it is what we refer to as non-European Union students in urban suburbs. In our work together we have come to agree that we both use language as a proxy for race and socio-economic class (see also Setati & Moschkovich, 2010). Nevertheless, our work does not directly address issues of how race and socio-economic class impact mathematics education in multilingual contexts. This is an important matter for future research to consider. Our own position is that race and class are highly interconnected with issues of language, but the connection is much more complex than is sometimes assumed (see Lubienski, 2001, for a discussion on the isolated examination of race, class and gender in the field of mathematics education research and some of the implications of this tradition in the analysis of students' mathematics achievement in the United States).

Rather than viewing race and class as fixed categories that determine the use and learning of the language of instruction among particular racial and socioeconomic groups, poststructuralist theories (see the work by Gutiérrez, 2010) explore how race and class are shifting categories that get constructed by discourses. The big issue then is how people with certain racial and socioeconomic status get positioned or position themselves in relation to mathematics and the language used for learning and teaching it. For non-European Union students in the area of Barcelona, who come from Bolivia for instance, race may be somewhat fixed – especially if these students have "indigenous" features – but the construction of a racial identity is variable and part of a complex process of socialization with implications for the possibilities of learning. This poststructuralist approach avoids reinforcing the classical dichotomy between middle-class and working class, or white people and those of African descent, to move towards a more sophisticated knowledge of how and why certain groups learn mathematics in a classroom with a language that is not their own.

In our view, race and socioeconomic class need to be unpacked according to diverse power relationships. All language, race, and class discourses are socially related to discourses on "difference" that have partially substituted previous discourses on "deficit". One of the questions is whether these discourses on difference have been also incorporated in the research without explicitly clarifying the political dimension that the notion of difference has. By not paying enough attention to power, researchers can easily refer to language, race, and class as if they were unique conditions for specific groups. Quite often, social class is only mentioned in research in which the students are thought of as working-class and race is only mentioned when the students are, for example, of African descent. It is similar to what happens with studies on gender, in which the notion of gender has practically become synonymous with "groups of girls". Discourses on difference do not necessarily need to have reductionist consequences in research; in fact, they can help put the emphasis on the socio-political construction of differences for the purpose of either reproducing or opposing power relationships. The problem, however, is that the emphasis on difference turns into a way of only marking concrete differences.

Further research in multilingual mathematics classrooms needs to seek frameworks that investigate issues of language, race, class, and power from the perspective of how differences (of language, race, and class) are constructed, and what is the dynamic role of power in the orchestration of these differences. Therefore, the focus of research needs to evolve towards students who learn mathematics in a language that is not their home language and become marginalized due to the dominant perspectives on various parts of their language, racial, and class identities. Possible research directions might consist of stimulating dialogue about interactions among language, race, and class differences as well as issues of power in relation to these differences, rather than examining on its own the topic of language, or that of race, as if they had a pure and fixed existence in the social arena of the mathematics classroom and the institutions. From the very beginning, simple and common expressions such as "native language" and "indigenous people" are strongly charged with issues of power by suggesting that only formerly colonised people are indigenous and their languages native. In our view this discourse of naming "the other" needs critique. Much remains to be done in that direction.

NOTES

1. Setswana is one of the 11 official languages in South Africa. The other official languages are: isiZulu, IsiXhosa, TshiVenda, Xitsonga, Sesotho, Isindebele, Siswati, Sepedi, Afrikaans, and English. According to the 2001 census, Setswana is the primary, or main, language of 8.2% of the population in South Africa, which is the same as the percentage of those for whom English is the primary language.
2. The existence of a parallel system of special classes for groups of late arrival immigrant students introduces the physical dimension of "access", together with the social and epistemological dimensions of this notion.
3. We use the word "marks" to mean negative stereotypes about people that become substitutes for experience and reduce our understanding.

REFERENCES

Adler, J. (2001). *Teaching mathematics in multilingual classrooms.* Dordrecht, The Netherlands: Kluwer.

Barwell, R. (2005). Empowerment, EAL and the National Numeracy Strategy. *International Journal of Bilingual Education and Bilingualism, 8*(4), 313–327.

Barwell, R. (Ed.) (2009). *Mathematics in multilingual classrooms: Global perspectives.* Bristol, England: Multilingual Matters.

Bourdieu, P. (1991). *Language and symbolic power.* Cambridge, MA: Polity Press.

Bourdieu, P. (1998). *Practical reason.* London: Polity Press.

Cummins, J. (2000). *Language, power and pedagogy: Bilingual children in the crossfire.* Clevedon, England: Multilingual Matters.

Gee, J. P. (1999). *An introduction to discourse analysis: Theory and method.* New York: Routledge.

Gutiérrez, R. (2002). Enabling the practice of mathematics teachers in context: Toward a new equity research agenda. *Mathematical Thinking and Learning, 4*(2&3), 145–187.

Gutiérrez, R. (2010). The sociopolitical turn in mathematics education. In R. Gutiérrez (Ed.), *Journal for Research in Mathematics Education, Special Equity Issue,* 1–32. Reston, VA: National Council of Teachers of Mathematics.

Gutstein, E. (2003). Teaching and learning mathematics for social justice in an urban, Latino school. *Journal for Research in Mathematics Education, 34*(1), 37–73.

Gutstein, E. (2006). *Reading and writing the world with mathematics: Toward a pedagogy for social justice.* New York: Routledge.

Hartshorne, K. B. (1987). Language policy in African education in South Africa, 1910–1985, with particular reference to the issue of medium of instruction. In D. Young & L. W. Lanham (Eds.), *Bridging the gap between theory and practice in English Second Language Teaching: Essays in honour of L. W. Lanham* (pp. 62–81). Pretoria, South Africa: Maskew Miller Longman.

Khisty, L. L. (1995). Making inequality: Issues of language and meanings in mathematics teaching with Hispanic students. In W. G. Secada, E. Fennema, & L. B. Adajian (Eds.), *New directions for equity in mathematics education* (pp. 279–297). Cambridge: Cambridge University Press.

Khisty, L. L. (2006). Language and mathematics: Toward social justice for linguistically diverse students. In J. Novotná, H. Moraová, M. Krátká, & N. Stehlíková (Eds.), *Proceedings of the 30th Conference of the International Group for the Psychology of Mathematics Education* (vol. 3, pp. 433–440). Prague, Czech Republic: PME.

Lubienski, S. T. (2001, April). *A second look at mathematics achievement gaps: Intersections of race, class, and gender in NAEP data.* Paper presented at the annual meeting of the American Educational Research Association, Seattle, Washington.

Mar-Molinero, C. (2000). *The politics of language in the Spanish-speaking world: from colonisation to globalisation.* London: Routledge.

Morrow, W. (1994). Entitlement and achievement in education. *Studies in Philosophy and Education, 13*, 33–37

Moschkovich, J. N. (1999). Supporting the participation of English language learners in mathematical discussions. *For the Learning of Mathematics, 19*(1), 11–19.

Moschkovich, J. N. (2002). A situated and sociocultural perspective on bilingual mathematics learners. *Mathematical Thinking and Learning, 4*(2&3), 189–212.

Phillipson, R. (1992). *Linguistic imperialism.* Oxford: Oxford University Press.

Phillipson, R. (2003). *English-only Europe? Challenging language policy.* London: Routledge.

Planas, N. (in press). Bilingual students acting as monolingual in two different contexts of the mathematics classroom. In G. H. Gunnarsdóttir et al., (Eds.), *Proceedings of the 11th Nordic Conference on Mathematics Education.* Reykjavík, Iceland: University of Iceland.

Planas, N., & Civil, M. (2009). Working with mathematics teachers and immigrant students: An empowerment perspective. *Journal of Mathematics Teacher Education, 12*(6), 391–409.

Planas, N., & Civil, M. (2010a). Discourse processes in critical mathematics education. In H. Alrø, O. Ravn & P. Valero (Eds.), *Critical mathematics education: Past, present and future* (pp. 145–160). Rotterdam, The Netherlands: Sense Publishers.

Planas, N., & Civil, M. (2010b). El aprendizaje matemático de alumnos bilingües en Barcelona y Tucson [Bilingual students learning mathematics in Barcelona and Tucson]. *Quadrante, 29*(1), 5–28.

Planas, N., Iranzo, N., & Setati, M. (2009). Language switching with a group of bilingual students in the mathematics classroom. In M. Tzekaki, M. Kladrimidou, & H. Sakonidis (Eds.), *Proceedings of the 33rd Conference of the International Group for the Psychology of Mathematics Education* (vol. 4, pp. 393–400). Thessaloniki, Greece: PME.

Planas, N., & Setati, M. (2009). Bilingual students using their languages in their learning of mathematics. *Mathematics Education Research Journal, 21*(3), 36–59.

Reagan, T. G., & Ntshoe, I. (1992). Language policy and black education in South Africa. In I. A. Stevenson (Ed.), *Sociolinguistics for applied linguistics* (pp. 25–52). Pretoria, South Africa: University of Pretoria Press.

Setati, M. (2002a) *Language practices in intermediate multilingual mathematics classrooms.* Unpublished doctoral dissertation. University of the Witwatersrand, Johannesburg.

Setati, M. (2002b) Researching mathematics education and language in multilingual South Africa. *The Mathematics Educator, 12*(2), 6–20.

Setati, M. (2005). Learning and teaching mathematics in a primary multilingual classroom. *Journal for Research in Mathematics Education, 36*(5), 447–466.

Setati, M. (2008). Access to mathematics versus access to the language of power: The struggle in multilingual classrooms. *South African Journal of Education, 28*(1), 103–116.

Setati, M., & Adler, J. (2000). Between languages and discourses: Language practices in primary mathematics classrooms in South Africa. *Educational Studies in Mathematics, 43*(3), 243–269.

Setati, M., Adler, J., Reed, Y., & Bapoo, A. (2002). Incomplete journeys: Code-switching and other language practices in mathematics, science and English language classrooms in South Africa. *Language and Education, 16*(2), 128–149.

Setati, M., & Barwell, R. (2006). Discursive practices in two multilingual mathematics classrooms: An international comparison. *African Journal of Research in Mathematics, Science and Technology Education, 10*(2), 27–38.

Setati, M., Chitera, N., & Essien, A. (2009). Research on multilingualism in mathematics education in South Africa: 2000–2007. *African Journal for Research in Mathematics, Science and Technology Education, 13* (Special Issue), 64–79.

Setati, M., Molefe, T., & Langa, M. (2008) Using language as a transparent resource in the teaching and learning of mathematics in a Grade 11 multilingual classroom. *Pythagoras, 67*, 14–25.

Setati, M., & Moschkovich, J. N. (2010). Mathematics education and language diversity: A dialogue across settings. *Journal for Research in Mathematics Education. Special Equity Issue.* Reston, VA: National Council of Teachers of Mathematics.

Strubell, M. (2006). A language policy overview with special reference to Catalan. In J. Lainio (Ed.), *Spain and its languages: A comparative view on the regional and minority languages policies of Spain and Sweden* (pp. 65–95). Mälardalen, Sweden: Mälardalen University.

Taylor, N., & Vinjevold, P. (1999). *Getting learning right.* Johannesburg, South Africa: Joint Education Trust.

Vithal, R., & Skovsmose, O. (1997). The end of innocence: A critique of "ethnomathematics". *Educational Studies in Mathematics, 34*(2), 131–157.

Zentella, A. C. (1997). *Growing up bilingual: Puerto Rican children in New York.* Malden, MA: Blackwell.

Mamokgethi Setati
University of South Africa

Núria Planas
Universidad Autònoma de Barcelona

CHAPTER 8

GENEALOGY OF MATHEMATICS EDUCATION IN TWO BRAZILIAN RURAL FORMS OF LIFE

GELSA KNIJNIK AND FERNANDA WANDERER

INTRODUCTION

In this chapter, we aim to develop a genealogical analysis of Brazilian geopolitics and mathematics education[1]. We consider that mathematics education must be understood in a broader sense, which means not to restrict it to what happens at institutions of formal education. In particular, it goes beyond school mathematics. For example, as discussed before (Knijnik 2007b), the educational process developed by the Brazilian Landless Movement[2] (in Portuguese, *Movimento Sem Terra* – MST) throughout its history must be seen beyond schooling, since each Landless subject educates her/himself through her/his participation in the everyday life of their communities and also through the wide range of political activities developed by the Movement. In this enculturation process to which the Landless people are subjected, they learn how to use the language games that constitute their mathematics.

Based on Later Wittgenstein's ideas, we assume that there are many different mathematics, like the Brazilian Landless peasant mathematics described by Knijnik (2007a, 2007b) and the *Costão* German-descendent settlers' mathematics studied by Wanderer (2007)[3]. The first section of the chapter briefly clarifies the philosophical thinking that gives support to that assumption. The section ends with empirical data from Knijnik and Wanderer's above-mentioned studies.

The second section of the chapter presents a genealogical analysis of mathematics education in two Brazilian time-space forms of life. Although both are situated in the country's southernmost state, one occurred during the mid-20th century, while the other is connected to a national social movement which emerged about 25 years ago in this part of the country, and is now considered the most important social movement in Latin America. The genealogical analysis makes explicit how school mathematics in those two different time-space forms of life works as a gear in the production of what Hardt and Negri (2003) call "differential inclusion". Taking genealogy as an analytical approach means following Michel Foucault's understanding about this notion. For him, genealogy "rejects the metahistorical deployment of ideal significations and indefinite teleologies. It opposes itself to the search for 'origins'" (Foucault 1977, p. 140). Inspired by Nietzsche, Foucault argues in favour of effective history, which considers "knowledge as a perspective" (p. 156). Effective history "deals with events in terms of their most unique characteristics, their most acute manifestations… The forces operating in history are not controlled

O. Skovsmose and B. Greer (eds.), Opening the Cage: Critique and Politics of Mathematics Education, 187–202.

by destiny or regulative mechanisms, but respond to haphazard conflicts" (p. 154). The philosopher considers that "it is necessary to master history so as to turn it to genealogical uses, that is, strictly anti-Platonic purposes. Only then will the historical sense free itself from the demands of a suprahistorical history" (p. 160).

DIFFERENT FORMS OF LIFE, DIFFERENT MATHEMATICS

Postmodern times have been characterized by the proliferation of multiple interpretations of the social world, at the same time as there begins a "sort of suspicion of the place from which these interpretations are constructed, i.e., of the idea of reason itself" (Condé, 2004, p. 16). According to this author, from the second half of the 19th century and the beginning of the 20th, with the crisis in mathematics, the theory of evolution, the rise of human sciences, relativity theory in Physics, and other movements, the rejection of the idea of a universal scientific rationality based on ultimate and true foundations was triggered. Using the ideas of Wittgenstein, Condé (2004, p. 29) says that "… we need friction. Back to the rough ground (PI §107)[4] of the social practices, and there to establish the criteria of our rationality". Returning to the rough ground drives us to regard the Modern Project and, consequently, Modern Science with suspicion. In particular, it allows us to problematise the existence of a unique and totalizing mathematics language, sustained by a specific rationality with its marks of asepsis, order, and abstraction.

In his later work, Wittgenstein repudiates the notion of an ontological foundation for language. Language takes on a contingent, particular character, acquiring meaning through its different *uses*. "The meaning of a work is its use in language", explains the philosopher (PI §43). In this way, since the meaning of a word is generated by its use, the possibility of essences or fixed guarantees for language is problematised, leading us to also question the existence of a single mathematical language with fixed meanings.

Highlighting the generation of diversified languages that gain meanings by their uses, Wittgenstein (1995) introduces the notion of language games as being the "whole, consisting of language and the actions into which it is woven" (PI §7). Hence, processes such as describing objects, reporting events, building hypotheses and analyzing them, telling stories, solving calculations, and others, are exemplified by Wittgenstein as language games.

According to Wittgenstein's interpreters such as Moreno (2000, p. 56), with the expression "language game" the philosopher points to the relevance of the praxis of language, emphasizing that understanding the meaning is not a matter of seeking a logical and definitive determination which could apprehend it "once and for all", but the purpose is to analyze the criteria "supplied by the use we make of language in many different games, i.e., in the different forms of life".

In aphorism 23, Wittgenstein states that language games are part of a form of life, which leads Glock (1996, p. 124) to highlight that the notion of form of life emphasizes the "intertwining of culture, world-view, and language" or, as Condé (1998, p. 104) writes: "The form of life is the last mooring place of language", i.e., the meaning of the language games that constitute the different mathematics and the rationality criteria embedded in them are constituted in the materiality of the forms of life in culture. Thus, academic mathematics, school mathematics, peasant mathematics, indigenous mathematics, in brief, the mathematics generated by specific cultural groups, can be understood as networks of language games engendered in different forms of life. However, these different games do not have an invariable essence that maintains them completely incommunicado from each other, nor a property common to all of them, but some analogies or relationships – what Wittgenstein (1995) calls "family resemblances". According to the philosopher, language games have "a complicated network of similarities overlapping and criss-crossing: sometimes overall similarities, sometimes similarities of detail" (PI §66). And, he explains: "I can think of no better expression to characterize these similarities than 'family resemblances'; for the various resemblances between members of a family: build, features, colour of eyes, gait, temperament, etc. etc. overlap and criss-cross in the same way – And I shall say: 'games' form a family" (PI §67).

These Wittgensteinian formulations allow us to consider the mathematics produced in different forms of life as networks of language games that have family resemblances between them. In Condé's words: "What exists are the different aspects of language that are expressed through language games which are multiple, varied and, mainly, particular" (1998, p. 124)[5]. Hence, it can be said that there are no universal superconcepts that can serve as parameters for others. Distinct mathematics language games have analogies, similarities that permeate them and allow for engendering different criteria of rationality. Different games resemble each other, they have analogies, similarities that penetrate them and afford engendering different rationality criteria.

Later Wittgenstein's ideas briefly presented here are at the kernel of the discussion undertaken in this chapter. Knijnik (2007a, 2007b) and Knijnik, Wanderer, and Oliviera (2005), taking support from empirical studies performed with the MST, have shown that the language games that shape the Landless Peasant mathematics bear the marks of the orality of that peasant culture. Thus, for instance, to find how much could be made available monthly, during one year, with the 900 reais[6] obtained from the sale of thirty sacks of ecological rice, Seu Otílio – a 64-year old peasant who only had 4 years of schooling – explained:

> … we tried to know how much would be over to spend every month. For instance: nine hundred reais divided by twelve. Out of one thousand two hundred to get it to nine hundred reais, you had to take a quarter out of one hundred, which would become seventy-five reais. Because you take it out of ten, you have two and a half, making up the logic of ten ... So, as I reckon it, in this case there would be seventy-five reais a month to buy the other things. Any person who wants to use the machine or the pen will reach this value, I'm sure … When I reckon it in my head I always have to look for the best path. I always have to round it out, to look for the large numbers. The closest, simplest way is to bring it to one thousand and two hundred reais. According to this logic it would be one hundred, but it could not be one thousand and two hundred because it is nine hundred. The twelve have numbers the same size as those that form nine. The nine can be formed by three times three, and twelve, four times three. So you have to take the total and see that twelve has

> a quarter more than the nine as a difference. This one quarter more is what I added, so I have to take it off the hundred (Knijnik & Wanderer, 2008).

As shown by the excerpt above, the language game played by Seu Otílio considers the orders that are most relevant to find the final value. When he was asked about the ways in which he performed such language games, Seu Otílio said:

> I always tried to get to know and practice the three kinds of ways to do mathematical sums. I always used my memory, which I place first. I have also always used the pen. I use the pen a lot to calculate and check large sums, in which one becomes very tired and have to record it. And another thing I have also used is the little machine. What I learned today [one of the mathematics classes of the Course] was to operate those memories [of the calculator] that I had never managed to get explained, so I was treading water. One would buy the little machine and only use it to add and divide. And one has to know all of them, and realizes what does not fit … But, actually, I can reckon very well in my head, right. I can reckon very well in my head. And I even like to. But my logical reasoning about the numbers is always in my head. I always approach; I can't switch off the reasoning for reckoning the idea, with the machine sum or of the pen. I can reckon with a pen, but I always project so many bags will give, more or less so much. I have always practiced this and I think it is very good. Because one manages to see if the sum is wrong, one can realize that it is wrong. When you reckon it by pen, or even on the machine, I can see it immediately, OK. But this is not right. Because I have already projected it this way. So what I was trying to find out is how to theorize this. (Knijnik & Wanderer, 2008)

In Seu Otílio's description, different language games can be identified: those of peasant mathematics – which use the "reasoning of sums through ideas", the language games connected to school mathematics in which he had been socialized – in his words, "pen sums", and the games involved in using the calculator – whose further learning had occurred during the pedagogical work that we were developing. However, despite the specificities of such games, Seu Otílio shows that he knows they bear a family resemblance, in the sense given to it by Wittgenstein.

Wanderer's study (2007) about the *Costão* settlers' form of life at the time of the Nationalization Campaign also shows how such a rural form of life was marked by orality. Converging with the empirical findings of Knijnik, Wanderer describes the language games of that rural culture, showing that they involved rules such as decomposition, estimating, and rounding. These rules were different from those that engendered the language games of school mathematics produced in that time-space, marked by writing, formalism, and abstraction.

When the participants of the study were asked about their oral mathematics practices they said that they did not learn them at school. As Seu Ivo said: "This was in practical life. In class, I learned nothing, in class only the multiplication tables we learned, only the multiplication tables". Another participant, Seu Seno, told that, at the time he attended school, the mathematical calculations "had to be done on the blackboard. Then, in the last years, in the third and fourth grades, one already had to write in the copybook". Besides positioning school mathematics as a knowledge marked by writing, Seu Seno also highlighted the requirement to "show how one does it". "One had to reckon it out. If you knew it in your head, you couldn't just write only the value there, you had to do the calculation, he [the teacher] wanted to see it."

In summary, Knijnik and Wanderer's empirical data (coming from two distinct Brazilian time-spaces) brought evidence of different mathematics language games practiced in rural forms of life in the south of Brazil: school mathematics and the peasant mathematics. So, based on empirical studies and using Later Wittgenstein's theoretical tools, it can be argued that there are different mathematics, each of which shapes and is shaped, inside a form of life, by a network of language games, which have specificities but also family resemblances. This argument drives us to the statement that there is more than a single and unique rationality, a single and unique mathematics, a single and unique set of rules by which individuals and cultures deal with space, time, and quantification processes – what in Western civilization is associated with the notion of mathematics. As shown elsewhere (Knijnik 2007a), it is important to understand how a single rationality among other rationalities – with its Eurocentric bias and its rules marked by abstraction and formalism – enters into the order of the pedagogical discourse – in Foucault's words, positioning the others as "wild exteriorities" (Foucault 2001, p. 35). In the next section, based on studies developed in two different time-space forms of life in the south of Brazil, it will be shown how school and, in particular, school mathematics works through what Hardt and Negri (2003) called "differential inclusion".

DIFFERENTIAL INCLUSION AND MATHEMATICS EDUCATION IN TWO BRAZILIAN TIME-SPACE FORMS OF LIFE

In his class of March 17, 1976, at Collège de France, Foucault (2002) goes further in the discussions about biopower, showing its connections with the mechanisms of racism. He highlights first that it can be considered as a means to insert a cutoff in life, "the cutoff between the one who is to live and the one who is to die" (Foucault, 2002, p. 304). Secondly, racism allows maintaining a relationship of the kind "to make people live, you must massacre your enemy", i.e. the "death[7] of the other, the death of the bad race, of the inferior race (or of the degenerate, or of the abnormal) is what will make life in general healthier; healthier and purer" (p. 305). Thus, "taking life … tends not to victory over political adversaries, but to the elimination of the biological danger and to the strengthening, directly linked to this elimination, of the species or race itself" (p. 306).

The arguments presented by Foucault converge with the analysis undertaken by Hardt and Negri (2003) on imperial racism. For them, even with the end of slavery and of the apartheid laws, it cannot be said that racist practices have diminished. On the contrary, they continue as intensely as ever, but now present themselves under different forms in our society. Étienne Balibar (cited by Hardt & Negri, 2003, p. 192) considers these new forms of racism as "a racism without race, or more precisely a racism that does not rest on a biological concept of race". As discussed by Hardt and Negri (p. 213), based on theorizations of Deleuze and Gattari, the imperial racist practice is not sustained by a theory of racial superiority in which there would be a binary division between races and exclusion processes, but by mechanisms that act as differentiated inclusion. Thus, for the authors, there is not, as a point of departure, a difference among races that can generate antagonistic blocks that separate those "inside" and "outside", but processes that act by inclusion and subordination. In their words:

> White supremacy functions rather through first engaging alterity and then subordinating differences according to degrees of deviance from whiteness. This has nothing to do with the hatred and fear of the strange, unknown Other. It is a hatred born in proximity and elaborated through the degrees of difference of the neighbor. (Hardt & Negri, 2003, p. 194)

In constructing their argument, Hardt and Negri also emphasize the impossibility of saying that there are no racial exclusions, but that it must be understood that this type of exclusion "arises generally as a result of differential inclusion" (p. 194). For them, it would be a mistake to consider even the apartheid laws as "the paradigm of racial hierarchy" (p. 194), since the racial differences would not be absolute or of nature, but differences in degree. "Imperial racism, or differential racism, integrates others with its order and then orchestrates those differences in a system of control" (p. 195).

The theoretical tools briefly presented here will be useful to perform the genealogical analysis about how the mathematics education of two Brazilian time-space forms of life works as a gear in mechanisms of differential inclusion.

Costão rural community during the Nationalization Campaign and mathematics education

The settlers of Costão, the rural community studied by Wanderer, were descendents of German immigrants from Rheinland, Saxony, and Westphalia, who came to the south of Brazil in 1824[8]. According to Dreher (1994) this wave of immigration, like those from other regions in Europe, was triggered by the need to people the Brazilian territory, constantly threatened by invasions from the Platine countries, to stimulate economic development, especially supply agriculture, as well as by the "population whitening" policy. The strong connection between this "whitening" policy and encouraging European immigration was sustained by the argument that "assumed the superiority of whites and the inferiority of other races, especially the black one, and sought its scientific legitimacy in the racial theories in vogue in Europe and the United States" (Seyferth, 1990, p. 18).

The worldwide geopolitical reconfigurations during the first half of the 20th century not only interrupted these immigration movements to Brazil, they also positioned the Germans who had immigrated and their descendants as "foreigners" to be effectively integrated into the country. Thus, as the *Estado Novo* (1937–1945) began, the Nationalization Campaign was implemented, constituting a mechanism that engendered power technologies to manage the population.

During the *Estado Novo*, forms of racism were gradually implemented to ensure the supposed integrity and "purity" of the Brazilian race compared to those groups that meant a kind of "danger" to the biological and political order of the nation, like the Germans and their descendents. However, it was not a racism marked by complete repulsion or expulsion of such groups from the country, but a kind of racism that acted through differential inclusion (Hardt & Negri, 2003), which allowed a putting together and, at the same time, a subordination. This differential inclusion was produced by the implementation of the Nationalization Campaign decrees, which began to disseminate technologies of population control by imposing the use of the Portuguese language in church, at school, and in other spaces of society; by incorporating civics lessons into the pedagogical processes; by allowing only teachers appointed by the State Education Offices to teach Portuguese, Brazilian History, and Geography by establishing public schools in the German and Italian colonized regions; by using only teaching materials published in Portuguese, among other decrees.

Hence, it was not the exclusion of Germans and their descendants from society, but a differential inclusion, since these groups continued to participate in the activities of church, school, and other public spaces. However, they were constantly positioned as men and women who did not communicate in the "right" language, who did not understand church services (which began to be held only in Portuguese), who did not know historical and geographical aspects of Brazil, or did not use the "appropriate" teaching materials. In summary, it was a process that produced a putting together, and, at the same time, a subordination.

In the *Costão* school, the imposition of Portuguese as the only form of communication produced breaks also in the way of dealing with their own mathematical knowledge, especially concerning the difficulty of doing mathematical calculations in Portuguese, and the need to "change the way of thinking", to perform the calculations. It is important to highlight that the imposition of the Portuguese language at school produced a cultural reconfiguration there. Before that time, the few black people who lived in the community did not attend school, which was private and charged monthly fees that could not be paid by the children of families that did not own land to farm, and also did not have any other paid work. However, during the implementation of the Campaign decrees, the black children were allowed to participate in the school.

The presence of black children in school at that time shows a putting together of the two cultural groups, but, at the same time a subordination of the new arrivals, since racist practices could be identified operating in the curriculum and, particularly, in school mathematics. Thus, blacks began to be positioned not only as a group that "survives from small thefts", but as those who were responsible for their “own death”, as in the “Story of the 10 little niggers” in the textbook used for mathematics lessons:

The Story of the 10 Little Niggers

Once there were 10 little niggers. They were brothers.
1, 2, 3 – 4, 5, 6 – 7, 8, 9 – and 1 more are 10.
That is great. Now hear ye!
One day they went for a walk. Do you know what the oldest did?
He hanged himself from a head of cabbage – and only nine were left.

Then there were 9 little niggers. They were brothers
1, 2, 3 – 4, 5, 6 – 7, 8, 9 – 3 are missing for them to be twelve.
One day they went for another walk
Do you know what the thinnest did?
He ate a corn cob – and died from a tummy ache.

Then there were 8 little black boys. They were brothers.
1 – 2 – 3 – 4 – 5 – 6 – 7 – 8.
Do you know what they did?
They went to play skittles – the fattest – his heart stopped.

Then there were 7 little niggers. They were brothers.
Do you know where they went?
They went to the witch’s house – she was evil, she stuck one in the cauldron.

Then there were 6 little niggers. They were brothers.
1 – 2 – 3 – 4 – 5 – 6
Do you know what they did?
They went swimming in the lake – one died of lack of breath.

Then there were 5 little niggers. They were brothers.
1 – 2 – 3 – 4 – 5
Do you know what they did?
They drank a lot of beer.
For the smallest – beer was poison.
But he was not convinced – he drank too much and died.

Then there were 4 little niggers. They were brothers.
1 – 2 – 3 – 4
Do you know what they did?
They played cops and robbers – and one fell dead to the ground.

Then there were 3 little niggers. They were brothers.
1 and 2 and 3.
Do you know what they did?
They wanted to make porridge
The greediest went to see – and oops! He fell into the pot.

Then there were 2 little niggers. They were brothers.
1 and 1 makes 2

> Do you know what they drank?
> They drank wine and more wine – one died, and one remained alone.
>
> The last little nigger did not want to remain alone, poor thing.
> He married a little black girl, her hair was very curly.
> They had ten little children. See how they have already grown.
> (Nast & Tochtrop, 1933, pp. 20–21)

When discussing the above text with the participants of her investigation, Wanderer observed the multiple meanings assigned by them to it, which allowed her to analyze how putting together of the black children with the others occurred in that school during the Nationalization Campaign. One of the interviewees told that at the *Costão* school during that time, the students memorized all the verses of "The story of the 10 little niggers", to be sung and recited – including the black children who, at the time were "employed" by the teacher to speak with those who did not know Portuguese during the school break (because in class only the teacher spoke). However, during the implementation of the Campaign decrees "the teacher let the *caboclos* [attend school] without paying, because they taught us Portuguese". Another participant in Wanderer's study explained:

> These students were caboclos, descendents from slaves. They generally lived on the river banks, because they had no land to live on. They survived from small thefts, and sometimes helping a neighbor on his farm. But we did not mix much with them. Because, in those days, we were German, superior, we felt superior. So, a black, a little black boy in class … that we didn't like, nobody liked these people, and they also did not like it.

As Wanderer (2007) discusses, although the blacks were identified as a group that held very highly relevant knowledge for the school institution at the time – the Portuguese language – which allowed them to attend classes, none of this erased the discrimination to which they were submitted. Further, the school mathematics practiced there was directly implicated in the processes of racism which created "racial hierarchies that are nonetheless stable and brutal" in school and in society (Hardt & Negri, 2003, p. 194).

Hence, as during the period of the Nationalization Campaign, nowadays the closing of the itinerant schools of MST and the incorporation of the children into the urban public schools – which we will discuss below – also constitutes a differential inclusion process in which school mathematics is a gear in its production.

Brazilian Landless Movement and mathematics education

The schooling processes performed by the Landless Movement comprise specificities that have been studied by scholars of important international research centres (Kane, 2000). Among these specificities it should be emphasized that their schools of Infant Education, Primary Education, Secondary School, and, more recently, Higher Education, belong to the public system of education (at the municipal, state, or federal levels), i.e., they are subject to official guidelines and regulations. However, due to the relative autonomy given by the Brazilian Educational System to its institutions, the MST has organized the curriculum of its schools based on pedagogical and philosophical principles (Knijnik, Alekseev, & Barton, 2006), which fulfil the purposes of a schooling that will serve the interests of their struggle for land reform.

As explicitly stated by Caldart (2003, p. 62), "under pressure from the mobilization of families and teachers, the movement decided to take on the task of organizing and articulating, inside its organicity, this mobilization [for the schooling of its members], to produce a specific pedagogical proposal for the schools achieved and to educate people who are capable of working from this perspective". Thus, the MST considers it a key issue that their schools should not only be located in the camps and settlements, but, most importantly, that they should develop pedagogical work closely connected with the Landless peasant culture (Knijnik 2007b), with its marks of the Brazilian rural culture in its crossings with the specificities of the struggle practices developed by the movement.

This educational perspective is followed by the work developed by MST schools and its teacher training courses (Knijnik 2007a; Lucas de Oliveira 2004; Monteiro, 2004) in the sphere of teaching and learning mathematics. The pedagogical practice described below (Knijnik & Wanderer, 2008) very clearly exemplifies this approach.

It was centred on a report written by a woman student who belonged to the Landless Movement National Committee (the group of elected peasants who coordinate the movement at the national level). Her report was about a march that the Landless Movement was doing at the time in a specific region of the state of Rio Grande do Sul. The march involved hundreds of peasant families who were walking along the main roads of that region, to press the state authorities to expropriate an unproductive large-holding whose owner had been in debt to the State for a long time, to the tune of about 32 million reais. When the discussion about her report started, she interrupted what was going on in the class, stood up, and, moving from a student subject-position to a leader subject-position, in a stentorian voice, as if she was in front of thousands of her comrades, explained:

> This is what is going on. We have eleven thousand and six hundred families settled in [the state of] Rio Grande do Sul. Following the data given by our Production Sector, the total amount of the State debts is seventy million reais, its total, counting everybody's debts. What is the point? The point is that Senhor Sotal, the landowner himself, has a debt of thirty-two million reais. In fact, it is not thirty-two, it is thirty-seven, but let's assume thirty-two million. Then, he alone has a debt of thirty-two million. And then I have a question because it is hard to debate about this in the schools, in the communities we are visiting during our march, in the media: What is the percentage that a single farmer took of public government money compared to our debts, to the "claims" that we are making? This question was asked on the first day of the march, already on the first day, when we sat down to prepare the people who were going to talk at the schools. This question came up and we looked at each other and couldn't [answer]. Then someone said: it can't [the answer to the question] be more or less … we become insecure and afraid to speak.

> I never managed to explain this part, then … reckon what is the percentage that a single landowner took from the government. … To give you an idea we [in the report] only took economic data. So [if we were to take] the question, what is the social result [of this situation] certainly it would call much more attention even. But [what we wrote in the report] is an economic result. … Let us get into the economic issue, because if we get into the social issue, it can't even be compared (Knijnik & Wanderer, 2008).

Her talk was interrupted by demonstrations from the other students, applauding her. The continuity of pedagogical work had as its centre the analysis of her report. This analysis was performed using some of the Landless mathematics language games, which were briefly mentioned in the previous section. Initially, the group was interested in discussing mainly the economic dimension of the situation, even if its social and political dimensions were always present. As one of the peasants justified: "With very concrete data, [one] strengthens our debate, our militancy". Another student completed this, saying: "It is important to take this to the march". The group consensus was that it would be important to write a text with the results of the analysis of the situation we had carried out in our mathematics class. Thus, the next stage of pedagogical work involved writing a text, which not only showed the analysis results but also highlighted the reasoning developed by the group, marked by Landless mathematics language games associated to their form of life. From the following week onwards, the text was distributed in the communities through which the march moved.

As happens in all other marches, the Landless children who participated in that specific march with other member of their families did not stop their schooling, thanks to the itinerant school[9] to which they belonged.

In her major research, Camini (2009) highlights that approximately four thousand and six hundred students have already attended the itinerant schools of the MST camps in the State of Rio Grande do Sul, from the time they were instituted in 1996, to the end of 2008. Even following principles and guidelines that regulate the public schools in the state, such as the requirement of 200 days of school a year, the itinerant schools, ever since they were made official, had some specificities: the students and teachers were Landless people living in the movement camps; teacher training was performed in Secondary School courses and courses of Higher Education belonging to the Movement; the students entered at any time during the school year; the general organization of the schools and of pedagogical work was implemented by the teachers and by the camped community; the school curriculum was structured by stages that were the equivalent of the initial grades of Basic Education and supported by the principles of the Landless Movement Pedagogy. Besides, the teaching materials used by the students were prepared by members of the MST Sector of Education, comprised by researchers and educators connected to the Movement, according to their interests, needs, and purposes (Camini, 2009).

Although Rio Grande do Sul has been a pioneer in the organization and implementation of the itinerant schools in Brazil, it is in this state that, since March 2009, decrees of the State Government together with the State Attorney's Office were issued for the purpose of interrupting this educational process. Closing the itinerant schools was preceded (during 2008) by the implementation of measures that pointed to this, such as delays in paying the teachers' salaries and suspension of the delivery of teaching materials to the students, as happened in all of the MST itinerant schools around the country. Thus, in March 2009, in the State of Rio Grande do Sul, the government agencies ordered that the children be transferred from the itinerant schools to the urban public schools of the municipalities where the camps are located. It was threatened that if this transfer were not made, the students would not receive their certificates at the end of the year and their parents would be held legally responsible for their "negligence". The debate about this issue has been widely disseminated by the national (and also international) media. In one of the most important newspapers, the State Attorney said:

> The [Itinerant] schools perform brainwashing. We have to guide the children about the possibility of becoming part of the world that is there, of the productive world. ... In a civil enquiry during which several things connected to MST were investigated, one of the proposals was an *Agreement on Conduct Adjustment* with the State Department of Education, for the public school system to absorb the students from these schools. This should be done so that they will have access to the knowledge imparted to all people.[10]

Statements like the one above indicate that the *Agreement on Conduct Adjustment* involves sending the Landless children to urban public schools of the Brazilian educational system, enabling a more effective control of their presence and attendance at school. Thus, the members of the Landless Movement become a target of the technology of power Foucault called biopolitics. Such technology, exercised through biopower, "takes the population as its object, as a large living body, so as to manage to govern this population in the best way possible" (Veiga-Neto, 2006, p. 35). This government puts into action control mechanisms also on the knowledges that will be taught to the Landless children. As a government authority emphasized, "Mandatory public teaching must be the same everywhere. It simply aims at ensuring that these children will have a right which is to be in an equal situation to the others".[11]

In brief, the elements of pedagogical work presented here and the discussion performed in the previous section about different mathematics, in particular about language games that constitute the peasant mathematics and its family resemblances to school mathematics, point to the specificities of teaching and learning mathematics at the MST schools and in its teacher training courses. This is a schooling process that is strongly linked to the Landless form of peasant life.

The closing of the itinerant schools in Brazil's southernmost state was an event with unique characteristics. Hundreds of children were forced to move to regular schools, most of them situated in towns linked to the camps in which they live by precarious roads. The access to those towns must be done using school buses in a state of disrepair. In these schools, they will be guided "about the possibility of becoming part of the world that is there, of the productive world" so that "they will have access to the knowledge imparted to all people": a teaching that "will ensure that the children will be in an equal situation with the others". In other words, closing the itinerant schools would be favouring school inclusion and, consequently, social inclusion.

However, with the support of Hardt and Negri's theorizations, we are led to state that this inclusion will be, above all, a differential inclusion. When the Landless children are obliged to attend urban schools, they will in fact not be excluded from the official educational processes. However, this inclusion will be permanently marked by a differentiation that will produce hierarchies and subordination. The MST struggles, its history, the Landless peasant culture, the language games that constitute what we have called Landless peasant mathematics, all this will be distant (not only geographically) from the urban school. The teaching materials used at urban public schools, as well as the training of their teachers, are also very distant from the Landless peasant form of life. The State Attorney said that the school curriculum would also enable the Landless children "to have access to the knowledge imparted to all people". It can easily be deduced that this knowledge is not the one from the Landless peasant form of life.

In particular, it does not include the language games that constitute Landless peasant mathematics, that are possibly unknown to the teachers of urban public schools. Thus, the language games of their mathematics will be considered spurious and, therefore, absent from the school curriculum, "repel[led], out of its borders" (Foucault, 2001, p. 33). So, the school mathematics will work as a gear in the mechanism of differential inclusion; the Landless children will be in the official school, they will learn "the knowledge offered to everybody" and, at the same time, they will position their own mathematical knowledge at a lower level. This is how differential inclusion functions; it attracts alterity, but subordinates and hierarchises the differences. This is how the post-modern racism discussed by Hardt and Negri functions.

FINAL REMARKS

The purpose of this chapter was to discuss Brazilian geopolitics in its relationship to mathematics education from a Foucaultian genealogical perspective. We took as empirical data two events that happened in different time-space schooling processes in the south of the country. They were analyzed in "terms of their singularity, the interrelations that define them and the conditions that make them possible" (Veyne, cited in Miller & Rose, 2008, p. 6). Following Johannesson (1998: 304), we considered the productivity of doing this type of history proposed by Foucault since genealogy traces how discursive themes – like educational issues presented here – generate ruptures and breaks in social practices and identify the formation of new historical complex events.

The two events analyzed here – imposing the Portuguese language on the German colonisation schools during the time of the Nationalization Campaign, and the closing of the itinerant schools of the Landless Movement that recently occurred – generated ruptures and breaks in the established order as regards life in society and, in particular, in the sphere of school. More specifically, the chapter problematises these breaks in the practices of school mathematics.

Based on Later Wittgenstein ideas, we argued about the existence of more than a single mathematics – the one produced by a specific form of life, namely the academic form of life, in which work those who are legitimized in Western society as scientists. We consider that there are different mathematics – different networks of language games associated with different forms of life which have family resemblances. In the chapter, when analyzing two distinct rural forms of life, we showed how the language games that constitute their peasant mathematics have family resemblances, like the rules that shape the grammar of their oral mathematics practices. It is important to say that what allows us to identify such practices as mathematics is that they maintain family resemblances with the language games of academic mathematics in which we were schooled.

The articulation of this argument with the discussion above using Hardt and Negri's notion of differential inclusion drives us to the main point of the chapter's rationale: the inclusion process of the black children in Costão school at that time and, nowadays, the inclusion process of Landless children in public urban schools were considered as differential inclusion. Both processes, in their specific ways, included children in school but at the same time subordinated them. In particular, we showed how school mathematics in those two time-space forms of life acted as a gear in the production of such differential inclusion. So, we can say that in the past, and also in the present, differences are established, hierarchies are built, and identities are produced in the interior of school mathematics practices.

ACKNOWLEDGMENTS

This chapter is a slightly edited version of our paper (Knijnik & Wanderer, 2010), which was originally published in *Zentralblatt für Didaktik der Mathematik*. We thank Springer for the permission to include it in this book.

We also thank Professor Paola Valero whose contribution for the first version of the paper was very significant.

NOTES

1 In recent years, a Foucaultian genealogical approach has been increasingly used in the field of Education. Studies developed by Miller and Rose (2008), Popkewitz and Brennan (1998), Walshaw (2004) and Bridge (2002) are examples of relevant works performed from such a theoretical perspective.

2 The Landless Movement website http://www.mst.org.br/mst/index.html presents a complete overview of its history and current struggles for land reform and education.

3 We aimed to analyze the discourses about school and school mathematics of a group of German-descendant, Evangelic-Lutheran settlers who attended the school of *Costão* (a rural community situated in the Southernmost state of Brazil), during the Nationalization Campaign – one of the actions taken by the *Estado Novo* (1937–1945), implemented by the dictator Getúlio Vargas. This historical period is considered a discursive space from which emerge statements about national consciousness, protection of the family, work and country, seeking a national identity in the cause of a Modern State united, homogeneous, and strong.

4 The author refers to aphorism 107 of Wittgenstein's book *Philosophical Investigations*. Following him, throughout the chapter similar notation will be used to mention Wittgenstein's aphorisms in that book.

5 Glock (1996, pp. 120–121) says that one can understand the notion of family resemblances not as a single line that permeates all the language games, but as threads that are interwoven, as in a rope,

constituting these games. According to him, "when we 'look and see' whether all games have something in common, we notice that they are united not by a single common defining feature, but by a complex network of overlapping and criss-crossing similarities, just as the different members of a family resemble each other in different respects (build, features, colour of eyes, etc.)".

6 In March, 2009, 1 Real was equivalent to approximately 0.3 Euro.

7 According to Foucault (2002:306), when using the word "death" he is not considering "simply direct murder, but also everything that may be indirect murder: the fact of exposing to death, of multiplying the risk of death for some, or, pure and simply, political death, expulsion, rejection, etc."

8 The immigrants were not a homogeneous group. Many were peasants, others marginalized urban people and people excluded from the industrialization process, besides those who could be considered politically exiled intellectuals. Some came to serve in the Brazilian Imperial army, others to work in the coffee plantations in the Southeast, and most of them came to the South because of the land occupancy policy for the development of family agriculture on small properties (Meyer, 1999, p. 34).

9 According to Camini (2009, p. 135), official MST documents indicate that the *Itinerant School* received this name because it means "a school that follows the camp itinerary until the time when the families have achieved land ownership, the settlement. Then comes another stage of the process, obviously connected to the previous one. It is the time to take the legal measures to establish the Peasant School for the sons/daughters of those who, returning to the rural area, wish to continue studying, working, living. The name Itinerant also means a pedagogical position of walking with the Landless, which is a great advance in the sense of affinity between the formal schooling processes and the educational experiences and practices of an organized social movement, such as MST."

10 Source:http://www.unisinos.br/ihu/index.php?option=com_noticias&Itemid=18&task=detalhe&id=20077

11 Source:http://www.jusbrasil.com.br/noticias/884592/mp-manda-fechar-escolas-itinerantes-do-st-no-rs-decisao-provoca-protestos

REFERENCES

Bridge, D. J. (2002). *Developing understandings of "inclusion" and "inclusive schooling"*. Doctoral Dissertation. Perth, Australia: Curtin University of Technology, Science and Mathematics Education Centre.

Caldart, R. S. (2003). A escola do campo em movimento. *Currículo Sem Fronteiras*, 3(1), 60–81.

Camini, I. (2009). *Escola Itinerante dos acampamentos do MST – Um contraponto à escola capitalista?* Unpublished doctoral dissertation. Universidade Federal do Rio Grande do Sul, Porto Alegre.

Condé, M. L. L. (1998). *Wittgenstein linguagem e mundo*. São Paulo: Annablume.

Condé, M. L. L. (2004). *As teias da razão. Wittgenstein e a crise da racionalidade moderna*. Belo Horizonte: Argvmentvm Editora.

Dreher, M. (1994). O Estado Novo e a Igreja Evangélica Luterana. In T. Muller (Ed.), *Nacionalização e imigração alemã* (pp. 87–110). São Leopoldo: Editora UNISINOS.

Foucault, M. (1977). *Language, counter-memory, practice: Selected essays and interviews*. Ithaca, NY: Cornell University Press.

Foucault, M. (2001). *A ordem do discurso*. São Paulo: Loyola.

Foucault, M. (2002). *Em defesa da sociedade: curso no Collège de France (1975–1976)*. São Paulo: Martins Fontes.

Glock, J. H. (1996). *A Wittgenstein dictionary*. Oxford: Blackwell.

Hardt, M., & Negri, A. (2003). *Empire*. Cambridge, MA: Harvard University Press.

Johannesson, I. A. (1998). Genealogy and progressive politics: Reflections on the notion of usefulness. In T. S. Popkewitz & M. Brennan (Eds.). *Foucault's challenge: Discourse, knowledge, and power in education* (pp. 297–315). New York: Teachers College Press.

Kane, L. (2000). Popular education and the Landless People's Movement in Brazil (MST). *Studies in the Education of Adults*, 32(1), 36–50.

Knijnik, G. (2007a). Brazilian peasant mathematics, school mathematics and adult education. *Adults Learning Mathematics – an International Research Forum*, 2(1), 54–62.

Knijnik, G. (2007b). Mathematics education and the Brazilian Landless Movement: three different mathematics in the context of the struggle for social justice. *Philosophy of Mathematic Education Journal*, 21(1), 1–18.

Knijnik, G., Alekseev, V., Barton, B. (2006). Other conventions in mathematics and mathematics education. In F. Leung, K. Graf, & F. Lopez-Real (Eds.), *Mathematics education in different cultural traditions: A comparative study of East Asia and the West* (pp. 567–580). New York: Springer.

Knijnik, G., & Wanderer, F. (2008). *Adult education and ethnomathematics: an analysis of a pedagogical experience with Brazilian Landless Movement leaders*. Paper presented in Topic Study Group 8 at 11th International Congress on Mathematics Education. Monterrey, Mexico.

Knijnik, G., & Wanderer, F. (2010). Mathematics education and differential inclusion: A study about two Brazilian time-space forms of life. *Zentralblatt für Didaktik der Mathematik, 42*(3–4), 349–360.

Knijnik, G., Wanderer, F., Oliveira, C. J. (2005). Cultural differences, oral mathematics and calculators in a teacher training course of the Brazilian Landless Movement. *Zentralblatt für Didaktik der Mathematik, 37*(2), 101–108.

Lucas de Oliveira, H. D. (2004). Atividades produtivas do campo no currículo: reflexões a partir da Etnomatemática. In G. Knijnik, F. Wanderer, & C. J. Oliveira (eds.), *Etnomatemática, currículo e formação de professores* (pp. 305–322). Santa Cruz do Sul: EDUNISC.

Meyer, D. (1999). *Identidades traduzidas. Cultura e docência teuto-brasileiro-evangélica no Rio Grande do Sul.* Unpublished doctoral dissertation. Universidade Federal do Rio Grande do Sul, Porto Alegre.

Miller, P., & Rose, N. (2008). *Governing the present: Administering economic, social and personal life.* Cambridge: Polity Press.

Monteiro, A. (2004). A etnomatemática em cenários de escolarização: alguns elementos de reflexão. In G. Knijnik, F. Wanderer & C. J. Oliveira (Eds.), *Etnomatemática, currículo e formação de professores* (pp. 432–446). Santa Cruz do Sul: EDUNISC.

Moreno, A. R. (2000). *Wittgenstein: os labirintos da linguagem. Ensaio introdutório.* São Paulo: Moderna.

Nast, W., & Tochtrop, L. (1933). *Meu livro de contas* (1). São Leopoldo: Editora Rotermund.

Popkewitz, T. S., & Brennan, M. (1998). *Foucault's challenge: Discourse, knowledge, and power in education.* New York: Teachers College Press.

Seyferth, G. (1990). *Imigração e cultura no Brasil.* Brasília: Editora Universidade de Brasília.

Veiga-Neto, A. (2006). Biopolítica, estado moderno e a inclusão na escola. *IHU On-line, 144* (5), 34–37.

Walshaw, M. (2004). (Ed.), *Mathematics education within the postmodern.* Greenwich, CT: Information Age Publishing.

Wanderer, F. (2007). *Escola e matemática escolar: mecanismos de regulação sobre sujeitos escolares de uma localidade rural de colonização alemã do Rio Grande do Sul.* Unpublished doctoral dissertation. Universidade do Vale do Rio dos Sinos, São Leopoldo.

Wittgenstein, L. (1995). *Philosophical investigations.* Oxford: Blackwell.

Gelsa Knijnik
Programa de Pósgraduação em Educação
Universidade do Vale do Rio dos Sinos
Brazil

Fernanda Wanderer
Faculdade de Educação
Universidade Federal do Rio Grande do Sul
Brazil

CHAPTER 9

ON BECOMING AND BEING A CRITICAL BLACK SCHOLAR IN MATHEMATICS EDUCATION: THE POLITICS OF RACE AND IDENTITY

DANNY BERNARD MARTIN AND MAISIE GHOLSON

Fifteen years ago, William F. Tate (1994) authored a paper titled, *From Inner City to Ivory Tower: Does My Voice Matter in the Academy?* Building on the work of critical race scholars (Delgado, 1989, 1990; Williams, 1991), and reflecting on his own early schooling and later experiences as a professor in the academy, Tate echoed the call for *voice* scholarship as one way to explain the experiences of minority scholars.

In this chapter, we revisit Tate's earlier discussion and continue to reflect, in a manner consistent with critical race counterstorytelling (e.g., Solórzano and Yosso, 2002), on the politics of race and identity in mathematics education. We do so by sharing our own experiences of becoming and being critical Black scholars in a field dominated by White scholars. We are not suggesting a singular conception of what it means, or should mean, to be a Black scholar, as our individual experiences and trajectories in the field will attest. Nor are we suggesting that our voices receive special privilege simply because we identify as critical Black scholars.

As a result of our attempts to alter research, policy, and practice with respect to Black children and mathematics, we claim that asserting and developing an identity of a critical Black scholar is not just a voluntary assertion of identity but, in our view, a necessary one. It is necessary in order to challenge the masternarrative and discursive, representational practices that continue to construct Black children as mathematically illiterate and intellectually inferior to children from other social groups. Moreover, we are well aware of a disturbing trend in society that attempts to strip Black children of their childlike and human qualities altogether by using such labels as "thugs", "urban terrorists", and "endangered species". These identities are supposedly the result of genetic, cultural, and intellectual inferiority (D'Souza, 1991; McWhorter, 2001; Steele, 1990; S. Thernstrom & A. Thernstrom, 1997; A. Thernstrom & S. Thernstrom, 2004).

In using our voices to foreground issues of race and identity and to centre Black children in our discussion, we realize that there are certain risks associated with doing so. One risk concerns our intentional blending and blurring of the personal, political, and scholarly. Scholarly work is supposed to be neutral and apolitical. However, we believe that *all* scholarship is political and shaped by our personal experiences. To deny this would be intellectually dishonest.

O. Skovsmose and B. Greer (eds.), Opening the Cage: Critique and Politics of Mathematics Education, 203–222.

Some colleagues and readers might also suggest that our focus on race and processes of racialisation are unnecessary diversions from focusing on teaching, learning, curriculum, and assessment; an attempt to inject these issues where they have no place. They may believe that race- and identity-centric research borders on advocacy and lacks rigor. In fact, some may believe that we now live in a post-racial society in which race and racism are no longer relevant and that Black scholars should stop "playing the race card" (Bonilla-Silva, 2003, 2005). However, we resist this line of thinking. The first author has made compelling arguments that mathematics learning and participation, in addition to being conceptualised as cognitive, sociocultural, and situated activities, can be conceptualised as *racialised forms of experience* (Martin, 2006, 2009c). This perspective reveals the salience of race and racism not only in structuring the ways that learning and participation unfold but also in shaping mathematics identities and the beliefs that people develop about who can and cannot do mathematics.

Some readers might also resist our efforts by suggesting that White scholars in mathematics education do not write about whiteness or explicitly advocate for White children. In our view, this critique is blinded by the ubiquity and normalization of whiteness – represented numerically, ideologically, epistemologically, and in material power (Ladson-Billings, 2000; Smith 1999) – which characterizes mathematics education research and policy contexts. The lack of scholarly interrogation of whiteness in mathematics education – in relation to learning, teaching, assessment, knowledge production, and power – is entirely consistent with the norms of White institutional spaces (Martin, 2008; Moore, 2008). Moreover, this lack of interrogation has the potential to render mathematics education, as an enterprise, fundamentally different in character than the racialised contexts that characterize most of USA society.

TALKING B(L)ACK

It is in the contexts of our characterization of mathematics education as a highly racialised domain and our commitment to meaningful mathematics education for Black children that we share our stories and voices in the dialogue below. We reflect on the processes of becoming aware of, and making meaning for, what it means to be a critical Black scholar and how our evolving awareness has shaped our present sensibilities on issues of race, identity, and mathematics education. We also reflect on the issue of positioning; how we attempt to position ourselves in the domain and, to the degree that it matters, how we might be positioned by others. For emerging scholars, this is a very real concern due to fears that they may be marginalized in the field. Finally, given our relationship as advisor and advisee, we address the dynamics involved in mentoring and being mentored for the purpose of engaging in the discipline as a critical Black scholar.

Danny: I'll open the dialogue by sharing my own sense of what it means to be critical and how this identity is co-constructed with my evolving sense of my Black identity; then I will share an example from my own experience.

For me, being a *critical* Black scholar entails unapologetically challenging the common-sense understandings, routine practices, policies, and forms of scholarship that intentionally or unintentionally dehumanise, depersonalise, and oppress Black people in symbolic and material ways (Ladson-Billings, 2000). Borrowing from Robin Kelley (1997), I see my efforts as "a defense of black people's humanity and a condemnation of scholars and policymakers for their inability to see complexity" (p. 4). I see this as a floor in my efforts not the ceiling since remaining stuck in a defensive posture is neither desirable nor effective. Yet, given the pervasiveness of academic and everyday assaults on Black identity, particularly on Black children, I find myself compelled to speak up and act.

I also take the *scholarly* part of this identity seriously. Because I place high value on the power of the pen and the power of ideas, all of my writing in the field has sought to challenge mainstream and conventional thinking about Black children and their competencies. And while it may appear that I have a singular focus on those issues, I believe that my writing has offered both a direct challenge to scholarship addressing more conventional topics and made a contribution to understanding these topics and that race-centred analyses have helped me to do this. For example, rather than following the widely-used race-comparative paradigm that considers only the collective status of Black students in relation to students in other socially constructed racial groups and focusing on failure and so-called racial achievement gaps, I have asked student-cantred and agency-related questions such as, *what does it mean to be Black in the contexts of learning and doing mathematics*? and *what does it mean to be a learner of mathematics in the context of being Black*? These questions force researchers to consider the power of Black subjectivities and how those subjectivities confirm or deny the supposedly objective research conducted about Black learners.

I have also asked the question, *who should teach mathematics to Black children?* (Martin, 2007), leading to other important questions about the kind of knowledge, beyond content and pedagogy, that teachers must possess to teach these children effectively. These questions have, in turn, led me to raise fundamental questions regarding the aims and goals of mathematics education and knowledge production about Black children, including, *why should Black children learn mathematics?* and *what is the study of Black children the study of?* How one answers these questions has a profound effect on the ways that learning, teaching, curriculum, and assessment are organized for Black children. My critical analyses have shown that these questions have, for the most part, received shallow responses from the mainstream mathematics education community (Martin, 2009a, 2009c, in press-a).

Several months ago, I was reminded how the *Black* part of my identity can be used to position me in the field. In the year 2000 my book, *Mathematics Success and Failure Among African American Youth*, was published. That book was based on my dissertation, completed three years earlier. The dissertation and the book fleshed out my earliest thinking on issues of mathematics *socialization* and *mathematics identity* and laid out a multilevel framework for studying these issues. After nine years in circulation, one would think that my place in the literature on mathematics identity, particularly for Black children, would be well established and somewhat secure. Many students and fellow scholars have utilized and improved on those early ideas.

In early 2009, a prominent White scholar co-authored an article in a well-known mathematics education research journal where he laid out a framework for studying mathematics identity. The article made reference to my book and the work of other scholars who have been focusing on these issues. A few weeks after that journal article appeared, a prospective student for the graduate program in mathematics education at my university phoned me for information about entrance requirements. During the conversation, he indicated that he had read the journal article and suggested that I was probably excited to have my work referenced by such a well-known scholar. In the words of the prospective student "You know you have made it when someone like professor X cites your work, especially in a journal like Y." It took some time for me to process the conversation but when I did, I realized that despite the book being in circulation for nearly ten years and despite subsequent publications and dozens of presentations across the country, it was not until a White scholar validated my work that I was supposed to feel validated.

I am not recounting this experience for self-serving reasons or due to any diminished feelings about the importance of my work. My identity as a critical Black scholar is not equated with a sense of inferiority. People who know me best can attest to the fact that the validation of my work by White scholars, or any other scholar, is not what drives me. However, I think this example, in its own way, should force us to consider issues of power and voice within our domain.

Surely, the prospective student is not alone in his perception and one is left to wonder how pervasive such thinking might be in the field regarding the contributions of Black scholars to the conversations on mathematics teaching and learning. Richard Delgado does a brilliant job addressing the larger issue concerning the politics of citation and scholarly authority in his articles titled *The Imperial Scholar: Reflections on a Review of Civil Rights Literature* (1984) and *Imperial Scholar Revisited: How to Marginalize Outsider Writing, Ten Years Later* (1995).

Maisie: My trajectory in becoming a critical Black scholar thus far has been a bit different. In fact, my trajectory was rooted in my personal childhood experiences, but has expanded to my academic experiences, as I learn more about the structures that allow racism to persist (Bonilla-Silva, 2001). I would say that my critical orientation was born circa second, perhaps, third grade, when I realized that little girls with long ponytails received hugs and adulation and little girls with cornrows and plaits received pats on the back and half-baked smiles. And, as the story goes, more than choosing to be a critical Black scholar, critical Black scholarship chose me. I had no decision in being Black and, rather than loathe myself, my beautiful mother, and my family, generally, I unwittingly adopted a stance early on that challenged external structures that sought to characterize me (and those that I loved) in demeaning and deficit ways. Growing up in the suburbs of Houston, various slights and insults were heaped on the shoulders of little Black girls: "You have a big nose"; the absence of party invitations; "Aren't you embarrassed that all the Black people work in the cafeteria?" and, once a year, someone whispering "nigger" to test how those two syllables change the atmosphere. My parents, in their wisdom, guarded and armed my brother and me with knowledge about European versus

African aesthetics, stories of Civil Rights struggles, the reality of inequity: "Come tell me if anyone mistreats you." and "Don't let anyone call you that. Never." In other words, a critical orientation to my small world, even as a young Black child, was part of my socio-emotional development.

Interestingly enough, with all that "good home training", there was an internal tension when I broached the subject of *being a critical Black scholar* in mathematics education. I strongly believe that being an emotionally healthy Black person requires a critical filter on life. However, what does it mean to present yourself as a critical Black scholar to the academy and what does it mean to aspire to be critical Black scholar within the academy, particularly in mathematics education, which is heralded by many as socio-politically and culturally neutral (although scholars such as Valero & Zevenbergen (2004) have critiqued this perspective)? In other words, what does it mean to politically "tip your hand" as to how you see the world and how you protect and preserve Black culture and Black people? Further, where are the spaces within the field of mathematics education that are tractable, unexplored, or underexplored with respect to race critical analysis?

Consider a recent conversation in a research meeting where I posed a question to the team regarding Black and Latino/a students that are specifically targeted within the study. I asked, "How are we getting to know these students outside of their scores on these administered assessments?". Silence and, then, "What do you mean?". Additional conversation ensued that day, but no efforts have been made since that time to pursue an understanding of these students beyond assigning them a rank, percentile, or pseudonym. As a critical Black scholar, what is my responsibility beyond raising critical questions? Does this stance require activism? And, how do I reconcile my position as a mere graduate student within the reality of research and knowledge production that continues to frame students of colour in typical ways?

I didn't ask the question in the meeting to be combative or provocative. But after observing the students in their classrooms, I was compelled. To me, it felt irresponsible to make no effort to capture these students as children with personalities and lives that extend outside of the classroom doors. Thus, I can't help but wonder if I even have a choice in *who* I become as a Black scholar in mathematics education. Can a Black scholar in mathematics education be anything other than *critical* and be whole? And, what are the permutations and striations of critical Black scholarship? And, finally, without a critical orientation can I even dare to maintain hope for the future of Black children in mathematics education?

In his essay, *A Talk to Teachers* (from a speech delivered in 1963), James Baldwin captures the perpetual reality for many Black children and, perhaps, sets the stage for the work of the critical Black scholar.

> As adults, we are easily fooled because we are so anxious to be fooled. But children are very different. Children, not yet aware that it is dangerous to look too deeply at anything, look at everything, look at each other, and draw their own conclusions. They don't have the vocabulary to express what they see, and we, their elders, know how to intimidate them very easily and very soon. But a black child, looking at the world around him, though he cannot know quite what to make of it, is aware that there is a reason why his mother

> works so hard, why his father is always on edge. He is aware that there is some reason why, if he sits down in front of the bus, his father or mother slaps him and drags him to the back of the bus. He is aware that there is some terrible weight on his parents' shoulders which menaces him. And it isn't long – in fact it begins when he is in school – before he discovers *the shape of his oppression* (emphasis added) (Baldwin, 1985, pp. 326–327).

Is our work as critical Black scholars to discover the shape of Black children's oppression or does this work entail more?

Danny: Maisie, I am moved by your early memories and the questions you ask about becoming and being a critical Black scholar, especially as you begin the journey you have embarked on as a graduate student. I am also moved by the example with the research team. Earlier, I mentioned the power of the pen and invoking activism and advocacy in writing. But you raise an important question about what a critical Black scholar should *do* beyond being critical. I wholeheartedly agree that principled action is the key. The "simple" act of you raising that important question about students in a meeting where others on the project presumably had more material power served to disrupt the path characterized by what I see as an unfortunate backgrounding, on one hand, or a conceptually flawed foregrounding, on the other, of race and identity (Martin, in press-b). I am reluctant to attribute intentionality to those present but my guess is that they knew what you meant when you asked the question. So, the response of "What do you mean?" was not one of ignorance. The subtext of the response was probably more along the lines of "We don't want to deal with issues of race and identity" even though nearly 90% of the students in the District for whom the project is intended are African American and Latino. Only resistance – although colour-blindness might be more appropriate – could explain ignoring of this fact. If so, one has to question if these folks can truly intervene in ways that are meaningful. So, the implications of your question are profound. As an emerging scholar, you will, of course, have to pick your spots; when to write, when to question, and so on. But if you are truly committed to challenging research, policy, and practice that dehumanise and simplify Black children, you are likely to find yourself acting more than you might have imagined.

Of course, as a mid-career scholar with tenure I realize that I have a bit more space to be critical. I am simultaneously inside and outside of the enterprise and I realize that I have also been granted (as well as earned) a certain amount of privilege. But because my trajectory into the academy was atypical, I have never feared speaking truth to power and pursuing my own path. I do find it interesting, however, that I have been contacted by many graduate students and new scholars who tell me they want to pursue issues of race and identity in their work but who have been discouraged from doing so, if not overtly then implicitly, usually by White scholars and mentors. They are told to wait until the dissertation is done and then told to wait until they have tenure. My advice has usually been to maintain their sense of purpose and, if necessary, find allies. As a result, I have been asked to sit on dissertation committees at many universities outside of my own. My point here is that issues of status, rank, and hierarchy are real in the field, just like they are outside of it. But to say nothing, write nothing, and do nothing only leaves those structures in place.

So, a partial answer to your question about the permutations and striations of critical Black scholarship would be that you will exert your identity and voice, and manifest them, in many different ways. In one instance, it might be to change the direction of a conversation by pointing out colour-blindness. In another instance it might be to counter or halt the inhumane representation of Black people that is being implied in research.

IN-BETWEEN A ROCK AND WHITE/BLACK PLACE

Maisie: Danny, it is interesting that you use the phrases "defend the humanity of Black people" and "dehumanise and simplify Black children" when discussing the likely obligation of critical Black scholars to act as a *humanistic defender*. This leads quite naturally to our positioning as Black scholars in mathematics education. Cornel West takes up this issue more generally in an article entitled *The Dilemma of the Black Intellectual* (1985). In this piece, West harkens back to Harold Cruse's seminal work *The Crisis of the Negro Intellectual* (1967) and quotes in the opening:

> The peculiarities of the American social structure, and the position of the intellectual class within it make the functional role of the negro intellectual a special one. The negro intellectual must deal intimately with the white power structure and cultural apparatus, and the inner realities of the black world at one and the same time. But in order to function successfully in this role he has to be acutely aware of the nature of American social dynamic and how it monitors the ingredients of class stratification in American society…Therefore, the functional role of the negro intellectual demands that he cannot be absolutely separated from either black or white world. (p. 451)

West notes that the precarious position that we find ourselves, as Black scholars is a "self-imposed marginality". Being positioned between the White academy and the Black community results in Black scholars typically functioning within four models of Black intellectual activity. It is beyond the scope of my reflection to exhaustively inspect all four of these models, but I find it worthwhile to explore the bourgeois model (or what West calls the "Black Intellectual as a Humanist"). As intimated by the name, the hallmark of this model is providing a defence of Black humanity. West problematises intellectual activity exclusively situated in this model by stating that "The basic problem with the bourgeois model is that it is existentially and intellectually stultifying for black intellectuals. It is existentially debilitating because it not only generates anxieties of defensiveness on the part of black intellectuals; it also thrives on them" (p. 116). I find navigating this minefield particularly hazardous in mathematics education scholarship, as you noted earlier, where hegemony surrounding Black children in mathematics classrooms maintains such grossly deficit orientations. Simply said, there is an overwhelming need to set the record straight, but there is a question as to how one eventually transcends this need of constantly defending our community.

If this were the only complexity, we could count ourselves lucky. Yet, what remains equally hardening is the strained relationship between Black scholars and the Black community itself. While the work in critical Black scholarship certainly comes at a high premium in terms of "academic legitimation", ironically Black scholars are also often marginalized from the Black community. This marginalisation is the result of a historical legacy that has bred an understandable mistrust of the research community. Among other issues, Carruthers (1994) notes that some of the *first* Black scholars conducted research to subdue the Black community, for example, during the 1919 during race riots in Chicago and during the Civil Rights Movement. The consequences of this mistrust has been the further marginalisation of the Black community – under-researched and demonised by studies conducted at an altitude of 20,000 feet.

In my brief experience as a graduate researcher, I feel a tremendous strain between the interests of the research endeavour and the mathematics education needs of Black children. There is a looming question regarding whom I ultimately serve that comes with every classroom observation within the project. To the point, a distrust of Black scholars is grounded in the fact that Black scholars often do not own their own research agendas and methods (Smith, 1999). The internal strain that I feel is commensurate with my frequent sense of disconnect from the philosophy and aims of the research project itself, over which I have no control.

Danny, what do you make of this Du Boisian double consciousness of knowing two worlds and being "at home" in neither place, particularly as a Black scholar in mathematics education?

Danny: Maisie, I appreciate you raising these issues and problematising the roles, identities, and quite frankly, the relevance of Black scholars. I am sure that many White scholars and Black scholars alike raise similar points and question the very notion of some Black scholars choosing to identify as critical and not just as a scholar. One point to be taken from your comments is that history is a good teacher. West, Carruthers, and others, remind us that Black scholars and self-proclaimed black intellectuals – for me, these are not synonymous – often find themselves in precarious positions not only with respect to the academy but also to Black communities. Personal reflection on this positioning is a good thing. Am I doing work that matters and that is relevant? Am I being complicit in negative constructions of children and their identities and competencies? Am I willing to speak truth to power? Am I being faithful to my core beliefs and values? Am I instantiating my Black identity in ways that make White scholars, and some Black scholars, comfortable with my presence and scholarly perspective?

My own reading of West's analysis suggests that his typology of Black scholars – Black intellectual as humanist; Black intellectual as revolutionary; Black intellectual as postmodern sceptic; and Black intellectual as critical organic catalyst – simply points out the complexity of that identity. As I stated earlier, there is, and should be, no singular conception of what it means to be a Black scholar. Depending on where they are and what their role happens to be, a given Black scholar will be more or less connected to Black communities and more or less critical in their orientation. In my view, West's categories are an oversimplification of this complexity. Pushed a bit further, one could view West's typology as an attempt to impose levels of authenticity on Black scholars. My own view is that such discussions are not constructive and that they have the potential to devalue the kinds of contributions that do not fall in line with the categories he places at the top of his hierarchy. Moreover, like all typologies, West's categories fail to deal adequately with the agency of Black scholars who strategically move in and across these categories in ways that do not foster self-imposed marginality. In this way, Black scholars are not situated exclusively in one category or another. Their movement across these categories is political for the very purposes of Black empowerment, insurgency, and emancipation. Moreover, the self-imposed marginality within the academy cited by West fails to consider the structural arrangements in place work to marginalize Black scholars no matter what their status and standing. One only has to consider social networks, composition of editorial panels, members of edited volumes, and so on.

In my opening comments, one point I failed to mention is the following: developing an identity as a critical Black scholar is not a destination. To *be critical* and to resist assaults on Black humanity and identity should not be the end goals of one's efforts. One doesn't become a critical Black scholar on a particular day or time or as a result of a particular act. In my opinion, it is a lifetime of work. The PhD doesn't signal this nor does entrée into the academy. And certainly it is not signalled by achieving a distanced and so-called objective disconnect from Black communities. So, my own view is to be mindful, but not deterred, by West's commentary. It serves to point out the dangers of a limited vision on Black scholarly work. No scholar should seek to conform to, or remain trapped in, West's typological categories. Moreover, one can, in fact, do meaningful work on behalf of Black children and communities from within the confines of the academy, if that is where life happens to find us. For example, your work on a research project that gives minimal attention to issues of race and student identity does not define your personal commitment to Black children. However, you can achieve multiple aims and goals on behalf of Black children via participation in such an effort. And if that effort is not compatible with your values, then you always have the option of joining projects and efforts that are more compatible.

A second point that you raise concerns the need to constantly defend the humanity of Black people, a role that West (1993) says is characteristic of the Black Humanist scholar whose actions border on bourgeois behaviour. I think you are absolutely right in raising the question of why this is even necessary. As you know, I have written about examples where it was claimed that Black children and poor children lacked the capacity to engage in abstraction and formal mathematical thinking (Martin, 2009a). Unless questioned, these views are allowed to persist and rise to the level of accepted truth. In constructing arguments against such views, one can simultaneously point out the flaws of the scholarly arguments that support these viewpoints and point out how these arguments contribute to a further dehumanisation of Black learners. Black scholars, working within the context of the academy, are uniquely positioned to do this. If the scholars working down the hall from you are consistently publishing articles that imply Black cultural and intellectual inferiority, fail to report on Black student success, advocate militaristic discipline for Black children in schools, minimize the diminishing status of Black boys in schools, and so on, it is much harder for someone "outside" the academy to challenge this and do so in the very same forums where this work appears. And considering the fact that such claims can make their way into print is an indication of the willingness of so many other scholars to say nothing.

My own view is that in contexts where claims of Black inferiority become normalized and taken as truth, to say nothing and resist assaults on Black humanity are not options, particularly for Black scholars. The historical record shows that many Black scholars, representing many different traditions, have made such arguments. Beyond the academy, Black people continue to fight for their humanity everyday. We must remain vigilant outside the academy as well as *inside*.

I want to raise another point. And this, again, may be a function of my particular experiences and trajectory. But, just as I indicated that one's identity as a critical Black scholar should not be a destination, I will say the same thing about the academy. It is an important context but it is not a spiritual home, for example. It is a context rife with politics but it will not be a place that will break my spirit. For me, carrying out my work in the academy is just a means to a larger end. While West (1993) gives primacy to the academy, there are many other contexts where one can engage in critical work and take a scholarly, principled approach to that work. My own decision to remain in the community college context, working directly with students who had often been underserved in public schools represented a statement of my commitments. In doing that work, I considered myself to be no less of a critical scholar than I do now. But I also realized that my efforts would forever be limited if I confined my efforts to localized practice alone. I also knew that I wanted to make a contribution to the scholarly debates about Black children and mathematics. When I made the decision to enter the university, I realized that it would provide me with the opportunity to do different work but still consistent with my fundamental beliefs and goals. While my work in the community college was channelled into teaching and working on parent and community math projects, my administrative work in the university context, for example, has allowed me to shape the College mission and vision and its programs in ways that can be favourable to Black students.

Because I find myself working in an administrative capacity, my focus has been on institutional change. My individual work up to now has been less engaged in community contexts. However, this administrative work is just another piece in my overall efforts. For example, when I came to University of Illinois, Chicago (UIC) a few years ago, we had one Black doctoral student in mathematics education. Since then, that one student has graduated and we currently have seven Black students in the doctoral pipeline. None of them came here to shelter themselves from Black communities and the needs of Black children. To be certain, they will learn mainstream methods and be exposed to conventional forms of thinking and problem formulation but they will also help to transform the academy in ways that would not be possible without their presence. I am confident they will raise new questions, choose contexts for research that are important to them, and produce scholarship that is relevant.

I am going to assume that, in your own case, the teaching you did before coming to UIC was a particular instantiation of your commitment to Black students and that the work you do here at UIC will further that commitment. While your efforts on behalf of, and in concert with, Black students may have taken on particular forms as a classroom teacher, they will take on different forms as a graduate student. My role, in an advising capacity, is to support you in your pursuits. If that means helping you situate your research in Black communities, focused squarely on Black children, then that will be the case. If it means facilitating other opportunities to help you learn valuable research skills, that will be done also.

My overall point here is to say that that there are many different points of leverage for critical Black scholars and while the efforts at these points may seem more or less connected to Black communities, we need work at many different levels based on many different configurations of working in the academy and community. Community efforts that help to empower Black children in mathematics are important. However, if university structures are not favourable to Black students and we do not have Black scholars inside the academy who can alter those structures, the progress of those Black students who enter can be limited.

This does not mean that Black scholars have to shoulder the full responsibility of furthering the cause of the Black community. Moreover, this new generation of Black scholars, yourself included, will face challenges in not only taking up particular subject positions relative to Black communities but also in being assigned various positions within a rapidly changing academy. For example, as the academy moves from a social project to a market force, from producing and disseminating knowledge as a public resource toward the privatisation of knowledge (Newson, 1998), there is a risk, and it is already true to some extent, of Black scholars being commodified based on their identities as Black scholars. As is true in society, the politics of representation will require that those in power appoint some Black scholars as leaders. The price of admission for these appointments should not be a lessening of one's critical perspective or commitment to Black children.

Maisie: This gives me quite a bit to think about. It seems that you are challenging the process of meaning making and assessing value of Black scholarship, regardless of whether these views and metrics are imposed by another critical Black scholar or the academy at large. Also, the point that you raise regarding the commodification of Black scholarly identity is well taken. The lure to "brand" oneself in the academy must be grounded in intellectual integrity. This seems to be a call for epistemologies that allow for Black scholars to exercise the authenticity of *their experiences*, but also an infrastructure that anchors us in academic discourse and accountability. I believe this was West's ultimate point, but his approach was somewhat prescriptive. Mohanty (1989) outlines a thought-provoking epistemological imperative of creating space for marginalized groups by stating:

> This issue of subjectivity represents a realization of the fact that who we are, how we act, what we think, and what stories we tell become more intelligible within an epistemological framework that begins by recognizing existing hegemonic histories. The issue of subjectivity and voice thus concerns the effort to understand our specific locations in the educational process and in the institutions through which we are constituted. Resistance lies in self-conscious engagement with dominant, normative discourses and representations and in the active creation of oppositional analytic and cultural spaces. Resistance that is random and isolated is clearly not as effective as that which is mobilized through systematic politicised practices of teaching and learning. Uncovering and reclaiming subjugated knowledge is one way to lay claim to alternative histories. But these knowledges need to be understood and defined *pedagogically*, as questions of strategy and practice as well as of scholarship, in order to transform educational institutions radically. And this, in turn, requires taking the questions of experience seriously. (p. 185)

This need for relevant and focused epistemologies and taking questions of experience seriously immediately invokes two frameworks with which I am becoming acquainted—*Black Feminist Thought* and *Critical Race Theory* (CRT). While I agree that the academy cannot be a spiritual home, the four "contours" of Black Feminist Thought, as defined by Patricia Hill Collins, provide a welcome sanctuary and contrast to the dominant epistemologies within mainstream mathematics education. Collins explicated the four contours as *concrete experience as a criterion of meaning, the use of dialogue in assessing knowledge claims, the ethic of caring*, and *the ethic of personal accountability* (Scheurich & Young, 1997). Cynthia Dillard (2000) expanded this list through her *Endarkened Feminist Epistemology*, wherein she adds, "research is both an intellectual and spiritual pursuit, a pursuit of purpose" (p. 674). These ways of knowing resonate especially well with me. CRT also provides a compelling epistemological framework, given its attention to counterstorytelling, the permanence of racism, whiteness as property, interest convergence, and a critique of liberalism (DeCuir & Dixon, 2004). I also wonder if these anchoring epistemologies can also be the ties that connect critical Black scholars throughout the African Diaspora.

How do you understand these epistemologies and how have these epistemologies influenced or shaped your work? To what extent do you think that these epistemologies will be embraced and seen as normative, rather than *other*, as cautioned by Beverly Gordon (1993)?

Danny: What you say in your last set of comments is key. First, it is not just that the academy or mathematics education, in particular, should change to create an infrastructure for alternative epistemologies. The academy has typically expanded itself to encompass a number of epistemologies but many are marginalized and regarded as too political and less rigorous by many in the mainstream. Rather than settling for the typical choices of assimilation or accommodation, I believe that critical Black scholars must create and claim spaces for themselves. They should not do so in ways that exclude or re-inscribe new hierarchies and oppressions but in ways that de-centre White logic and White perspectives and methods (see Zuberi & Bonilla-Silva (2008) for thorough discussion of these terms), and shattering notions of what is normative. I should say that my use of the terms *White logic* and *White methods* is not flippant or meant to essentialise White scholars. I would encourage other scholars in the field to read Linda Tuhiwai Smith's (1999) *Decolonizing Methodologies: Research and Indigenous Peoples* or Zuberi and Bonilla-Silva's (2008) recent book entitled *White Logic, White Methods* for historical accountings that justify the use of these terms. As noted by Zuberi and Bonilla-Silva (2008):

> Some readers will argue that the logic of social science, like mathematics and physics, is without racial biases, and can be applied regardless of racial and other individual considerations. However, as we have argued, all scientific endeavors transpire in a world where race, gender, and class are important not only as subjects for investigation, but as structural factors that partly shape researchers and their scientific gaze... Hence, whereas the knowledge/experience basis of Whites, as a group, leads them to produce racial knowledge that tends to reproduce the racial order, the knowledge/experience of non-Whites, as groups, leads them to produce racial knowledge that uncovers social relations of domination, practices of exclusion, and the like... (p. 18)

Part of my concern with mainstream mathematics education is the frequent narrowness of research. I understand issues of grain size and units of analysis for particular kinds of studies. However, your example pointing out how the research team failed to ask questions about student identities implies that curriculum design can be done somewhat independent of who the learners are as persons in the world. In my view, this merely reduces students to objects and consumers of the curriculum that we design for them. Their subjectivities as learners and doers of mathematics with emerging identities often do not inform the process. In my view, Black feminist thought, CRT, poststructuralist, Freirian, and other perspectives can certainly inform research on mathematics teaching, learning, curriculum, and assessment.

I should point out that I do not utilize scholarship outside of mathematics education, particularly sociology, as a way to be novel. In relation to the issues that concern me, I do so out of necessity. I also do it based on a lifetime of personal experience and in relation to the experiences of those who have been the focus of my research. The complexity of these experiences – often shaped by race and the negotiation of racial identity – often requires more than traditional, mainstream theories of cognition and being in the world.

Maisie: Danny, this raises a whole host of issues that I will eventually navigate under your guidance, which relate globally to graduate students who are interested in race critical work in mathematics education. For example, in the event that students can avoid the pitfalls of physical and cultural isolation within their graduate programs, to what extent can students avoid intellectual isolation when adopting frameworks that stand in opposition to Western[1] positivist epistemologies

(Gay, 2004; Smith 1999)? Further, while my experience at UIC may be supportive and nurturing and pushes away from Eurocentric, male biased curriculum, how does this milieu prepare a critical Black female scholar for the academy at large (Gay, 2004)?

Danny: I think part of the answer to your question is that context matters. Where you study, with whom you study, and what you choose to study matters. Some environments and the people in those environments are likely to be more supportive of an emerging identity as a critical Black scholar and the choice to utilize alternative epistemologies. But I also think you, and other graduate students, might consider the possibility that in environments where there is great support for your work and ideas, even if they seem far-removed from mainstream areas of focus, it is still wise to "master the master's tools". Critique and selective use of these tools and ideas can only come from knowing them and understanding the logic behind them. For example, it is often said in colloquial talk among many Black people that in order to function sanely in the world not only do we have understand what it means to be Black but we also have to understand what it means to be White. Applying this to your development as a graduate student, I believe that part of my role as a mentor and advisor is to help you develop deep understanding of the dominant epistemologies and modes of research in addition to fostering your development as a critical Black scholar. I do not say this to imply the old adage that you have to work twice as hard as someone who is White. But if I only fostered your development by promoting only race-critical methods and perspective, for example, I would be engaging in the very same exclusion of ideas and perspectives that I noted when discussing the experiences of graduate students at other institutions.

On the heels of these last points, I want to loop back to something I mentioned earlier, having to do with the increased number of Black mathematics education doctoral students at our institution. We now have eight Black students in our pipeline and seven of them, including you, are my advisees. And my assessment is that all of you have some critical orientation as it has been talked about in this chapter. Despite the fact that many of you have come to UIC specifically to work with me, you all brought that critical orientation with you. And of course, you will all continue to develop your own individual voices as a result of your experiences outside of working with me. However, what I want to point out is that your collective presence in this program, and eventually in the field, has far-reaching implications. I know this because, as I go out into the field, one of the questions that I often get in response to my critiques of mainstream theory and methods and my characterization of mathematics education research and policy contexts as instantiations of White institutional space (Martin, 2008, in press-b) is *who can do the work of researching Black children's mathematical experiences?*

At the conclusion of a recent keynote address (Martin, 2009b), where I foregrounded Black children, issues of race, racism, and racialisation, and offered a critique of what I characterized as mainstream mathematics education research, I was asked this very question by a White scholar in the audience. If that scholar were to look at me, you, and your classmates as part of an emerging critical mass, our collective presence could be interpreted as sending a message that only critical Black scholars can study Black children.

Yet, I want to point out how the audience member's question further highlights the politics of race and identity in mathematics education. In my view, such a question can have the effect of momentarily re-centring the discussion to focus on the needs or sensitivities of White scholars, even critical White scholars whose work might focus on Black children.

Clearly, anyone can study the mathematical experiences of Black children, and the history of research in the field will verify this. Moreover, in the same way that I have argued against ineffective teachers, Black or White, I support the work of White scholars whose research facilitates mathematical, social, and epistemological empowerment (Ernest, 2002) for Black children.

Maisie: Danny, I agree whole-heartedly that race is not *the* primary factor as to who can do this work. Using a recent example, I would like to be more specific of what empowerment means in conducting research of Black students and teachers in a mathematics classroom.

In discussions among the research staff, several teachers' names were tossed about for in-depth study; that is, consecutive days of videotaping and field notes. An African-American teacher's name was suggested. I will call her Janice (a pseudonym). Although she was immediately positioned as "mathematically weak", she was selected as a teacher of interest based primarily on her race. While I had already been assigned to observe another teacher, I was concerned about how Janice was being talked about, so I also volunteered to observe Janice as a means to "protect" her from potentially deficit-oriented reports of her classroom practice throughout the academic year. As mentioned before, being from the neither-world – between the academy and the Black community – I was able to see her in what I believed was a different light.

Janice's classroom was 100% Black. During class, Janice was often abrasive in tone and academically demanding of her students. There was always a constant stream of students coming by to visit, to get a hug, or to be verbally and lovingly chastised. It was clear that she was well-liked by her students, despite her "tough love." She did not take excuses and often launched into mini-speeches regarding her background within the same community as her students (Clark, Johnson, & Chazan, 2009). Her style was a sharp contrast to White teachers in the project. At one point during the academic year, Janice's conversations became increasingly didactic and authoritative, more so than usual. This was antithetical to the intended design of the reform-based program, which called for open discourse and mathematical argumentation among the students.

During this time of Janice's heightened authority, there was also a series of physical fights among students in the hallways. In one of these fights, a teacher was struck. One day after class, Janice told me quite directly that the kids were not going to "punk" her. It was clear that she felt that she had to reassert her authority within the classroom or lose her status (as this other teacher had lost their authority). This had a definite impact on classroom discourse. Janice often employed an Initiation-Reply-Evaluation (I-R-E) model in order to maintain classroom order and her status as the authority figure.

During research meetings, the only information requested from me about Janice was how the conversation went in classrooms or whether I had captured any "good" video. It was clear to me that the design of the reform-based curriculum had normalized open discourse and mathematical argumentation, which was not always a viable, or desirable, part of Janice's school and classroom culture. It was also clear that the research team privileged mathematical knowledge and practices over classroom culture and practices that provided a stable environment in Janice's eyes.

Perhaps because of my own background as a Black female mathematics teacher, I was able to understand her perspective whether I agreed with it or not. However, transferring what I learned about Janice's practice into the research team discussions did not fit into the overarching goals of the research study at large. From the point of view of many on the research team, the lack of mathematics discourse, as called for in the curriculum, worked to further marginalize Janice within the research and solidified framings of her mathematical inadequacy. This is not to demonise the research team, but to recognize that their narrow aims often helped to instantiate and perpetuate deficit constructions of Black students and teachers.

From this experience, I found that one challenge for Black scholars is to interject oneself in the construction of standards (i.e., successes, failures, and models) (McDermott and Varenne, 2006) because, without perspectives that call attention to the limitations in these standards, they can work to harm Black students and teachers alike. So, you asked earlier, who can do this work? I humbly say it is those scholars who can actively and unrelentingly exercise reflexivity, who see Black children (and teachers) as objectified subjects in research projects but as informants whose life experiences and voices can give great insight into their needs as learners (and teachers).

Danny: The example that you provide above is powerful and I hope that its significance is not lost on readers. Clearly, you are not attempting to vilify the members of the research team. However, what you have pointed out is a kind of conceptual blind-spot that that many scholars, of all backgrounds, can have with respect to developing their research or implementing their projects. I do think a responsibility of all scholars is to think about these blind-spots and the implications of moving forward in their work without giving them attention. Although it might be an unfair demand, I think is a requirement for any critical Black scholar. Again, the price of admission to the field or to a particular project should not be the suspension of one's willingness to demand full consideration of Black children's (and teachers') humanity even in the context of research on curriculum.

Maisie: As we begin drawing to a close in this conversation, I do want to make a few final points. I could quite conveniently choose not to study Black children in mathematics education, particularly in the haze of post-racial delusion, yet I have decided to do so. Being a Black scholar in and of itself *is* a political act, being a

Black scholar studying Black children qualifies as further politicisation, and being a *critical* Black scholar can only be understood as political activism.

Danny, in your work, you often repeat several provocative questions that have been raised by other Black scholars (Perry, Steele, & Hilliard, 2003, p. 19):

- Why should African-American youth take school seriously if they cannot predict when and under what circumstances their intellect or intellectual work is likely to be taken seriously?
- Why should African-American youth commit themselves to doing outstanding intellectual work if – because of the color of their skin – this work is likely to be undervalued, evaluated differently, or ignored?
- Why work hard at school, or anything else for that mater if these activities are not inextricably linked to and address one's status as a member of a historically oppressed people?

While these questions pertain to African-American youth, critical Black scholars can easily turn these questions on their head and inquire simply: *Why do this work?* Every critical Black scholar will have to answer that question for him or herself. Earlier, you mentioned the defence of Black people's humanity. I will endeavour to answer this question poetically, but nonetheless honestly. I believe that the great understandings of mathematics education, American education for all children, will and must be found in the classroom of Black children. I take a page from the great poet, Nikki Giovanni (2002), who makes a similar argument in *The Quilting of the Black-Eyed Pea.* Speaking of future astronauts who set out for Mars, Giovanna wrote:

> So let me slow this down:
> Mars is 1 year of travel to get there.....
> plus 1 year of living on Mars......
> plus 1 year to return to Earth......
> = 3 years of Earthlings being in a tight
> space going to an unknown place with an
> unsure welcome awaiting them...
> tired muscles...unknown and unusual
> foods....harsh conditions...and no known
> landmarks to keep them human...
> only a hope and a prayer that they will be
> shadowed beneath a benign hand and there
> is no historical precedent for that except this:
> The trip to Mars can only be understood
> through Black Americans
> I say, the trip to Mars can only be understood
> through Black Americans
> ...
> and that is why NASA needs to call Black America
> They need to ask us: How did you calm your
> fears...How were you able to decide you

were human even when everything said you
were not...How did you find comfort in
the face of the improbable to make the
world you came to your world... How was
your soul able to look back and wonder

RAPPING UP

By engaging in this dialogue and sharing our voices, it is our hope that we have successfully revisited, and extended, Tate's earlier reflections on the experiences of Black scholars in mathematics education. Clearly, our stories are our own and we do not profess to speak for others. As a mid-career scholar and a new graduate student, we enact our identities as critical Black scholars in similar and different ways. However, we share a concern for mathematics education that is meaningful, relevant, and responsive to the needs of Black children as *Black* children. We share this concern while carrying out our work in a domain numerically dominated by White scholars and in a society where Black children continue to be devalued. Our critical engagement with the field is driven by the need and necessity for confronting this devaluation. In doing so, we see our scholarly work as deeply personal and political.

NOTE

[1] I use "Western" as delimited by Zuberi and Bonilla-Silva (2008). They stated, "White logic assumes a historical posture that grants eternal objectivity to the view of elite Whites and condemns the views of non-Whites to perpetual subjectivity; it is the anchor of the Western imagination, which grants centrality to the knowledge, history, science, and culture of elite White men and classifies 'other' people without knowledge, history, or science, as people with folklore but not culture." (Emphasis added, p. 17).

REFERENCES

Baldwin, J. (1963). *A talk to teachers. The price of a ticket, collected non-fiction: 1948–1985*. New York: Saint Martin's Press.

Bonilla-Silva, E. (2001). *White supremacy & racism in the post-civil rights era*. London: Lynne Rienner Publishers.

Bonilla-Silva, E. (2003). *Racism without racists: Color-blind racism and the persistence of racial inequality in the United States*. Lanham, MD: Rowman and Littlefield.

Bonilla-Silva, E. (2005). "New racism," color-blind racism, and the future of Whiteness in America. In A. W. Doane & E. Bonilla-Silva (Eds.), *White out: The continuing significance of race* (pp. 271–284). New York: Routledge.

Carruthers, J. H. (1994). Black intellectuals and the crisis in Black education. In M. Shujaa (Ed.), *Too much schooling, too little education: A paradox of Black life in white societies* (pp. 37–55). Trenton, NJ: Africa World Press.

Clark, L., Johnson, W., & Chazan, D. (2009). Researching African American mathematics teachers of African American students: Conceptual and methodological considerations. In D. Martin (Ed.), *Mathematics teaching, learning, and liberation in the lives of black children* (pp. 75–102). New York: Routledge.

Cruse, H. (1967). *The crisis of the Negro intellectual*. New York: Morrow.

DeCuir, J. T., & Dixon, A. D. (2004). So when it comes out, they aren't that surprised that it is there: Using critical race theory as a tool of analysis of race and racism in education. *Educational Researcher, 33*, 26–31.

Delgado, R. (1984). The imperial scholar: Reflections on a review of civil rights literature. *University of Pennsylvania Law Review, 132*, 561–578.

Delgado, R. (1989). Storytelling for oppositionists and others: A plea for narrative. *Michigan Law Review, 87*, 2411–2441.

Delgado, R. (1990). When a story is just a story: Does voice really matter? *Virginia Law Review, 76*, 95–111.

Delgado, R. (1995). The imperial scholar revisited: How to marginalize outsider writing, ten years later. In R. Delgado (Ed.), *Critical race theory: The cutting edge* (pp. 401–408). Philadelphia: Temple University Press.

Dillard, C. (2000). The substance of things hoped for, the evidence of things not seen: Examining an endarkened feminist epistemology in educational research and leadership. *Qualitative Studies in Education, 13*(6), 661–681.

D'Souza, D. (1991). *The end of racism*. New York: Free Press Paperbacks.

Ernest, P. (2002). Empowerment in mathematics education. *Philosophy of Mathematics Journal, 15*. Retrieved from http://www.ex.ac.uk/~PErnest/pome15/contents.htm

Gay, G. (2004). Navigating marginality en route to the professoriate: Graduate students of color learning and living in academia. *International Journal of Qualitative Studies in Education, 17*(2), 265–288.

Giovanni, N. (2002). *Quilting the black-eyed pea*. New York: HarperCollins.

Gordon, B. M. (1993). African-American cultural knowledge and liberatory education: Dilemmas, problems, and potentials in a postmodern society. *Urban Education, 27*(4), 448–470.

Kelley, R. (1997). *Yo' mama's disfunktional! Fighting the culture wars in urban America*. Boston, MA: Beacon Press.

Ladson-Billings, G. (2000). Racialized discourses and ethnic epistemologies. In N. Denzin & Y. Lincoln (Eds.), *Handbook of qualitative research* (pp. 257–277). Thousand Oaks, CA: Sage.

Martin, D. (2000). *Mathematics success and failure among African American youth: The roles of sociohistorical context, community forces, school influence, and individual agency*. Mahwah, NJ: Lawrence Erlbaum.

Martin, D. (2006). Mathematics learning and participation as racialized forms of experience: African American parents speak on the struggle for mathematics literacy. *Mathematical Thinking and Learning, 8*(3), 197–229.

Martin, D. (2007). Beyond missionaries or cannibals: Who should teach mathematics to African American children? *The High School Journal, 91*(1), 6–28.

Martin, D. (2008). E(race)ing race from a national conversation on mathematics teaching and learning: The national mathematics advisory panel as white institutional space. *The Montana Mathematics Enthusiast, 5*(2&3), 387–398.

Martin, D. (2009a). Liberating the production of knowledge about African American children and mathematics. In D. Martin (Ed.), *Mathematics teaching, learning, and liberation in the lives of Black children* (pp. 3–38). New York: Routledge.

Martin, D. (2009b). Little Black boys and little Black girls: How do mathematics education and research treat them? In Swars, S. L., Stinson, D. W. & Lemons-Smith, S. (Eds.), *Proceedings of the 31st Annual Meeting of the North American Chapter of the International Group for the Psychology of Mathematics Education* (pp. 22–41). Atlanta, GA: Georgia State University.

Martin, D. (2009c). Researching race in mathematics education. *Teachers College Record, 111*(2), 295–338.

Martin, D. (in press-a). Learning mathematics while Black. *The Journal of Educational Foundations*.

Martin, D. (in press-b). Race, racial projects, and mathematics education. *Journal for Research in Mathematics Education*.

McDermott, R., & Varenne, H. (2006). Reconstructing culture in educational research. In G. Spindler & L. Hammond (Eds.), *Innovations in educational ethnography: Theory, methods, and results*, (pp. 3–31). Mahwah, NJ: Lawrence Erlbaum.

McWhorter, J. (2001). *Losing the race: Self-sabotage in Black America*. New York: Harper Perennial.

Mohanty, C. T. (1989–90, winter). On race and voice: Challenges for liberal education. *Cultural Critique*, *14*, 179–208.

Moore, W. (2008). *Reproducing racism: White space, elite law schools, and racial inequality*. New York: Rowman & Littlefield.

Newson, J. (1998). The corporate linked university: From social project to market force. *Canadian Journal of Communication*, *23*, 107–124.

Perry, T., Steele, C., & Hilliard, A. (Eds.) (2003). *Young, gifted and black: Promoting high achievement among African-American students*. Boston: Beacon Press.

Scheurich, J. J., & Young, M. D. (1997). Coloring epistemologies: Are our epistemologies racially biased? *Educational Researcher*, *26*(4), 4–16.

Smith, L. T. (1999). *Decolonizing methodologies: Research and indigenous peoples*. New York and London: Zed Books.

Solorzano, D., & Yosso, T. (2002). Critical race methodology: Counter-storytelling as an analytic framework for education research. *Qualitative Inquiry*, *8*(1), 23–44.

Steele, S. (1990). *The content of our character: A new vision of race in America*. New York: St. Martin's Press.

Tate, W. F. (1994) From inner city to ivory tower: Does my voice matter in the academy? *Urban Education*, *29*(3), 245–269.

Thernstrom, S., & Thernstrom, A. (1997). *America in black and white: One nation indivisible*. New York: Simon & Schuster.

Thernstrom, A. & Thernstrom, S. (2004). *No excuses: Closing the racial gap in learning*. New York: Simon & Schuster.

Valero, P., & Zevenbergen, R. (Eds.). (2004). *Researching the socio-political dimensions of mathematics education*. Dordrecht, The Netherlands: Kluwer.

West, C. (1985, autumn). The dilemma of the Black intellectual. *Cultural Critique*, *1*, 109–124.

Williams, P. J. (1991). *The alchemy of race and rights: Diary of a law professor*. Cambridge, MA: Harvard University Press.

Zuberi, T., & Bonilla-Silva, E. (Eds.). (2008). *White logic, white methods: Racism and methodology*. New York: Rowman & Littlefield.

Danny Bernard Martin
College of Education
University of Illinois at Chicago

Maisie Gholson
College of Education
University of Illinois at Chicago

TOTAKAHINI: THE TALE OF THE PARROT

RABINDRANATH TAGORE
TRANSLATED BY SWAPNA MUKHOPADHYAY

Once there was a bird. It was ignorant. It sang all day, but it never read a thing. It hopped and flapped its wings, it flew here, there, and everywhere but it never showed the slightest sign of good manners. In other words, the bird did not know how to behave.

The King thought to himself "What's the use of such a bird to our society? It devours our orchards and causes such a havoc in our fruit market." He called his cabinet ministers together and ordered them to take action. "Go and educate that Bird!"

The primary responsibility of educating the Bird was bestowed upon the Royal Nephew. The Scholars came from far and wide and deeply investigated the matter. The question they pondered was: What are the reasons for this creature's failure to learn? They came to a consensus that the Bird's nest of twigs and straw was a structure too simple and flimsy to hold any knowledge. The key recommendation was that the Bird should have a properly constructed cage. The King's Scholars were endowed with handsome rewards and went home satisfied.

The Goldsmith started building a golden cage. It was such a magnificent construction that people poured in from every quarter to view it. Some exclaimed, "This is Education at its very best". Others said "Education or not, it's a grand cage. What a lucky Bird!" The Goldsmith was laden with rewards. Likewise contented, he started towards his home with his sack full of baksheesh.

The Scholar sat down to educate the Bird. He took a pinch of snuff and declared: "It is not a matter of just a few books."

The Nephew immediately called in the Scribes. They rushed to copy the texts and copy the copies of the texts till the pile grew as high as a mountain. Awestruck bystanders declared "Bravo! Education is breaking all bounds!" The Scribes loaded their trophies on their oxen and hurried homewards. Henceforth, their families would want for nothing.

Looking after the expensive cage became the Nephew's endless task; it needed constant maintenance. The perpetual dusting and sweeping, cleaning and polishing, prompted all those who witnessed it to agree: "It certainly is improving."

Many were hired to begin with; still more were hired to supervise them. Each month, faster and faster, they filled their chests with fistfuls of gold. Their

O. Skovsmose and B. Greer (eds.), Opening the Cage: Critique and Politics of Mathematics Education, 223–226.

cousins and brothers enthusiastically crammed the upper levels of the King's Office.

In every family, there is always a shortage of something or other; critics are the only commodity that we have in plenty. They kept saying, "Yes, indeed, the cage is impressive, but nobody's bothered to ask about the Bird."

The word got to the King. Annoyed, he called for his Nephew and demanded an explanation, "Nephew, what is all this gossip about?"

The Nephew replied, "Dear King, Your Highness, if you really care to know the truth, ask the Scholars, the Scribes, the Engineers, the Designers, the Managers, and the Supervisors. Those Critics don't get enough to eat – that's why their outlook is gloomy."

This reply made everything clear to the King and immediately the Nephew's neck was wrapped with gold.

The King wanted to inspect for himself the neck-breaking speed of the great educational enterprise. So one day he showed up at the School with his Ministers, their Advisers, and all the rest.

As they approached the main gate of the School, bells and horns, cymbals and drums, flutes and kettledrums, started up. The Scholars shook their sacred locks high in the air and began loudly chanting mantras. The Engineers, the Workers, the Goldsmiths, the Scribes, the Overseers, the Managers and all the Royal Cousins sang in chorus a hearty greeting for the King.

The Nephew said, "Oh King, Your Highness, behold the spectacle." The King said, "A magnificent feast for the eyes! And the sound is no less impressive." The Nephew added, "It is not just the sound, Sire, it is a lavish extravaganza. Nothing has been spared."

Pleased by the performance, the King walked through the main gate, ready to leave the place. Just as he was getting on his royal elephant, the Critic, hiding behind a bush, whispered, "But Sire, have you seen the Bird yet?" That startled the King. "Oh yes, it completely slipped my mind – I haven't seen the Bird yet."

He went back to the Scholar and asked him for a demonstration of the teaching technique: "Show me the methods you use to teach the Bird." What a delightful process to observe! The technique was far more important than the Bird itself, which scarcely needed to be visible. That made the King realize that the system was totally flawless. There was no food or drink in the cage, nothing but the tip of a pen stuffing page after page shredded from piles of books into the Bird's mouth. Its song has stopped, of course; there was not even a moment to scream. It's enough to make one tingle all over!

As he climbed onto his elephant, the King ordered the Royal Ear-twister to administer a good dose to the Critic.

Day by day, the Bird became half-dead and docile. The Guardians saw this as an encouraging development. Yet, true to its old habits, it would still occasionally see the morning light and flap its wings in an unacceptable manner. Moreover, it was caught a few times trying to bend the bars of the cage with its feeble beak.

The Chief of Police howled, "What an impertinent act!"

The Blacksmith appeared immediately at the School with his hammers, bellows, and fire. With resounding pounding, clanging and banging, an iron chain was forged. The Bird's wings were also cut off.

The King's relatives, deeply disturbed, glumly shook their heads. "In this Kingdom, the birds have no intelligence or sense of gratitude!"

Then the Scholars, pen in one hand and rod in the other, resumed the process that is called Education.

The Blacksmith's reputation went up to the extent that his wife became festooned with gold, and a turban of honour was bestowed on the Chief of Police for his conscientious and careful service.

The Bird died. No one even noticed it. Again, it was the god-forsaken Critic who started the rumour, "The Bird is dead."

The King called the Nephew and asked: "Nephew, what's this I hear now?" The Nephew replied: "Oh King, the Bird has completed its education."

The King inquired, "Does he still jump?" "No, Your Highness!"

"Does he still sing?" "Oh Lord, no!"

"Does he still scream when he is hungry?" "No!"

The King wanted to see the Bird. "Go get the Bird at once." The Bird was brought to the King. It was followed by the Chief of Police and his bodyguards and horsemen. The King poked and prodded the bird, but never a squawk or chirp did it utter. All that was heard was the rustle of the dry shredded paper inside its belly.

Outside, a sighing southerly wind began to stir the forest air.

REFERENCE

Tagore, R. (1961). Totakahini. In *Lipika (Vol. 7 of Rabindra Rachanabali [Collected works])*, pp. 812–814. Calcutta: West Bengal Government Centenary Collection. (First published in *Sabuj Patra*, a Bengali magazine, Magh, 1324 [January–February, 1918]).

Swapna Mukhopadhyay
Portland State University
Oregon, USA

PART III

MATHEMATICS AND POWER

CHAPTER 10

THE HEGEMONY OF MATHEMATICS

BRIAN GREER AND SWAPNA MUKHOPADHYAY

In this chapter, we use the term "hegemony" with specific reference to how mathematics and mathematics education are implicated in various forms of interpersonal dominance and in ideological struggles, in particular cases of dominance and suppression of some cultural groups by others, and the attempts by dominated groups to find voice and agency. We analyse symbolic violence of several kinds; we do not attempt to deal with the extensive uses of mathematics in dominance by means of military or economic force.

The most pernicious form in which hegemony operates was characterized thus by Foucault (1980, p. 133):

> ... hegemony is a state within society whereby those who are dominated by others take on board the values and ideologies of those in power and accept them as their own; this leads to them accepting their position within the hierarchy as natural for their own good.

Often these processes operate at a level below consciousness; they remain unexamined or even unnoticed, in which case the task at hand is to render them visible and expose them to critique. In other cases, notably the "culture wars" (most obviously, but by no means exclusively, in the United States), the clash between value systems and world-views is extremely visible.

The title of this chapter echoes that of the book *Hegemony of English* (Macedo, Dendrinos, & Gounari, 2003). In Greer and Mukhopadhyay (in press), in suggesting parallels between English and mathematics as two instruments of hegemony, we show how passages from that book can be directly transposed to refer to mathematics. For example, one task of unmasking that is necessary in bringing ideological conflict into the open is to clarify that:

> ... given the social and ideological nature of different functions and uses of language **mathematics**, the proposition that language **mathematics** is neutral or non-ideological constitutes, in reality, an ideological position itself. (Adapted as indicated from Macedo et al., 2003, p. 27)

Being necessarily selective, we discuss three contexts in which aspects of hegemonic manifestations of mathematics and mathematics education are manifest, namely:

1. Mathematics and cultural imperialism

O. Skovsmose and B. Greer (eds.), Opening the Cage: Critique and Politics of Mathematics Education, 229–248.

2. Hegemony of mathematics in society
3. Hegemonic aspects of mathematics education

Finally, we discuss forms of resistance seen in each of these contexts.

MATHEMATICS AND CULTURAL IMPERIALISM

Eurocentric narrative of history of mathematics

At the most general level, as expressed by Dias (2002, p. 205):

> The understanding of what is "human" and what should be regarded as "development" has taken ... a specific turn since the coming up of the "Invention of Man" toward the end of the sixteenth century, and of the subsequent hegemonic construction of the alien, subaltern (non-European) other. The consequence has been the steady marginalisation, separation, and sub-ordination of difference and diversity of the world-wide existing human and cultural experience and its multifaceted expressions. This domination structure has as its correlates the privileging and selective imposition of reduced cognitive structures, of one-sided interpretation patterns, of restricted scientific and technical solutions and of monolinguistic habits.

This characterization by Dias finds clear expression in relation to mathematics in the words of Kline (1953, p. 27):

> Mathematics is a living plant [that] finally secured a firm grip on life in the highly congenial soil of Greece and waxed strong for a brief period. In this period it produced one perfect flower, Euclidean geometry. The buds of other flowers opened slightly and with close inspection the outlines of trigonometry and algebra could be discerned; but these flowers withered with the decline of Greek civilization, and the plant remained dormant for one thousand years. Such was the state of mathematics when the plant was transported to Europe proper and once more embedded in fertile soil.

Cracks in the Eurocentric narrative have increasingly appeared in recent years through the work of many scholars (for overviews, see Joseph, 1992; Powell & Frankenstein, 1997). At a level of detail, many examples have been documented of intellectual precedence by non-European mathematicians in relation to pieces of mathematics, including many that traditionally bear the names of Europeans. In the collection of original sources compiled by Swetz (1994), examples include a Chinese version of the pattern usually known (within academic mathematics) as "Pascal's Triangle" predating Pascal by several centuries, and several versions of the theorem conventionally associated with the name of Pythagoras.

Beyond these cases of provenance of particular pieces of mathematics, there are more substantive general issues. A particularly significant example is the analysis of achievements in calculus in Kerala in the two centuries preceding its "discovery" by Leibnitz and Newton and the possibility that this knowledge could have been conveyed to Europe by Jesuit missionaries or otherwise (Joseph, 2009). Even if it is not possible to establish that such transmission took place, parallel development of significant ideas of calculus further deconstruct the narrative of academic mathematics being solely a European achievement. In a talk to the Mathematical Association of America, Joseph (2008) asked: "Why is there such difficulty for new evidence on Non-Western contributions to become accepted and then percolate into standard histories of science?" and contended that the standard of evidence required to establish transmission from East to West was generally much higher than that required for transmission in the opposite direction, because of Eurocentric ideology (and see Ernest, 2009). Indeed, Eurocentric accounts of mathematics typically fail to acknowledge the degree of cultural cross-fertilization of mathematical knowledge, as discussed, for example, with reference to India at various points by Amartya Sen (2005), and in terms of the major contributions to, and influences on, "Greek" mathematics from North Africa and beyond. Sen (e.g. 2005, pp. 146–149) documents how the mindset of British imperialists in relation to India was such that, on *a priori* grounds that a primitive culture could not be capable of such, they refused to acknowledge intellectual achievements of Indian scholars in the face of irrefutable documentary evidence.

If we consider the contemporary state of mathematics as a discipline, it is now an enterprise with global institutions and fast and comprehensive communication. The great mass of known mathematics, as a codified system, is recent, and, at the frontiers of research, specialized to such a degree that it is typically the case that a mathematician can only discuss her or his work with a few colleagues in the world. Why, then – beyond the obvious interest for students of humankind – dwell on the history of mathematics? It is certainly a matter of cultural justice, particularly when it comes to teaching non-Western students. Beyond that, as many mathematicians and others have pointed out, the study of diverse conceptions of mathematics related to different epistemologies and forms of life could be very productive in terms of opening up new mathematics – the history of mathematics has not come to an end.

Cultural imperialism in action

Gramsci (2009, p. 416) commented on the European culture as the "only historically and concretely universal culture" dependent, of course, on European (and later, US) military and/or economic dominance of much of the rest of the world, perpetuated in various forms. As put by Said (1994, p. 300): "… if direct political control has disappeared, economic, political, and sometimes military domination, accompanied by cultural hegemony – the force of ruling and, as Gramsci calls them … directive ideas – emanating from the West and exerting power over the peripheral world, has sustained it."

What part does mathematics play? We are not proposing to address the confluence of scientific and technological developments in Europe and the era of colonization through military force, in which mathematics was centrally implicated, but rather consider its place within cultural hegemony. In a seminal paper entitled "Western mathematics: The secret weapon of cultural imperialism", Bishop (1990, p. 51) declared that:

> [Mathematics] had, in colonial times, and for most people, it continues to have today, the status of a culturally neutral phenomenon in the otherwise

> turbulent waters of education and imperialism. This article challenges that myth, and places what many now call "Western mathematics" in its rightful position in the arguments – namely, as one of the most powerful weapons in the imposition of western culture.

Bishop challenged (p. 63): "Should there not be more resistance to this cultural hegemony?" and concluded that "Resistance is growing, critical debate is informing theoretical development, and research is increasing, particularly in educational situations where culture-conflict is recognized. The secret weapon is secret no longer". Gary Urton (2009, p. 27) commented that the title of Bishop's paper "must surely be one of the most provocative in the recent literature concerning the history of mathematics and the nature and status of mathematical practice." Illustrating Bishop's argument, Urton provided a very detailed analysis in relation to the European invasion of the New World, and in particular to the imposition of European accounting methods on peoples of South America, displacing the highly-developed culturally embedded systems already in place there (2009; see also Urton, in press).

Said (1994) analysed the roles of literature in both reflecting and creating the internalisation of the imperialist relationship by both oppressed and oppressors. We are not aware of any work of comparable scope that deals with mathematics in the same way.

HEGEMONY OF MATHEMATICS IN SOCIETY

There is nothing modern about the intimate relationships between mathematics, governance, and the structuring of cultural life through mathematical abstractions such as kinship systems and, most notably, money and capital. However, those relationships have complexified greatly in terms of scientific, industrial, and technological developments, including the massive growth in computational and information-processing power and sociopolitical developments, such as post-colonial globalisation. Applications of mathematics, science, and technology have brought great benefits but also occasioned great harm – "wonders and horrors", as D'Ambrosio (1994) put it.

A prominent development within the uses of mathematics is the extent to which modelling is being applied, not just to physical phenomena, but also social phenomena. Already in 1986, Davis and Hersh (p. xv) warned that: "The social and physical worlds are being mathematized at an increasing rate" and "We'd better watch it, because too much of it may not be good for us". Industrialized societies have not developed means of solving human problems at the same rate as they have developed means of solving technical problems. Moreover, technical solutions are often substituted for the more complex human solutions that are needed. Improving mathematics education, for example, is a human problem, not a technical problem (Kilpatrick, 1981). Yet recent educational policy in the USA, and well beyond, treats the system, and the students, teachers, and schools within it, as a black box that can be controlled by external levers of test data, financial incentives and disincentives, punitive measures, and market forces (Ravitch, 2010; Spring, 2010).

We focus here on two main points. The first is that many aspects of contemporary life, both beneficial and harmful, are mediated by mathematical constructions that are often inaccessible to most people they affect – if they are even aware of what is going on. The second is that people generally are ill-prepared to react critically, and with agency, to these circumstances and are underserved in this regard by their education, and by forms of discourse within society. As a consequence of this lack of critical agency, people are subject to many forms of control, resulting in a combination of powerlessness and uncritical compliance.

Mathematics in action in society

Skovsmose (2005, p. 86) uses the term "mathematics in action" to refer to "how mathematical conceptions are projected into reality". As an example, he discusses the computerized models for the overbooking of seats on planes that are intended to maximize profit by taking into account how many passengers do not show up, how much it costs to "bump" passengers, and so on. Since the passengers so affected modify their behaviour in response to their experience, and these reactions in turn lead to adjustments to the models, an important aspect that this example illustrates is that "when part of reality becomes modelled and remodelled, then this process also influences reality itself" (Skovsmose, 2000, p. 5). Similar feedback effects have been analysed by Hacking (1995) in relation to the classification of people. The creation of categories and the associated social constructions (discourse, legal frameworks, and so on) not only affect the people so categorized but – since those people are aware of being so categorized – they change their behaviour which, in turn, changes the nature of the category. A manifestation of this phenomenon of great importance within the US currently is the ethnic categorization of students, and in particular the related discourse of "achievement gaps", i.e. group differences on standardized tests (see below).

A related formulation of relevance to our discussion is the principle enunciated by Campbell (1975) that "the more any quantitative social indicator is used for social decision-making, the more subject it will be to corruption pressures and the more apt it will be to distort and corrupt the social processes it was intended to monitor". The effects of Campbell's Law in the context of high-stakes testing in the USA have been fully analysed by Nichols and Berliner (2007). All of these examples illustrate Skovsmose's point that the modelling of situations involving people are neither neutral nor static descriptions but change the behaviour being modelled and are themselves changed.

In comparison with the modelling of overbooking for air travel, other examples have much deeper impact on people's lives, such as the very complex models used in governmental economic planning (see Skovsmose, 2005, pp. 82–85 for a discussion of a Danish macro-economic model). In another domain, for some time computer models have been used extensively in making military strategic decisions. Mayer (2010), in an article about the use of unmanned aircraft (drones) in Afghanistan and Pakistan, reported the following:

> Though the C.I.A.'s methodology remains unknown, the Pentagon has created elaborate formulas to help the military make such lethal calculations. A top military expert, who declined to be named, spoke of the military's system, saying, "There's a whole taxonomy of targets." Some people are approved for killing on sight. For others, additional permission is needed. A target's location enters the equation, too. If a school, hospital, or mosque is within the likely blast radius of a missile, that, too, is weighed by a computer algorithm before a lethal strike is authorized.

The last sentence in the quotation prompts a myriad of questions: What is the nature of the algorithm? On what assumptions is it based? Is its application evaluated in any way? How is the value of a non-American life relative to an American life quantified in it? Who constructed the algorithm and what were their qualifications? How much consideration was given to ethical, as opposed to strategic, issues? Could "I was just obeying algorithms" become a new defence in war crime trials?

Lack of societal support for critical mathematical agency

Skovsmose (2005, p. 140) suggests that there are four main categories of people in relation to how mathematical constructions are created, used, and experienced. "Constructors" are those who "develop and maintain the apparatus of reason", "Operators" are those who use them, often with little understanding, "Consumers" are those who are effected, and the fourth group are those that capitalist society is prepared to discard as "disposables".

In our opinion, people, including even many of the constructors, are not equipped with enough insight, criticality, and agency to respond to the multiple ways in which uses of mathematics affect their lives. We attribute this form of disempowerment to both inadequacies of mathematics education (addressed further below), and to lack of societal support.

A major contributory factor in this disempowerment is the supreme irony of the "Information Age", namely that the amount and accessibility of information swamp the intellectual tools available for evaluating and interpreting it. In a society in which those with the most money often can choose what will become accepted truth, we often do not know what information is to be trusted. Societal discourse is dominated by the values and strategies of advertising, in which truth is irrelevant and what matters is what people can be persuaded to believe. In these circumstances, it would certainly be useful to have more groups of unbiased and trustworthy experts, at the societal or community level, to act as "interpreters" mediating among the constructors, operators, and customers – and indeed "disposables" – of whom Skovsmose speaks. In this spirit, Evans and Rappaport (1999, cited by Yasukawa and Brown, Chapter 11, this volume) introduced the notions of "parastatistician" and of "barefoot statistician". Yasukawa and Brown explain (p. 250) that "the role of the barefoot statistician in community settings (and the parastatistician in the business settings) is to be a resource person who does research in their local community (or workplace) in order to promote critical citizenship, that is, to enable people to engage in policy debates and advocacy". (A useful function is also performed by experts in statistics and modelling who write accessibly for non-experts).

In a better world, the media would operate as honest brokers in this respect. Only a few journalists, particularly in the alternative media, display that kind of integrity. By way of example, media shortcomings are revealed in an analysis of how the estimation of civilian deaths in the Iraq war has been addressed by politicians, pundits, polemicists, the mainstream press, and the public (Greer, 2009a), drawing attention to lack of understanding or misrepresentation of the most basic statistical concepts, and distortion or rejection of the evidence on ideological grounds.

A further major impediment to being able to make sense of situations and issues important in their lives is the image of mathematics among the general population. According to Ernest (1996, p. 803), mathematics:

> ... has a bad press, with a widespread public image of being difficult, cold, abstract, and in many cultures, largely masculine. It also has the image of being remote and inaccessible to all but a few super-intelligent beings with "mathematical minds".

Indeed, Freire made precisely this comment (see Introduction, p. 3).

Such a widespread conception facilitates "intimidation by mathematics". An example that remains very telling was discussed by Koblitz (1981). It concerns an appearance on a popular television show in the USA by the author of a best-seller called "The Population Bomb". At one point, the author wrote in large letters:

$$D = N X I$$

and explained that D stood for damage to the environment, N for the number of people, and I for the impact of each person on the environment. He claimed that he had demonstrated that the more people, the more pollution. Koblitz (p. 111) commented:

> Who can argue with an equation? An equation is always exact, indisputable. Challenging someone who can support his claims with an equation is as pointless as arguing with your high school math teacher. How many of [the] viewers had the sophistication necessary to question [the] equation? Is [the author] saying that the "I" for the president of [an irresponsible chemical company] is the same as the "I" for you and me? Preposterous, isn't it? But what if the viewer is too intimidated by a mathematical equation to apply some common sense?

Several other examples illustrating the same phenomena were analysed by Schoenfeld (1991, p. 315), who referred to:

> ... the totalitarian power of mathematics. Throw enough formalism or appeal to mathematical authority in front of people, and they'll back down. For some reason ... people don't expect heavily mathematical statements to make sense to them. They assume they make sense to *somebody* (the "experts"), abdicate responsibility for understanding, and (at least sometimes) accept the statements on face value.

HEGEMONIC ASPECTS OF MATHEMATICS EDUCATION

Here we consider three aspects of hegemonic relationships as they are enacted in mathematics education, namely the ways in which the experience of school mathematics helps to shape the adult's worldview, the ongoing hegemonic struggle about how mathematics should be taught *and why*, and the use of mathematics education as a weapon of cultural violence. There are many other aspects that we could have chosen to discuss, such as those related to gender.

School mathematics experience as foundational

The ways in which experience of school mathematics is foundational in the formation of a worldview are numerous and complex, and here we can only sketch a few. Illustratively, then, we discuss the potential of mathematics schooling for intellectual violence against children, in particular through denial of the child's right of sense making, the related lack of the modelling perspective essential to an understanding of how mathematics formats society as discussed above, and the role of mathematics education as training in submission to objectification through assessment, reduction to values of a few statistical variables of dubious validity, and other dehumanising devices.

School mathematics can be a personally damaging experience. When the first author was teaching future elementary teachers about mathematics, he began each class by asking the students to think back to their own experience in mathematics classes and briefly describe the first memory that came to mind. Many times, what these (almost all female) students remembered and described was that at some stage a teacher in elementary school publicly mocked them for getting an answer wrong. Both of us have had many conversations with people we meet in all kinds of situations in which they recount similar experiences and their emotional reactions often extending well into adult life. It is not determined, but mathematics education affords opportunities for many kinds of symbolic violence, including abuse of sense making, suppression of creativity and intellectual diversity, and authoritarianism. In a strong critique of schooling in general, in the form of a reaction to the contradictions he experienced as a school mathematics teacher, Pais (2009, p. 62) declared that:

> In school we learn to be governed. Which behaviours are right, which things can and can't be said? This is how school appears in modernity, as an apparatus to govern the population by fabricating the kinds of subjects that hegemonic society stipulates as normal…

Here we consider, in particular, how children's natural inclination to make sense may be abused. In 1854, the novelist Gustave Flaubert wrote in a letter to a young relative:

> Since you are now studying geometry and trigonometry, I will give you a problem. A ship sails the ocean. It left Boston with a cargo of wool. It grosses 200 tons. It is bound for Le Havre. The mainmast is broken, the cabin boy is on deck, there are 12 passengers aboard, the wind is blowing East-North-East, the clock points to a quarter past three in the afternoon. It is the month of May. How old is the captain?

The intention was satirical, but researchers in France and elsewhere later posed a stripped-down version of Flaubert's question to schoolchildren: "On a boat there are 26 sheep and 6 goats. How old is the captain?". The reactions of the children, many of whom answered "26" without comment, became a *cause célèbre* (Baruk, 1985). How could children behave so irrationally? One answer was given by Schoenfeld (1991, p. 340) who commented that, by conforming with the expectations of the mathematics classroom, the children were making sense of a different kind.

Other researchers posed questions that are not "nonsensical" in the same way, but require judgment if the real-world implications of the text are taken seriously. An example: "John's best time to run 100 metres is 17 seconds. How long would it take him to run 1 kilometre?". Questions of this general kind have been given to children in countries all over the world under various experimental conditions. With remarkable uniformity, they predominantly answer in a way that appears to negate their real-world knowledge. For example, even when the question explicitly asks for comments on possible interpretations, almost all children asked the question above answered "170 seconds" without qualification (for details, see Verschaffel, Greer, & De Corte, 2000; for further discussion, Verschaffel, Greer, Van Dooren, & Mukhopadhyay, 2009). The implication is that, after some years at school, children learn to play a game in which one of the rules is that when asked a mathematical word problem, what they know about the world should be ignored. As Freudenthal (1991, p. 70) asked "Wouldn't it be worthwhile investigating whether and how this didactic breeds an antimathematical attitude and why the children's immunity against this mental deformation is so varied?"

In Verschaffel et al., (2000), the proposition was argued that word problems should be characterized as exercises in mathematical modelling. A major shortcoming of mathematics education in the United States (and elsewhere) is the lack of attention to the relationship between mathematics as a system of abstract structures and mathematics as a means for modelling aspects of the real and social worlds. One reason for this state of affairs is the privileging within mathematics as a discipline of mathematics that is "pure" over mathematics that is "applied". It is true that mathematical models are often very complex and require considerable knowledge of technical mathematics (e.g., see the discussion by Gutstein (Chapter 1, this volume) on the complexity of modelling the spread of HIV/AIDS). Nevertheless, there are some very fundamental ideas about the nature of mathematical modelling that can be conveyed through simple examples – for example, that a given calculation that appears appropriate at first sight may not be, that any model rests on assumptions, that a model is always a simplification, and that the nature of the model (e.g., its degree of precision), and how the results of the modelling exercise should be communicated, will be dependent on the tools available and on the purposes for which it is constructed. As many straightforward yet extremely powerful examples in the work of Frankenstein (1989, 2009) show, there are important issues that can be illuminated by clear thinking requiring relatively limited technical mathematics.

The most rudimentary of word problems can be characterized as exercises in mathematical modelling, and the four basic arithmetic operations can be interpreted as putative models that may or may not fit a given situation (Usiskin, 2007). Yet, as mentioned above, research that has been replicated in many countries shows that most students and teachers appear to accept a didactical contract that requires them to answer word problems without reference to their real-world knowledge. It does not seem too far-fetched to conjecture that this training in school mathematics may contribute to an attitude in later life whereby a veneer of mathematics inhibits critical application of real-world knowledge, as in the examples of intimidation described above.

Indeed, Skovsmose (2005, pp. 164–165) argued that mathematics education fulfils a societal role:

> The school mathematics tradition may provide qualities, like obedience, trust in numbers, exaggerated belief in authority etc. These aspects are considered problematic outcomes of mathematics education. But ... in many jobs, it is essential that people follow manuals and prescriptions.

In many circumstances in the USA, such prescriptions take the form of "zero tolerance" policies, which result in appalling decisions based on obeying the rules instead of applying judgment to particular circumstances. In the extreme case, what Skovsmose (2006, p. 324) termed "the banality of expertise" occurs when technical expertise is exercised without reflection or ethical control, as "experience might ignore its own humanity and become mechanised".

One way in which school, and mathematics education in particular, may mould people for such roles is through objectification by assessment or other forms of creditation. Assessment (or "accessment" as McDermott & Hall (2007, p. 10) put it) is a crude but powerful tool for control, as is currently being demonstrated most clearly, but by no means exclusively, in the USA. One way of looking at assessment is that it affords communication of various kinds (Miller-Jones & Greer, 2009). As indicated above, it is the instrument *par excellence* for communicating intellectual inferiority of individuals, groups, and even, through international comparison studies, nations. Typically, assessment items also communicate, in subtle and not so subtle ways, that mathematics has nothing to do with reality, even when a question appears to be framed in a context.

Kilpatrick (1993, p. 44) proposed that:

> The challenge for the 21st century, as far as mathematics educators are concerned, is to produce an assessment practice that does more than measure a person's mind and then assign that mind a treatment. We need to understand how people, not apart from but embedded in their cultures, come to use mathematics in different social settings and how we can create a mathematics education that helps them use it better, more rewardingly, and more responsibly. To do that will require us to transcend the crippling visions of mind as a hierarchy, school as a machine, and assessment as engineering.

Some twenty years later, there is no indication of this challenge being seriously accepted.

What mathematics in school, and why?

The "culture wars" in the United States may be regarded as a struggle for hegemonic dominance, in Gramsci's sense. A recent analysis by Schoenfeld and Pearson (2009) draws some very clear parallels between two fronts in this struggle from the mid-1990s onwards, namely the "reading wars" and the "math wars" (and there are other very clear parallels, in particular with the "history wars" (Nash, Crabtree, & Dunn, 2000), conflict over multicultural education, and language policies.) Schoenfeld and Pearson examine in some detail the history of the conflicts in relation to political developments in California and the USA in general and to pendular shifts in dominant educational theoretical positions (and see Spring, 2010).

Why should the opposing ideological stances in relations to these domains be so closely correlated? Schoenfeld and Pearson suggest (pp. 2–3) that underlying both is a tension between two views of the goals of education:

> Is education to serve as the path of opportunity for all citizens, especially the poor and powerless, to aspire to upward social mobility and political enfranchisement? Or, is the primary purpose of education to train workers just skilled enough to fill the jobs needed but not skilled enough to question the inequities of power and wealth in the society, perpetuating the status quo by reproducing the social and economic order in each generation?

Another deep-level proposed explanation for the linkages across domains in the culture wars is Lakoff's (1996) suggestion that there are two fundamental metaphors underlying the worldviews of conservatives and liberals – that of the authoritarian father, and that of the nurturing family, respectively.

"Mathematics for all" has become standard educational rhetoric in official discourse in the USA and beyond. In an essay setting out shared principles between mathematicians and mathematics educators (Ball, Ferrini-Mundy, Kilpatrick, Milgram, Schmid, & Schaar, 2005), the first "fundamental premise" is that: "All students must have a solid grounding in mathematics to function effectively in today's world." Really? Think about people you know. Aren't there many who do not have a solid grounding (assuming that means roughly what we take it to mean) in mathematics that are living full and productive lives? Isn't it offensive to tell such people that they are dysfunctional? "Mathematics for all" has a fine equitable sound to it, but neither theoretically, nor in terms of the actual situation for students, can it bear scrutiny (Gutstein, 2009; Martin, 2003).

In terms of the relationship between individual and societal needs, we may ask what school mathematics contributes to preparing people for life, work, and citizenship. D'Ambrosio (1999, p. 138) suggested that "we may be cheating our youth when we say that mathematics, as taught in our schools, opens good perspectives of employment for them". Instead, as discussed by Hersh and John-Steiner (2011, Chapter 9), mathematics too often operates as a filter unnecessarily restricting the choices for individuals.

Of course, the slogan "Mathematics for all" cannot be evaluated without at least a rough indication of what mathematics is intended. To make the discussion more grounded, let us take the premise of the National Mathematics Advisory Panel (Greer, Chapter 4, this volume) that school mathematics should result in every student learning algebra at least as far as the contents tabulated in its final report (United States Department of Education, 2008, p. 15). Greer (2008, p. 427) argued against the goal of "algebra for all" on the grounds that:

> … while collectively a cadre of mathematical and scientific specialists is needed for society to operate effectively, most individuals in our society do not need to have studied algebra. This by no means implies that anyone should be denied the opportunity to do so (in an intellectually stimulating way). Moreover, students should be encouraged to study algebra in the spirit of keeping options open, given its status as a gatekeeper to many educational and economic opportunities.

(A similar position is argued by Hersh and John-Steiner (2011, pp. 305–307).

Even within academic mathematics, there are alternatives to algebra as the focus. In particular, with obvious importance in contemporary life, there is the whole area of mathematical modelling, applications, and data analysis, as some prominent mathematicians and mathematics educators within the USA have argued (e.g., Shaughnessy, 2010; Steen, 2003).

A more radical alternative is "critical mathematics for all". Atweh (Chapter, 15, this volume) argues that "all students need a considerable amount of mathematics for effective citizenship in the increasingly mathematised world of today – albeit different types of mathematics depending on their interests, capacities and career choices". While this case can certainly be made, think of all the other desirable elements for an individual to be an effective citizen – in the domains of science (particularly environmental science), economics, history, and so on. Is it reasonable to expect every individual to be so well informed? One response to these heavy demands is to develop a more collective form of expertise whereby communities have access to people who can help explain and inform in relation to specific issues.

Beyond the question of "what mathematics" is the even more fundamental question "What is mathematics education *for*?" There are standard, partially valid, but eminently critiquable answers (Pais, 2009) – to reproduce the stock of mathematicians and carry forward the development of the discipline; the preparation of a workforce to enable the nation to compete successfully in the global economy; as part of cultural heritage as much as literature or music; as a training in reasoning and problem-solving; as preparation for practicalities of everyday life (Greer, 2009b). Currently, in the United States, the most prominent official reason is that strong mathematics and science education are necessary to maintain global military and economic dominance. Contrast that position with the plea by D'Ambrosio (2010) that mathematicians should collaboratively address the most universal problem of the world, namely survival with dignity.

Mathematics education and cultural violence

Earlier we discussed the role of mathematics education as a weapon of imperialism, a role that is by no means over. Cultural violence also happens internally. In the USA today, it is often argued that, while there were social and educational injustices in the past, the situation is now acceptable. However:

> The notion that ethnic minorities have achieved equity and social justice now, *even if it were true*, needs to be further qualified by taking into account what Ladson-Billings (2006) called "the education debt"... Whether talking about Native Americans, Latino/as, African Americans, and others, how do you characterize, let alone quantify, the results of decades or centuries of education as a tool of discrimination? (Miller-Jones & Greer, 2009, p. 175)

Mathematics education has been, and remains, part of the oppression, as shown, for example, in the work of Martin (e.g., 2010). Hersh and John-Steiner (2011, Chapter 8) describe a racist mathematics professor, Robert Lee Moore, who taught at the University of Texas until 1969 when he retired at the age of 87. The scandal is not so much that a mathematics professor could be a blatant racist (that is bad enough), but rather that the system within which he worked enabled him to behave in that way.

The teaching of mathematics is typically portrayed as uncontroversial, without any need for cultural responsiveness or considerations of diversity since "mathematics is the same for everyone", a position we reject. In particular, Greer, Verschaffel, and Mukhopadhyay (2007) argued that:

> If a decision is made to mathematise situations and issues that connect with students' lived experience, then it brings a further commitment to respect the diversity of that experience across genders, classes, and ethnicity (p. 96).

and illustrated the point with an example of a test item discussed by Tate (1995):

> It costs \$1.50 each way to ride the bus between home and work. A weekly pass is \$16.00. Which is the better deal, paying the daily fare or buying the weekly pass?

When African American students were interviewed about how they responded, it was discovered that they "transformed the 'neutral' assumptions of the problem – all people work 5 days a week and have one job – into their own realities and perspectives" (Tate, 1995, p. 440). In their experience, as opposed to white middle-class experience, a job might mean making several bus trips every day, not just two, and working more than five days a week. If items of this type are used for assessment, and assumptions are made about the "right" answers, the implications for inequity are clear, given that, as Tate (1995, p. 440) put it "the underpinnings of school mathematics curriculum, assessment, and pedagogy are often more closely aligned with the idealized experience of the White middle class" (see Martin (2010) on mathematics education as a white institutional space).

Martin (2009) argues that mathematics educators, instead of critiquing, generally take as natural a racial hierarchy of mathematical ability. This supposed hierarchy takes concrete form in differences in scores on standardized tests. McDermott and Hall (2007, p. 11) stated bluntly that: "Quantitative tests of aptitude and achievement have give U.S. education a way to sort children by race and social class, just like the old days, but without the words 'race' and 'class' front and center."

Gutierrez (2009, p. 9) commented on what she termed "gap-gazing":

> Although mainly concerned with the well-being of marginalized students … mathematics education researchers who focus on the achievement gap can unknowingly support practices that are against the best interests of those students. Some of the dangerous effects of gap gazing include: offering little more than a static picture of inequities; overlooking many assumptions embedded in measurement tools; supporting deficit thinking and negative narratives about marginalized students; accepting a static notion of student identity; relying upon a comparison group; dividing and categorizing students; offering a "safe" proxy for talking about students of color without naming them or acknowledging racism in society; perpetuating the myth that the problem (and therefore solution) is technical in nature; and relying upon narrow definitions of learning and equity.

FORMS OF RESISTANCE

> Remember that our basic message is: "We are allowed to think about alternatives" (Slavoj Žižek, speaking at Occupy Wall Street protest, 10 October, 2011)

For each of the contexts discussed above, namely cultural imperialism, mathematics in society, and mathematics education, we address some of the forms of resistance that critical mathematics educators and others are developing.

Diversity of mathematical practices

> The intellectual activity of those without power is always characterized as non-intellectual. (Freire and Macedo, 1987, p. 122)

As indicated above, the work of many scholars has been contributing to a counter-narrative to the dominant narrative of mathematics as an intellectual achievement of Europeans, with contributions from non-Europeans to be marginalized or appropriated. The resistance to this narrative takes place at two levels (Powell & Frankenstein, 1997) – first, in establishing that the contributions of non-European cultures to academic mathematics have been of fundamental importance (Joseph, 1992), and second in establishing that there are many forms of mathematical practice beyond that of academic mathematics, a form of argument particularly associated with the ethnomathematical perspective (D'Ambrosio, 2006).

Barton (2008, p. 10) drew a distinction between "near-universal, conventional mathematics" (NUC-mathematics) and "[systems] for dealing with quantitative, relational, or spatial aspects of human experience" (QRS-systems). Having argued that NUC-mathematics has been a contingent development, he asked: "If NUC-mathematics is not the only one possible, why does it have pre-eminence in curricula world-wide?" (p. 155). That leads into the perplexing dilemma of what and how to teach mathematics to indigenous groups who "learn mathematics in a distinct cultural-linguistic context – how can they study an international subject while retaining the integrity of a minority world view?" (p. 142). One approach (p. 170), namely to use mathematics to analyse culturally embedded activities, is exemplified by the long-term project of Jerry Lipka among the Yupik people of Alaska (e.g. Lipka, Yanez, Andrew-Ihrke, & Adam, 2009).

The complexity of this dilemma raised by Barton is illustrated by a number of contributions at a ICME Regional Conference when the president of the African Mathematical Union (Kuku, 1995) "warned against the overemphasis on culturally oriented curricula for developing countries that act against their ability to progress and compete in an increasingly globalized world" (Atweh & Clarkson, 2001, p. 87). Thus, there is a tension in that "an understanding of NUC-mathematics and a world-language such as English … [represent] access to communication, further educational opportunities, employment, and development" (Barton, 2008, pp. 167–168). (References to this tension, and parallels in language, will be found at several points in this book).

In a provocative paper which begins with a graphic narrative describing a Navajo child's complex navigation through difficult terrain, Pinxten and François (2011) put a different spin on the tension by asking, with respect to children, such as the one they described, who are already well adapted to their cultural milieux, what the benefits of school mathematics are – beyond keeping open the possibility of becoming a mathematician.

Teaching about mathematical modelling and its implications

Mathematics education could serve to prepare future citizens to be disposed to, and capable of, taking a critical view towards the formatting of their lives as described above. There are movements within many mathematical curricular documents, and within the academic field of mathematics education, to shift more attention to mathematical applications and modelling but most of these developments stop short of dealing with political issues or questions of ethics (Barbosa, 2006). Greer and Verschaffel (2007, p. 221) suggested three levels of modelling within school mathematics, namely implicit modelling (which is how we would characterize work done with word problems), explicit modelling of specific situations, and critical modelling "whereby the roles of modelling within mathematics and science, and within society, are critically examined". Elsewhere, we have argued for "modelling with purpose" (Mukhopadhay & Greer, 2001), and Frankenstein (2009) presents a comprehensive analysis of what she terms "*real* real-world word problems". Gutstein (2006, and Chapter 1, this volume) and Atweh (Chapter 15, this volume) present examples of curricula in action that integrate significant mathematics with the analysis of social and political issues.

Macedo et al., (2003, p. 1) refer to "the almost total absence of courses in the required [university School of Education] curriculum that would expose students to the body of literature dealing with the nature of ideology, language politics, and ethics". As a modest proposal, we suggest that mathematics students at university should have a course on the history of mathematics giving due prominence to the contributions of non-Europeans, the social history of mathematics, and the roles in society of mathematics in action, and the political implications of all those aspects.

Culturally responsive mathematics education

Against the background of the global interconnectedness of mathematics as a disciple, the struggle to counteract the rigid homogenisation of school mathematics worldwide is part of a larger struggle for cultural diversity, alongside other battles, in particular that against linguicide. It sits astride a fundamental fault line in worldviews and values.

What does it mean to aim for culturally responsive mathematics education (Greer, Mukhopadhyay, Nelson-Barber, & Powell, 2009)? The starting-point is the philosophical position that mathematics is "a human activity, a social phenomenon, part of human culture, historically evolved [and evolving], and intelligible only in a social context" (Hersh, 1997, p. xi). From this premise follows an openness to forms of mathematical practices that are aligned with forms of life, of which academic mathematics is just one, albeit very special, case. We then ask if it is appropriate (it certainly is not inevitable) that mathematics-as-a-school-subject be dominated by mathematics-as-a-discipline. And we suggest that the alienation that many children in school, and adults out of school, feel towards mathematics is partly the result of the lack of connexions between their experience in mathematics classrooms and their experiences out of school.

FINAL COMMENTS: HUMANIZING MATHEMATICS EDUCATION

Mathematics, in and out of school, is a powerful instrument for inflicting symbolic violence in the service of establishing hegemony. In the three contexts that we have discussed, it can be, and too often is, used to communicate to people that they are intellectually inferior and lack critical agency, and for these reasons should conform and submit to the banality of expertise.

Many of the points raised in this chapter, and elsewhere in this book, are about mathematics education having a dehumanising effect on people – refusing to acknowledge the intellectual achievements of others, demanding that people act in accordance with mathematical models they are unaware of and exert no control over, imposing rigid uniformity on content, methods, and interpretation of mathematics in schools.

Mathematics education could be an intellectual playground, a context for the celebration of diversity and creativity, a way to empower a critical disposition towards issues of social importance. Sometimes it is all of these. We *are* allowed to think of alternatives.

REFERENCES

Atweh, B., & Clarkson, P. (2001). Internationalization and globalization of mathematics education: Towards an agenda for research/action. In B. Atweh, H. Forgasz & B. Nebres (Eds.), *Sociocultural research on mathematics education: An international perspective* (pp. 77–94). Mahwah, NJ: Lawrence Erlbaum.

Ball, D. L., Ferrini-Mundy, J., Kilpatrick, J., Milgram, R. J., Schmid, W., & Schaar, R. (2005, October). Reaching for common ground in K-12 mathematics education. *Notices of the American Mathematical Society*, *52*(9), 1055–1058. Retrieved from http://smu.edu/education/ teachereducation/ graduate/Milgram_rdg_comm-schmid.pdf

Barbosa, J. C. (2006). Mathematical modelling in classroom: a socio-critical and discursive perspective. *Zentralblatt für Didaktik der Mathematik*, *38*(3), 293–301.

Barton, B. (2008). *The language of mathematics: Telling mathematical tales*. New York: Springer.

Baruk, S. (1985). *L'âge du capitaine: De l'erreur en mathematiques*. Paris: Seuil.

Bishop, A. J. (1990). Western mathematics: the secret weapon of cultural imperialism. *Race and Class*, *32*(2), 51–65.

Campbell (1975). Assessing the impact of planned social change. In G. Lyons & D. T. Campbell (Eds.), *Social research and public policies: The Dartmouth/OECD Conference*. Hanover, NH: Public Affairs Center, Dartmouth College.

D'Ambrosio, U. (1994). Cultural framing of mathematics teaching and learning. In R. Biehler, R. W. Scholz, R. Strasser, & B. Winkelmann (Eds.), *Didactics of mathematics as a scientific discipline* (pp. 443–455). Dordrecht, The Netherlands: Kluwer.

D'Ambrosio, U. (1999). Literacy, matheracy, and technoracy: A trivium for today. *Mathematical Thinking and Learning*, *1*, 131–154.

D'Ambrosio, U. (2006). *Ethnomathematics: Link between tradition and modernity*. Rotterdam: Sense Publishers.

D'Ambrosio, U. (2010) . Mathematics education and survival with dignity. In H. Alro, O. Ravn & P. Valero (Eds.), *Critical mathematics education: Past, present, and future* (pp. 51–63). Rotterdam: Sense Publishers.

Davis, P. J., & Hersh, R. (1986). *Descartes' dream: The world according to mathematics*. Brighton, England: Harvester.

Dias, P. V. (2002). Human development in history and society: Multi-perspectivity of perceptions and diversity of linguistic expressions. In S. C. Agarkar & V. D. Lale (Eds.), *Science, technology and mathematics education for human development, Volume 2* (pp. 205–214). Mumbai, India: Homi Bhabha Centre for Science Education/ Tata Institute of Fundamental Research.

Ernest, P. (1996). Popularization: Myths, massmedia, and modernism. In A. J. Bishop, K. Clements, C. Keitel, J. Kilpatrick, & C. Laborde (Eds.), *International handbook of mathematics education* (pp. 785–817). Dordrecht, The Netherlands: Kluwer.

Ernest, P. (2009). The philosophy of mathematics, values, and Keralese mathematics. In P. Ernest, B. Greer, & B. Sriraman (Eds.), *Critical issues in mathematics education* (pp. 189–204). Charlotte, NC: Information Age Publishing.

Evans, J., & Rappaport, I. 1999, Using statistics in everyday life: from barefoot statisticians to critical citizenship, In D. Dorling & S. Simpson (Eds.), *Statistics in society: The arithmetic of politics* (pp. 71–77). London: Arnold.

Foucault, M. (1980). Truth and power. In C. Gordon (Ed.), *Power/knowledge* (pp. 107–133). Brighton: Harvester.

Frankenstein, M. (1989). *Relearning mathematics – a different third R – radical maths.* London: Free Association Books.

Frankenstein, M. (2009). Developing a criticalmathematical numeracy through real real-world word problems. In L. Verschaffel, B. Greer, W. Van Dooren, & S. Mukhopadhyay (Eds), *Words and worlds: Modelling verbal descriptions of situations.* Rotterdam: Sense Publishing.

Freire, P., & Macedo, D. (1987). *Literacy: Reading the word and the world.* South Hadley, MA: Bergen & Garvey.

Freudenthal, H. (1991). *Revisiting mathematics education.* Dordrecht, The Netherlands: Kluwer.

Gramsci, A. (2009/1925–1930). *Selections from the prison notebooks.* [Edited and translated by Q. Hoare & G. N. Smith]. Hyderabad: Black Swan Orient.

Greer, B. (2008). Algebra for all? *The Montana Mathematics Enthusiast, 5*(2&3), 423–428.

Greer, B. (2009a). Estimating Iraqi deaths: A case study with implications for mathematics education. *Zentralblatt für Didaktik der Mathematik, 41*, 105–116.

Greer, B. (2009b). What is mathematics education for? In P. Ernest, B. Greer, & B. Sriraman (Eds.), *Critical issues in mathematics education* (pp. 3–6). Charlotte, NC: Information Age Publishing.

Greer, B., & Mukhopadhyay, S. (in press). The hegemony of ~~English~~ mathematics. In P. Ernest & B. Sriraman (Eds.), *Critical mathematics education: Theory and praxis.* Charlotte, NC: Information Age Publishing.

Greer, B., Mukhopadhyay, S., Nelson-Barber, S., & Powell, A. (Eds.), (2009). *Culturally responsive mathematics education.* New York: Routledge.

Greer, B., & Verschaffel, L. (2007). Modelling competencies – overview. In W. Blum, P. L. Galbraith, H.-W. Henne, & M. Niss (Eds.), *Modelling and applications in mathematics education. The 14th ICMI Study* (pp. 219–224). New York: Springer.

Greer, B., Verschaffel, L., & Mukhopadhyay, S. (2007). Modelling for life: Mathematics and children's experience. In W. Blum, P. Galbraith, H.-W. Henne, & M. Niss (Eds.), *Modelling and applications in mathematics education. The 14th ICMI Study* (pp. 89–98). New York: Springer.

Gutierrez, R. (2009). Embracing the inherent tensions in teaching mathematics from an equity stance. *Democracy and Education, 18*(3), 9–16.

Gutstein, E. (2006). *Reading and writing the world with mathematics: Toward a pedagogy for social justice.* New York: Routledge.

Gutstein, E. (2009). The politics of mathematics education in the US: Dominant and counter agendas. In B. Greer, S. Mukhopadhyay, S. Nelson-Barber, & A. B. Powell (Eds.), *Culturally responsive mathematics education* (pp. 137–164). New York: Routledge.

Hacking, I. (1995). The looping effect of human kinds. In D. Sperber, D. Premack, & A. Premack (Eds.), *Causal cognition: An interdisciplinary approach* (pp. 351–383). Oxford: Oxford University Press.

Hersh, R. (1997). *What is mathematics, really?* New York: Oxford University Press.

Hersh, R., & John-Steiner, V. (2011). *Loving and hating mathematics.* Princeton, NJ: Princeton University Press.

Joseph, G. G. (1992). *The crest of the peacock: Non-European roots of mathematics.* London: Penguin.

Joseph, G. G. (2008). George Gheverghese Joseph on the transmission to Europe of non-European mathematics [Talk to Mathematical Association of America]. Retrieved from http://www.maa.org/news/092608jglecture.html)

Joseph, G. G. (Ed.). (2009). *Kerala mathematics: History and its possible transmission to Europe.* Delhi: B. R. Publishing Corporation.

Kilpatrick, J. (1981). The reasonable ineffectiveness of research in mathematics education. *For the Learning of Mathematics, 2*(2), 22–29.

Kilpatrick, J. (1993). The chain and the arrow: From the history of mathematics assessment. In M. Niss (Ed.), *Investigations into assessment in mathematics education: An ICMI study* (pp. 31–46). Dordrecht, The Netherlands: Kluwer.

Kline, M. (1953). *Mathematics in western culture.* New York: Oxford.

Koblitz, N. (1981). Mathematics as propaganda. In L. A. Steen (Ed.), *Mathematics tomorrow,* (pp. 111–120). New York: Springer-Verlag.

Kuku, A. (1995). Mathematics education in Africa in relation to other countries. In R. Hunting, G. Fitzsimons, P. Clarkson, & A. J. Bishop (Eds.), *Regional collaboration in mathematics education* (pp. 403–423). Melbourne: Monash University.

Ladson-Billings, G. (2006). From the achievement gap to the education debt: Understanding achievement in U.S. schools. *Educational Researcher, 35*(7), 3–12.

Lakoff, G. (1996). *Moral politics.* Chicago: University of Chicago Press.

Lipka, J., Yanez, E., Andrew-Ihrke, D., & Adam, S. (2009). A two-way process for developing effective culturally based math. In B. Greer, S. Mukhopadhyay, A. B. Powell, & S. Nelson-Barber (Eds.), *Culturally responsive mathematics education* (pp. 257–280). New York: Routledge.

Macedo, D., Dendrinos, B., & & Gounari, P. (2003). *The hegemony of English.* Boulder, CA: Paradigm Publishers.

Martin, D. B. (2003). Hidden assumptions and unaddressed questions in Mathematics for all rhetoric. *The Mathematics Educator, 13*(2), 7–21.

Martin, D. B. (2009). Researching race in mathematics education. *Teachers College Record, 111*(2), 295–338.

Martin, D. B. (2010). Not-so-strange bedfellows: Racial projects and the mathematics education enterprise. In U. Gellert, E. Jablonka, & C. Morgan (Eds.), *Proceedings of the Sixth International Mathematics Education and Society Conference* (pp. 42–64). Berlin: Freie Universität.

Mayer, J. (2010). The predator war: What are the risks of the C.I.A.'s covert drone program? New Yorker. Retrieved from http://www.newyorker.com/reporting/2009/10/26/091026fa_fact_mayer?currentPage=all

McDermott, R., & Hall, K. D. (2007). Scientifically debased research on learning 1854–2006. *Anthropology & Education Quarterly, 38*(1), 9–15.

Miller-Jones, D., & Greer, B. (2009). Conceptions of assessment of mathematical proficiency and their implications for cultural diversity. In B. Greer, S. Mukhopadhyay, S. Nelson-Barber, & A. B. Powell (Eds.). *Culturally responsive mathematics education* (pp. 165–186). New York: Routledge..

Mukhopadhyay, S. & Greer, B. (2001). Modeling with purpose: Mathematics as a critical tool. In W. Atweh, H. Forgasz, & B. Nebres (Eds.). *Socio-cultural aspects in mathematics education: An international perspective* (295–311). Mahwah, NJ: Lawrence Erlbaum.

Nash, G. B., Crabtree, C., & Dunn, R. E. (2000). *History on trial: Culture wars and teaching of the past.* New York: Vintage.

Nichols, S. L., & Berliner, D. C. (2007). *Collateral damage: How high-stakes testing corrupts America's schools.* Cambridge, MA: Harvard Education Press.

Pais, A. (2009). The tension between what mathematics education should be for and what it is actually for. In P. Ernest, B. Greer, & B. Sriraman (Eds.), *Critical issues in mathematics education* (pp. 53–60). Charlotte, NC: Information Age Publishing.

Pinxten, R, & François, K. (2011). Politics in an Indian canyon? Some thoughts on the implications of ethnomathematics. Educational Studies in Mathematics, 78, 261–273.

Powell, A. B. & Frankenstein, M. (Eds.). (1997). *Ethnomathematics: Challenging Eurocentrism in mathematics education.* Albany, NY: SUNY Press.

Ravitch, D. (2010). *The death and life of the great American school system: How testing and choice are undermining education.* New York: Basic Books.

Said, E. W. (1994). *Culture and imperialism*. London: Vintage Books.

Schoenfeld, A. H. (1991). On mathematics as sense-making: An informal attack on the unfortunate divorce of formal and informal mathematics. In J. F. Voss, D. N. Perkins & J. W. Segal (Eds.), *Informal reasoning and education* (pp. 311–343). Hillsdale, NJ: Lawrence Erlbaum.

Schoenfeld, A. H., & Pearson, P. D. (2009). The reading and math wars. In G. Sykes, B. Schneider, & D. N. Plank (Eds.), *Handbook of education policy research* (pp. 560–580). New York: Routledge.

Sen, A. (2005). *The argumentative Indian.* London: Allen Lane.

Shaughnessy, M. (2010, August). Statistics for all – the flip side of quantitative reasoning. NCTM *Summing Up.*

Skovsmose, O. (2000). Aporism and critical mathematics education. *For the Learning of Mathematics, 20*(1), 2–8.

Skovsmose, O. (2005). *Travelling through education: Uncertainty, mathematics, responsibility.* Rotterdam: Sense Publishers.

Skovsmose, O. (2006). Reflections as a challenge. *Zentralblatt für Didaktik der Mathematik, 38*(4), 323–332.

Spring, J. (2010). *Political agendas for education (4th edition)*. New York: Routledge.

Steen, L. A. (2003). Data, shapes, symbols: Achieving balance in school mathematics. In B. Madison & L. A. Steen (Eds.), *Quantitative literacy: Why numeracy matters for schools and colleges.* Princeton, NJ: National Council on Education and the Disciplines.

Swetz, F. (1994). *From five fingers to infinity*. Chicago & La Salle, IL: Open Court.

Tate, W. F. (1995). School mathematics and African American students: Thinking seriously about opportunity-to-learn standards. *Educational Administration Quarterly, 31*, 424–448.

United States Department of Education (2008). *The final report of the National Mathematics Advisory Panel.* Washington, DC: Authors.

Urton, G. (2009). Mathematics and authority: a case study in Old and New World accounting. In E. Robson & J. Stedall (Eds.), *The Oxford Handbook of the history of mathematics* (pp. 27–56). Oxford: Oxford University Press.

Urton, G. (in press). Mathematics and accounting in the Andes before and after the Spanish conquest. In S. Mukhopadhyay & W-M. Roth (Eds.), *Alternative forms of knowing (in) mathematics: Celebrations of diversity of mathematical practices.* Rotterdam: Sense Publishers.

Usiskin, Z. (2007). The arithmetic operations as mathematical models. In W. Blum, P. L. Galbraith, H-W. Henne, & M. Niss (Eds.), *Applications and modelling in mathematics education The 14th ICMI Study* (pp. 257–264). New York: Springer.

Verschaffel, L., Greer, B., & De Corte, E. (2000). *Making sense of word problems.* Lisse, The Netherlands: Swets & Zeitlinger.

Verschaffel, L., Greer, B., Van Dooren, W., & Mukhopadhyay, S. (Eds.) (2009). *Words and worlds: Modelling verbal descriptions of situations.* Rotterdam: Sense Publications.

Brian Greer
Portland State University
Oregon, USA

Swapna Mukhopadhyay
Portland State University
Oregon, USA

CHAPTER 11

BRINGING CRITICAL MATHEMATICS TO WORK: BUT CAN NUMBERS MOBILISE?

KEIKO YASUKAWA AND TONY BROWN

WHY BRING CRITICAL MATHEMATICS TO WORK?

In Australia, as in many other OECD countries, employers are showing a keen interest in the literacy and numeracy levels of adults in the workforce. Much of this interest comes from the less than statistically impressive results of the International Adult Literacy and Life Skills (ALLS) Survey that was conducted in these countries. In Australia, the 2006 ALLS Survey results showed little change from the survey conducted 10 years prior, and showed that over fifty percent of adult Australians scored in the bottom two levels in numeracy. The results for literacy were not much better, and problem solving fared much worse (Australian Bureau of Statistics, 2007). In Australia, problems in the labour market, ranging from new entrants' employability skills, unemployment, and skills shortages in some industries, have been attributed to poor literacy and numeracy skills since the late 1990s, and while such attributions have been contested (see for example, Black, 2002) they continue to be made. The importance of literacy and numeracy for employment is signalled in policy documents that claim to describe the "core skills" that adults need and the skills needed for employability (Department of Education, Employment, and Workplace Relations, 2008; Department of Education, Science, and Technology, 2002), and one of the largest Federal Government programmes in adult literacy and numeracy is a labour market program for long-term unemployed people.

The discourses surrounding the labour market programmes, and documents such as *Employability Skills for the Future* (Department of Education, Science, and Technology, 2002), represent individuals as having a deficit that they need to take responsibility for addressing. These discourses found traction and were strengthened during the long period of conservative government in Australia, a time when precarious forms of employment increased, individually "negotiated" employment contracts grew, and the industrial relations environment became increasingly anti-union. While some of the anti-union sentiments have been moderated since the change of government in 2007, the focus on literacy and numeracy for employability has not waned. In 2009, the Federal Labor Government gave half a million dollars to the Australian Industry Group, a peak employer body, to research the literacy and numeracy skills of their workers (Australian Industry Group, 2010) and Skills Australia, the advisory body to the Federal Government on workforce skills and development, identified literacy and numeracy as a serious problem needing priority

O. Skovsmose and B. Greer (eds.), Opening the Cage: Critique and Politics of Mathematics Education, 249–264.

policy attention (Skills Australia, 2010). Not surprisingly, such employer-driven, human-capital-focussed, literacy and numeracy initiatives do not foreground critical literacy and numeracy, which is described by Anderson and Irvine (1993, p. 82) as "learning to read and write as part of the process of becoming conscious of one's experience as historically constructed within specific power relations". In other words, critical literacy and numeracy is learning that enables workers to challenge the politics of work and the workplace, and helps them to imagine alternative cultures and forms of work. Nor do these employer initiatives organise the learning of literacy and numeracy in ways that help workers develop collective power and voice in their workplaces or as workers in society.

In this chapter we examine the kind of numeracy – critical mathematics – that is missing from the dominant discourses of numeracy for work. We interpret what critical mathematics for work can mean, and illustrate its potential and limitations through a case study. The case study is a union-based campaign for casual workers. We will illustrate that while there was an abundance of statistics (numbers) that described the increasing numbers of these casual workers and their pay and conditions, these numbers did not reveal the inequities and the exploitation that these workers experienced powerfully enough to mobilise them, or their union, to actively challenge the state of affairs. However, this case study shows that revealing the human stories behind these numbers provided a breakthrough in the campaign to assist the workers to self-organise and thereby practise democracy; and, at that point, they needed to exercise numeracy skills in a very applied manner for themselves and as a means to counter the employer's claims.

In this chapter, we first introduce a taxonomy of workplace mathematics that helps to delineate the place of critical mathematics in the workplace. Secondly, we provide a very brief review of some of the relevant concepts and ideas from the critical mathematics literature that will be drawn upon in our analysis, in particular the notion of the "barefoot statistician" introduced by Evans and Rappaport (1999). Thirdly, we introduce and describe the case study and examine the ways in which critical mathematics was brought to work alongside workers' stories, and the roles that the union activists played as story tellers and "barefoot mathematicians". We conclude with some discussion and implications for ways in which critical mathematics learning can be understood and approached in the workplace.

HOW CAN WE THINK ABOUT WORKPLACE MATHS?

It's difficult to imagine someone who can function in the workforce without any knowledge or skills in mathematics. Even those who say "I'm no good at numbers" cannot say that mathematics plays no part in their working life. One way of thinking about the mathematics that a worker might encounter is to consider the different purposes that mathematics serves the worker.

One purpose is to enable a person to be able to learn the specialist skills and knowledge of their job, and/or to receive the qualifications and credentials that the job requires. So, for example, a tiler might need to know how to calculate the area of floors and walls, and so, in their tiling certificate course, they might be introduced to the concept of area and formulae for calculating the area of different shapes. Or, a person learning to become a commercial chef would need to know how the metric system works in order to read and interpret recipes that specify amounts of ingredients in grams, kilograms, and litres. An engineer would need to have knowledge of calculus to work with models of physical systems that are often developed from differential equations. In other jobs, there may not be any mathematical requirements specified for the job, but the job may still require an educational qualification at a level that involves a level of attainment in mathematics either directly related to, or leading up to, this qualification.

A second purpose that mathematical knowledge serves a worker is enabling the worker to do their job. The mathematics in this case may be quite narrowly specialised and focussed on the particular nature of the work. For example, a tiler may need to calculate the area of a bathroom floor. They may apply a formula they learnt in their tiling course, but could, in some cases, need to apply some additional skills of estimation and approximation for the little spaces that are not neatly calculable using the formulae they learned in their course. There are mathematical practices of estimation and approximation that the tiler develops through their apprenticeship and experience of watching more experienced tilers and actually doing the work. In the actual workplace, the tiler would also know that more tiles than what would exactly cover the floor would need to be purchased because of possible breakages, while ensuring that not too much surplus is purchased so that the cost stays within the budget. A commercial chef would need to know about measurements for measuring ingredients and temperatures that are relevant for their work. Knowledge that 1000 grams is the same as 1 kilogram would be needed, but knowledge about nanograms or gigalitres might not be needed for the chef. In addition to mathematical "facts" such as the decimal basis of the metric system and that angles are measured in degrees (or radians), workers use other mathematical knowledge that is learned and accepted within their local practice. Chefs would have a "feel" for how much a kilo of flour is – from working with different amounts of flour in their cooking practices. Experienced workers would also know when it is important to be "exact" or to go back to first principles, and when they could rely on what they could approximate. Workers develop mathematical practices that are specific for their work and which may look quite different to the mathematics that are taught in formal courses (Wedege, 2010).

Separate from the technical mathematics that is needed to "do the job" is the functional mathematics that a worker needs to negotiate as a consequence of being in paid work. Some of the mathematics involved would be "generic" for all workers in the workplace – for example, interpreting and checking the information that is shown on the payslip. There are other skills such as those involved in preparing a tax return at the end of each financial year (or knowing what their accountant needs to know in order to prepare their tax return), working out the train or bus timetable to get to work and home, and working out how to fill out time sheets and/or work to a workload formula of the workplace.

There is a fourth kind of mathematics that is less obvious and may not be actively used by all workers in all workplaces. This is the critical mathematics for "reading" the politics of work and the workplace – for example, what "counts" as work and what are the relationships of power and how do they limit or empower the workers. This might involve questioning why the definition of employment and unemployment can change in official public policy; analysing pay differentials among different groups of employees in the same workplace; comparing what are the hours they are supposed to be paid for, and the hours they are actually working; analysing what productivity increases the workplace has produced, and how these are reflected in any improvements in workers' conditions and pay; calculating the percentage of different categories of workers that exist in their workplace, and the level of political influence held by each; estimating the proportion of different categories of workers that are represented in their trade union, and the level of political influence exercised by each. We can ask: how do workers learn this kind of mathematics; who can assist them in this learning; how do workers learn to ask these kinds of critical questions; and what do/ can workers do when they have the answers?

Table 1 summarises the different purposes that mathematics can serve a worker.

Table 1: Different kinds of purposes for mathematics for, and in, the workplace

Enabling mathematics for accessing training and qualifications	**Technical mathematics for doing the job**	**Functional mathematics for being a paid worker**	**Critical mathematics for "reading" the politics of work and the workplace**
e.g. - concept of area, and how to calculate it - metric system for a commercial cookery course - calculus for engineers	e.g. - calculating area for a floor or wallspace - measuring ingredients/ materials - setting up mathematical models on a computer and interpreting the results	e.g. - filling out timesheets - checking your own payslips - banking your pay - accessing entitlements (leave, workers comp, overtime etc) - negotiating the transport system to get to work on time - doing the tax returns	e.g. - analysing the "logic" of the pay rates in the workplace - questioning data on workplace "productivity" and/ or link to pay - analysing the political representation of different groups of workers - analysing the political representations of different groups of workers in the union

WHAT IS CRITICAL MATHEMATICS?

Different terms exist, such as critical mathematics (Ernest, 1991), radical mathematics (Frankenstein, 1989), mathemacy (Skovsmose, 1994), matheracy (D'Ambrosio, 1999), mathematics/statistics for critical citizenship (Benn, 1997;

Evans & Rappaport, 1999), mathematical literacy (Gutstein, 2006), and (critical) numeracy (Johnston & Yasukawa, 2001), to describe and name the kind of mathematical knowledge that helps us to uncover and understand questions of power and social justice[1]. For the purposes of this chapter, we will name "critical mathematics" as both the approach and the knowledge that is underpinned by the belief that mathematics is political: mathematical knowledge can be used both to create and stabilise social inequities and injustices, and to expose these inequities and injustices in order to imagine and help to create alternative futures. What distinguishes "critical mathematics" from "mathematics" is this political dimension of questioning and engaging with questions of power.

Mathematics as a discipline is commonly viewed as "pure" and apolitical, a view firmly held by the number theorist G. H. Hardy and expressed in *A Mathematician's Apology* (1967), originally published in 1940. Hardy says mathematics is one science "whose very remoteness from ordinary human activities should keep it gentle and clean" (Hardy, 1967, p. 122). If the claim of remoteness of mathematics from "ordinary human activities" seems not to be very believable, this is because Hardy made a distinction between what he called "the real mathematics of real mathematicians" and "trivial" mathematics; the former is largely useless while the latter is "on the whole, useful"; it is the "real" mathematics that Hardy associates with remoteness from the lives of most people (1940, p. 139). Putting aside the debate about the suitability and value of the classification between "real" and "trivial" mathematics, the view of mathematics as, if not gentle, "clean" is compatible with the common perception that because mathematics is about numbers, mathematics is objective and universally understood (if only people were smart enough to learn it). The power of the perceived objectivity has been the subject of analysis by Porter (1995) in *Trust in numbers: The pursuit of objectivity in science and public life*. Porter argues that numbers, or more precisely quantification, is a "technology of distance"; but its perceived rigour and consistency in relation to a large set of rules, negates the need for "intimate knowledge" and trust (Porter, 1995, p. ix). Numbers speak with authority because they are perceived to sit above the petty political interests of mortal beings! But those who can see and use this common perception of mathematics are the ones with real power because they can exercise their will through the authority of numbers without necessarily exposing their self-interests.

How do "interests" get mathematised? Mathematics, and mathematical models in particular, can be understood as a kind of a "script" for writing the world according to the writer's worldview. Mathematics is a resource for description, inscription, prescription, and proscription (Skovsmose, Yasukawa, & Ravn, 2011). Mathematical models are designed for a purpose, to solve a problem *in a particular way for a particular interest*. So the reality of the problem is described only enough to be able to include and exclude aspects that are relevant to the end goal. In the selective process of determining those aspects that constitute the model, the values and ideologies associated with the interests of the modellers are inscribed into the mathematical version of reality. The model produces solutions that reflect these interests, and as these solutions get implemented, we end up subscribing to this way of understanding the original problem. We develop a habit of ignoring certain things, and focusing on other things when we meet the same kinds of problems. This is what Skovsmose (2005) calls "mathematics in action" and what MacKenzie (2005) calls the "performative dimension" of mathematical models.

Another description of the process of mathematising interests is to view mathematics or mathematical models as technological artifacts, and seeing how they are "mobilised" (Yasukawa, 2003). Using the lens of actor-network theory, mathematical models undergo four key stages of "translations" (Latour, 1987): 1. *problematisation*, where a problem is identified, and stakeholders or "actors" who hold some interest in the way the problem is resolved are identified; 2. *interressement* – where the interests of the different stakeholders are negotiated and some new relationships are formed while some old relationships might be dissolved; 3. *enrolment* – where stakeholders' interests are built into the model; and 4. *mobilisation* – where the model is put into operation and becomes part of the social and physical environment in which it is operating. Once the model is mobilised, the history of the model is no longer visible. This model might be a model of the financial market that MacKenzie writes about (2005), one of the many models about climate change currently being used to make predictions, or an encryption algorithm for network security (Skovsmose & Yasukawa, 2004). What makes mathematical models difficult to challenge is that they operate and are consumed as "black boxes", and the assumptions and interests on which the models were built are not readily accessible by the consumers or "victims" of the models.

The aim of critical mathematics is to expose what is in the black box because the solutions that mathematical models produce have political, economic, social, and environmental consequences. Frankenstein (1989), Gutstein (2006), Gutstein and Peterson (2005), among others, offer many examples of integrating critical social inquiry into mathematics teaching. For the non-formal and workplace settings, the notions of the "barefoot statistician" and the "parastatistician" described by Evans and Rappaport (1999) offer ways of thinking about how critical mathematical knowledge might be exercised. They describe the barefoot statistician as "a user and sometimes producer of information whose expertise lies in a balance of basic statistical skills and communication links with their community" (Evans & Rappaport, 1999, p. 71). The role of the barefoot statistician in community settings (and the parastatistician in the business settings) is to be a resource person who does research in their local community (or workplace) in order to promote critical citizenship, that is, to enable people to engage in policy debates and advocacy. The barefoot statistician, according to Evans and Rappaport (1999), needs to be able to demystify the relevant statistics, clarify and identify slippery uses of definitions (for example, definitions of labour market categories such as unemployed, full-time work), and hunt down data and information that are not readily available. They need to have effective communication links and strategies with the community/ members of the workplace to be able to disseminate the findings in an informative and comprehensible manner.

But will information on its own lead to anything apart from greater understanding and wider ownership of information that can explain an unsatisfactory situation? Will comprehension of the mathematics lead to the community members or workers taking action to change the situation? Is mathematics truly critical if it does not mobilise some form of collective action?

HOW DOES MATHEMATICS HELP A CRITICAL READING OF WORK AND THE WORKPLACE?

We will examine a case study in a workplace where a group of workers were, for a long time, very dissatisfied with their work situation. One of the sources of dissatisfaction was the way in which their work was constructed by a particular mathematical model that had been operating in the workplace and, indeed, across their industry for many years. In this case, there was some significant achievement in mobilising some action and effecting changes to policies. We will analyse how these actions and changes were achieved, and examine what role critical mathematics played, and what role another approach played.

The case study is based on what became known as the "casuals campaign" in the Australian higher education sector from around 2005. In 2004, the two of us and a colleague, James Goodman, who were, at the time, elected officers and delegates in our union, decided to tackle what until then was a largely ignored (some would argue strategically avoided) issue in the trade union for higher education workers, namely organising and increasing the density[2] of casual academic members[3] in their union.

Simple arithmetic tells us that casual academics have the numbers. In some universities casual academics outnumber the full-time academics by 500% (Brown, Goodman, & Yasukawa, 2010). In 1990, casuals delivered about a tenth of all university teaching, however by 2008 between a third and a half of university teaching was being delivered by casuals, and these proportions can be much higher in some academic areas (Brown et al., 2010). But their representation and voice in decision-making processes in the university are almost negligible. Their status as "casual" limits them politically as well as legally. Casuals are not represented and cannot stand for positions in many formal decision-making bodies. Casuals, because of the nature of their contracts, do not have security of employment or access to internal processes of promotion, paid study leave, and other conditions that could help them to develop their careers as academics. Casual academics are the majority underclass in the university sector.

Most studies of casualisation in Australia take a quantitative approach in an effort to present a picture of the scale of the issue (Buchanan, 2004; Campbell, 2004; Productivity Commission, 2006; Wooden, 1999), and this trend has been mirrored in research into academic casualisation. The most significant writings about the position of casual workers in Australia that are pertinent to research on casual academics had been conducted by Junor (2004) and Pocock, Prosser, and Bridge (2004). Junor's study involved quantitative surveying of casual university staff, with 2,494 casuals completing a survey out of the 9,000 that were distributed, while the study by Pocock and her colleagues (2004) was an exception to the large-scale survey approach as it centred on fifty-five in-depth interviews with casual staff in a number of different industries.

Junor's (2004) contributions were rich in data and relied on argument that sought to reveal the changing employment situation in Australian higher education as well as the predicament facing both casual and fixed-term staff and academics occupying continuing positions. The report outlined the shape of the national higher education sector, made comparisons over time, and provided important statistical information. It sought to influence policy at national and institutional levels through the weight of their arguments. The study by Pocock et al., (2004) challenged the dominant discourses around casual work through qualitative data that exposed the realities of casual work for those they interviewed; whatever benefits of flexibility that casual work provided were benefits to the employers, not to the workers.

The research conducted by us (Brown et al., 2010) at City University adopted a different approach to Junor's (2004). The research was initiated by union activists by applying for a small grant from the national union, which itself was eager to encourage research that would assist in increasing membership numbers. The Union, like many other unions, was experiencing a decline in membership numbers and membership density that could be explained by the changing age profile of its membership, a decade of funding cuts to higher education by the national government, reluctance on the part of university administrations to hire new staff, and difficulties faced by the union in attracting and recruiting new members. These difficulties occurred especially in areas where there was employment growth, such as among new younger staff, and, in particular, casual academic staff, where membership numbers and density were extremely low.

Taking the Junor (2004) and Pocock et al., (2004) reports as starting points and as a means of setting our research within a larger context, we focussed on one university that was well known to have high levels of casualisation, and then concentrated on twenty-five staff in two faculties at City University. The centrepiece of the research was a set of in-depth qualitative interviews with the twenty-five casual staff. The data contained in those interviews enabled us to change the discourse about casualisation. No longer was it primarily about numbers, about how many casuals there were, how the numbers had increased over a set time, what proportion of staff casuals made up, and how many classes casuals were co-ordinating and teaching. It didn't go into the cost savings of casual employment over continuing employment, as significant as that issue is. It began to focus on the life stories and expectations of the individuals who could be identified as colleagues, those staff a continuing academic would likely meet in the corridor, or work with, for example in preparing a subject. The data contained in the interviews provided new information, especially regarding relationships in the faculties, individual aspirations, the daily difficulties encountered by casuals both within the university (a lack of office space, printing access, privacy to meet students, etc.) and outside (the struggle to obtain a loan from a bank), and personal work histories. Reframing the nature of the discourse thereby enabled a counter-narrative to be created.

The personal stories that the individual respondents told helped to add a richer, more human, dimension to the picture of their individual situations than sets of numbers. They did more than simply invite the reader to respond; they called out for an appreciation of the challenges, pleasures, and obstacles they faced. Those individual accounts could not easily be put aside or dismissed, or countered with a competing set of statistical data. The real strength of this approach to the issue was to develop a shared story of the challenges these workers faced, why they faced them, and what others should do to assist, and why. It proved to be a means of going from articulating individual stories to one of articulating values – the values with which, in this case, union members and activists could identify, and then pursue means of addressing them.

The research added a dimension of stories, with the power to move and mobilise, to the pre-existing power of statistical information. Would this new data be useful in mobilising the casual staff, continuing staff who were the casuals' colleagues as well as supervisors, and the union to take up the issue in a way that was more than just rhetorical support?

UNION ACTIVISTS AS "STORYTELLERS"?

The personal stories of casual academics exposed the exploitation in ways that other casuals could relate to, and unsettled full-time academics. "Inconvenient truths" for the union as an organization, and the university, were told, such as:

> It's very difficult as somebody who's employed in casual work to commit to anything in the long-term, whether it's committing to a lease on a place, getting a loan. I mean, there's just no long-term planning so you just don't know where you are. (Paula, casual academic)

and from Marie, another casual academic:

Interviewer	How would you describe your access to training and development?
Marie	Well, I know it's there but I don't have time to do it. ... because it's always in my unpaid time. ... I just don't really see that I've got access to that training because I'm studying as well.
Interviewer	You're doing your PhD?
Marie	Yeah. So it's more that I feel that if I go to that training, I'm doing it in my own time. It's a sort of psychological thing in a way. ...
Interviewer	If you were being paid, would you feel happier?
Marie	Yeah. I mean, it's silly, it sounds mingy to say that but I've put a lot of extra time in. Even as a casual I've gone overseas to work [through] the faculty. And that's never been acknowledged. I mean, I loved it but after a while, I'm not going to do that any more, it's too expensive for me.
Interviewer	So a lot of those costs were not covered?

Marie Certainly the base costs were covered but not my, the other costs I incurred was childcare and all those extra costs.

Capturing the workers' experiences was a deliberate strategy, based on an understanding of social movement organising, to bring personal accounts to the fore. Detailed statistical data, while powerful in documenting trends and for making submissions, also tend to erect barriers to empathetic solidarity. The belief was that members would be more receptive to individual stories, and these stories would be a way of building a shared sense of solidarity among casuals themselves, rather than the common tendency of telling people the facts first. Heath and Heath (2007, p. 206) contend that "we need to open gaps before we close them… empathy emerges from the particular rather than the pattern." The story's power, then, is twofold. It provides simulation (knowledge about how to act) and inspiration (motivation to act).

In preparing for a campaign on the issue of academic casualisation we started with the idea that relying on individual stories would be a basis for developing dialogue and building relationships, especially among casual staff who are difficult to organise because of their varied working patterns. Spreading the stories through reports, articles, and meetings resulted in small networks and new relationships being formed. In those meetings new stories about who this community is, where it has been, where it is going, and how it could achieve change were elaborated. The expectation was that developing a self-organising network and campaign would be a way of interweaving and strengthening relationships, understanding and action so that each contributes to the others. As Ganz (2002, pp. 16–17) argues, the role of the organiser and activist in these situations (and ideally in an educational role) is to work with people to deepen their understanding of who they are and what they want. They work to help interpret why they should act to change their world – motivation, and how they can act to change it – strategy. Through developing feelings of hope, community, self-worth, agency, and urgency they challenge feelings of doubt, isolation, defeatism, and fear, which inhibit action.

These are strategic decisions about what is the most effective way to mobilise people to act in their interests and to change situations that confront them. Deepening understanding of how to act is further developed by creating opportunities for people to deliberate, to strategise how to use the resources they have in order to take advantage of opportunities to get what they want, and to motivate others on why they should help, that is, having to make choices and decisions. In the case of the casuals and the union these opportunities had to be created.

Having brought to the fore the casuals' stories, publicised them, and started developing networks, the need to supplement the qualitative data with quantitative data became evident.

UNION ACTIVISTS AS "BAREFOOT STATISTICIANS"?

As hope emerged among groups of casual academics that there were other workers in the academy who felt and recognised their dissatisfaction, and that perhaps there was some action they could take as members of the union, the number of casual academics joining and contacting the union started to increase. As union activists, we organised meetings of casual academics in their workplaces to provide space and time to share concerns about their employment conditions. As union delegates we had ready access not only to the pay schedule and how it was to be understood, but also the assumptions that were accepted by those who were party to the model (university managers and the union) that became the established way of paying casuals. One of the clauses in enterprise agreements of many Australian universities that has consistently caused confusion and misunderstanding is a critical one that describes how the pay structure works for casual academics works. Extracts from the enterprise agreement of City University say:

> The hourly rate of pay for lecturing, tutoring and undergraduate clinical nurse education contained in Schedule 2 will encompass the following activities in addition to the delivery of lectures and/or tutorials:

(a) preparing lectures or tutorials
(b) up to 20 minutes of marking for each hour of teaching
(c) administration of relevant records of the students for whom the casual employee is responsible and
(d) consultation with students.

Schedule 2 explains that

(c) The hourly rate of pay will be derived from the following formula – "(relevant full-time salary" / 52) / 37.5 + 23% (loading). Refer to the notes at the end of the schedule for the "relevant full-time rate".

and refers to Schedule 4 that says that the pay for one hour of lecture consists of pay for:

> one hour of delivery and two hours of associated non-contact duties.

There is much to get confused about in these explanations of the pay rates for casual lecturers, and many of the casual academics who were interviewed in our research responded in a way not dissimilar to the interviewee who said:

> I have no idea how to read my contract. The contract that I've received at the beginning of every semester is extremely vague. It's coded. We've never had a breakdown of what those codes mean. I do a rough calculation of what I know I should get paid for tutoring and for attending a lecture and then I see if there's some kind of equivalency. But that's about it. ... beyond that I don't understand what any of it means and there doesn't seem to be anybody able to explain it to me. I did try once to get clarification about a marking contract and I felt basically that I was fobbed off, I was basically told that it was a very complicated variable and it would take forever to explain it to me but my contract was correct, if I was concerned about its correctness, so go ahead and sign it. And that came straight from the Dean's office when I went to visit them. (Casual academic)

The pay structure for casual academics in the enterprise agreement was one that was negotiated by the University and the Union. In reality however, the low representation and consequently limited political power of the casual members of the union meant that they did not have a large part in the negotiations. The continuing full-time academics who formed the majority of the union membership directed the union negotiators to pursue conditions that impacted directly on themselves as a priority, leaving limited room to negotiate improvements for the casual members of the union.

In our meetings with the casual union members, we (as union activists) found that many of the casual academics knew "intrinsically" that they were being underpaid, but had little idea of how to make a case that could stand up in any industrial dispute. All the formulae and the clauses in the enterprise agreement were just a "black box" to them. We worked with these members, firstly to open up the "black box", and explained what the formula meant and its historical roots. We explained how the pay rate worked, and developed a table in which they could record the time they were spending on different aspects of their work: the actual classroom delivery of a lecture or tutorial, the preparation they did for each class, the time they spent in student consultation and administration, and the time they spent on marking. All members of one group of these academics came from the same department and took tutorials in the same subject. They collectively agreed to keep a log of their hours over a number of weeks. Comparing their logs, we were able to establish that the hours they spent on different aspects of the work were comparable, and the hours spent in some aspects of their work (for example, class preparation and marking) were much greater than the hours for which the university was paying them.

With the claim of underpayment created together with the casual academics, we presented the case of underpayment to the university, and were able to gain acknowledgement of underpayment and a partial remedy. More importantly, those casuals who were involved in their local campaign learned how to decipher the mathematical model that determined how they were being paid, and learned how the model needed to be conditioned in order to calculate the correct number of hours of pay for the casual academics in this area. This learning then led some of the casual academics to take part in a larger campaign organised by the union to gain recognition for the kinds of exploitation they were experiencing and to take an active role in improving their pay and conditions in the next round of negotiations with the university managements.

The role that we played as union activists in helping the casual academics collectively decipher the pay formula and structure could perhaps be understood as that of "barefoot mathematicians" – using our mathematical knowledge and communication networks within the union to encourage collective action in the workplace and their union.

SO, CAN NUMBERS MOBILISE?

The case study illustrates that there is critical learning – learning about the politics of work and the workplace – that is an important aspect of knowledge that critical mathematics education can address. However, in this case study, the workers could not be collectively mobilised as a result of deciphering the pay formula alone; the stories of the casual academics' lived experiences helped to build the sense of solidarity and collective imperative that preceded the meetings that opened up the black box of formulae.

There is a need for tapping into both the qualitative and quantitative sensibilities of the workers, and recognise the power of story alongside numerical / statistical data. To collectively mobilise workers to challenge injustices in the workplace, critical mathematics must be brought to work, but so, too, must the stories and voices of the workers that give human meaning to the numbers.

The case study has given us insights for critical mathematics education and union organisers. Critical mathematics education cannot be "truly critical" unless the process of learning itself is democratic and leads to collective action. The reluctance of people to engage with anything mathematical poses a challenge for any collective action that involves the exercise of numeracy skills. It is tempting for those who "know" the mathematics to provide the workers the "answers" with which to work. Not only would this do little to "empower" those workers who shift their subordination from the employer to their union leader (or whoever brings them the "answer"), but, as this case study has shown, would do little to mobilise and sustain a campaign by, and for, the workers who are experiencing the injustice. As Holland and Skinner (2008) showed in their study of literacy practices in social movements, unions as a social movement can also choose to encourage both inclusionary as well as exclusionary literacy activities. For union organisers and activists, Evans and Rappaport's idea of a "barefoot statistician/mathematician" provides a model of working inclusively with workers by using the authentic stories and voices of workers to organise them and to develop a collective understanding of, and challenge to, the numbers that create inequitable and exploitative situations in their workplaces.

In a recent keynote address at an adult education conference, policy activist and academic Joe Lo Bianco (2010) explained that both statistics and stories are critical in shifting dominant policy discourses, and they can interact in two ways. Statistics can emerge from capturing trends and averages from a large number of stories, but stories also emerge from interpreting the statistics in relation to lived experiences. In the case study that we have described, there was a conscious effort to move beyond overarching meta-statistics and instead capture the stories of the casual academics' lived experience. Articulating these stories and receiving a positive response from their colleagues and their union gave them the confidence to use afresh the mathematical formulae that shaped their pay, working hours, and conditions. Reframing their experiences helped others to develop a better understanding of their situation, and taking hold of the calculations that shaped their work enabled new authentic tools to be deployed to advocate their cause for recognition and better remuneration. This suggests that bringing together the affective, through the use of narrative accounts, along with the apparently objective, involved in the use of numbers, can make for a powerful partnership for mobilising.

ACKNOWLEDGEMENT

The authors would like to acknowledge the contributions of our colleague James Goodman who was a co-researcher in the original research project and comrade in the campaign that is cited in this paper.

NOTES

1. This is not intended to be an exhaustive list of people who have written in the area of critical mathematics – rather, a survey of the different ways in which people have described critical mathematics.
2. Density is another mathematical term used to represent the proportion of the workforce, or category within a workforce, who are members of the union. It is a term with increasing saliency in discussions of unionisation, especially in the context of declining rates of union density, as it is now widely recognised as being a more informative representation of union presence than a figure identifying raw membership numbers.
3. In Australia, the term casual academic refers to those academics who are employed by the university, usually for teaching related duties only, and are paid at an hourly rate. Typically, they would be employed for several hours a week for the duration of the teaching semester. They may or may not be given another casual contract for the following semester; however, they have no legal right for re-employment after each semester contract is completed.

REFERENCES

Anderson, G. L., & Irvine, P. (1993). Informing critical literacy with ethnography. In C. Lankshear & P. L. McLaren (Eds.), *Critical literacy: Politics, praxis, and the postmodern* (pp. 81–104). Albany, NY: SUNY Press

Australian Bureau of Statistics (2006). *Adult literacy and life skills survey: Summary results.* Canberra: Author.

Australian Industry Group (2010). *National workforce literacy project report: Report on employers [sic] views on workplace literacy and numeracy skills.* Sydney:Australian Industry Group, .

Benn, R. (1997). *Adults count too: Mathematics for empowerment.* Leicester: National Institute of Adult Continuing Education.

Black, S. (2002). *Whose economic wellbeing? A challenge to dominant discourses on the relationship between literacy and numeracy skills and (un)employment.* Retrieved from Adult Literacy and Numeracy Australian Research Consortium website: http://www.staff.vu.edu.au/alnarc/index.html.

Brown, T., Goodman, J., & Yasukawa, K. (2010). Academic casualisation in Australia: Class divisions in the university. *Journal of Industrial Relations, 52*(2), 169–182.

Buchanan J. (2004). *Paradoxes of significance: Australian casualisation and labour productivity.* Paper prepared for ACTU, RMIT and *The Age* Conference: Work Interrupted: Casual and Insecure Employment in Australia, Hotel Sofitel, Melbourne, 2 August.

D'Ambrosio, U. (1999). Literacy, matheracy, and technocracy: A trivium for today. *Mathematical Thinking and Learning, 1*(2), 131–153.

Department of Education, Employment and Workplace Relations. (2008). *Australian core skills framework, A summary.* Canberra: Commonwealth of Australia.

Department of Education, Science, and Technology. (2002). *Employability skills for the future*. Canberra: Commonwealth of Australia.

Ernest, P. (1991). *The philosophy of mathematics education*. London: Falmer.

Evans, J., & Rappaport, I. (1999). Using statistics in everyday life: From barefoot statisticians to critical citizenship. In D. Dorling & S. Simpson (Eds.), *Statistics in society: The arithmetic of politics*, (pp. 71–77). London: Arnold.

Frankenstein, M. (1989). *Relearning mathematics – A different third R – Radical Maths*. London: Free Association Books.

Ganz, M. (2002). What is Organizing? *Social Policy, Fall*, 16–17.

Gutstein, E. (2006). *Reading and writing the world with mathematics: Towards a pedagogy for social justice*. New York & London: Routledge.

Gutstein, E., & Peterson, B. (Eds.). (2005). *Rethinking mathematics: Teaching social justice by the numbers*. Milwaukee, WI: Rethinking Schools.

Hardy, G. H. (1967, originally 1940). *A mathematician's apology*. Cambridge: Cambridge University Press.

Heath, C., & Heath, D. (2007). *Made to stick: Why some ideas survive and others die*. New York: Random House.

Holland, D., & Skinner, D. (2008). Literacies of distinction: (Dis)empowerment in social movements. *Journal of Development Studies*, *44*(6), 849–862.

Johnston, B., & Yasukawa, K. (2001). Numeracy: Negotiating the world through mathematics. In B. Atweh, H. Forgasz & B. Nebres (Eds.), *Sociocultural research on mathematics education* (pp. 279–294). Mahwah, NJ: Lawrence Erlbaum.

Junor, A. (2004) Casual university work: Choice, risk, inequity and the case for regulation. *The Economic and Labour Relations Review*, *14*(2), 276–304.

Latour, B. (1987). *Science in action*. Cambridge, MA: Harvard University Press.

Lo Bianco, J. (2010). *Getting serious about adult literacy in Australia*. Keynote address presented at the 33rd Annual Conference of the Australian Council for Adult Literacy, Darwin, Australia.

MacKenzie, D. (2005). Mathematising risk: Models, arbitrage, and crises. In B. Hutter & M. Power (Eds.), *Oragnizational encounters with risk* (pp. 167–189). Cambridge: Cambridge University Press.

Pocock, B., Prosser R., & Bridge, K. (2004). *"Only a casual... ". How casual work affects employees, households, and communities in Australia*. Adelaide: University of Adelaide.

Porter, T. (1995). *Trust in numbers: The pursuit of objectivity in science and public life*. Princeton, NJ: Princeton University Press.

Productivity Commission (2006). *The role of non-traditional work in the Australian labour market*, Canberra: Author.

Skills Australia (2010). *Australian Workforce Futures: A National Workforce Development Strategy*, Canberra: Commonwealth of Australia.

Skovsmose, O. (1994). *Towards a philosophy of critical mathematics education*. Dordrecht, The Netherlands: Kluwer.

Skovsmose, O. (2005). *Travelling through education: Uncertainty, mathematics, responsibility*, Rotterdam: Sense Publishers.

Skovsmose, O., & Yasukawa K. (2004). Formatting power of "mathematics in a package": A challenge for social theorising? *Philosophy of Mathematics Education Journal*. Retrieved from http://www.ex.ac.uk/~PErnest/pome18/contents.htm

Skovsmose, O., Yasukawa, K., & Ravn, O. (2011). Scripting the world in mathematics and its ethical implications. *Philosophy of Mathematics Education Journal*, Retreived from http://people.exeter.ac.uk/PErnest/pome26/index.html

Wedege, T. (2010). People's mathematics in working life: Why is it invisible? *ALM International Journal*, *5*(1), 97–98.

Wooden, M. (1999). Job insecurity and instability: Getting the facts straight. *Business Council of Australia Papers*, *1*(1), 14–18.

Yasukawa, K. (2003). Towards a social studies of mathematics. In S. Kelly with B. Johnston and K. Yasukawa (Eds.), *The adult numeracy handbook: Reframing adult numeracy in Australia* (pp. 26–34). Sydney: NSW ALNARC and Language Australia.

Keiko Yasukawa
University of Technology, Sydney

Tony Brown
University of Technology, Sydney

CHAPTER 12

SHAPING AND BEING SHAPED BY MATHEMATICS: EXAMINING A TECHNOLOGY OF RATIONALITY

KEIKO YASUKAWA, OLE SKOVSMOSE AND OLE RAVN

INTRODUCTION

Since the mid 1980s, there has been increasing interest in understanding the nature of the relationship between mathematics and society. Some of this interest has led to the examination of different forms of mathematical practices using ethnographic methodologies (Gebre, Rogers, Street, & Openjuru, 2009; Lave, 1988; Nunes, Schliemann, & Carraher, 1993; Street, Baker, & Tomlin, 2008). Studies in this area show the cultural shaping of mathematics as social practices, and uncover the variety of forms of mathematical practices that can be found. Ethnomathematics, the study of the interactions of mathematics with cultures, has also attracted increasing interest, particularly among mathematics educators (see, for example, Ascher, 2002; D'Ambrosio, 1985; Joseph 1990; Zaslavsky 1999) interested in exploring possibilities of increasing cultural diversity and inclusion in the mathematics classroom. Although schooling is arguably the most powerful means of reproducing dominant social values and orientations, understanding the relationship between mathematics and society has implications in social spheres that are broader than the mathematics classroom. One only needs to look at the extent to which social debates and policies are influenced and shaped by statistical data that can be produced, analysed, and manipulated with increasing speed and ease through sophisticated information technologies.

Strangely, however, there has been relatively little attention in the field of sociology to trying to understand the nature of the relationship between mathematics and society.[1] In contrast to this, there has been an increasing interest and theoretical development in what have come to be known as Science and Technology Studies (STS). One concern of the STS scholars is to challenge the modernist assumptions about science and technology, for example that science operates according to its own quality criteria, and that technology can be considered simply as applied science. Instead, the focus of STS scholars is on the mutual shaping of science and technology, and society. A great variety of technologies has been studied by STS scholars – from large scale infrastructure and military systems (Hughes, 1998) to the more mundane technologies such as "bicycles, bakelite, and bulbs" (Bijker, 2002) and domestic technologies (Cowan, 1983), and, more recently, various forms of management technologies (Bloomfield, Coombs, Knights, & Littler, 1997; Grint & Woolgar, 1997). These studies have allowed for some radical reflections about the nature of scientific endeavours, the

O. Skovsmose and B. Greer (eds.), Opening the Cage: Critique and Politics of Mathematics Education, 265–284.

relationships between science, technology, and society, and the modernist thesis about "progress". An important rationale for these critical studies of technology, that challenge dominant theories of technology as autonomous and unstoppable, is expressed by MacKenzie and Wajcman (1999, p. 5):

> The view that technology just changes, either following science or of its own accord, promotes a passive attitude to technological change. It focuses our minds on how to *adapt* to technological change, not on how to *shape* it. It removes a vital aspect of how we live from the sphere of public discussion, choice, and politics. Precisely because technological determinism is partly right as a theory of society (technology matters not just physically and biologically, but also to our human relations to each other) its deficiency as a theory of technology impoverishes the political life of our societies.

MacKenzie and Wajcman go on to draw on Ulrich Beck's thesis of the risk society, which is that many of the risks facing us are manufactured by humans, to say that social change must not be seen as something that simply emerges upon us, but something that should become "a process that is actively, and democratically, shaped" (p. 5).

In many of these studies, references are made to the quantitative reasoning that is involved in the politics and cultural dynamics at the interface between the technological artefact (or system) and human actors (see, for example, the studies on standards in Star and Lampland, 2009). However, there has been little that has focused the gaze on the mathematical elements themselves in order to theorise mathematics as a distinctive element or actor in the broader socio-technical milieu.

In an attempt to theorise mathematics from a sociological perspective, several scholars, including authors of this paper, have done so from the starting point that mathematics can be considered a technology (Porter, 1995; Skovsmose 1994; Skovsmose & Yasukawa, 2009; Yasukawa, 2003). It is a tool for people to make meaning of the world around them. Porter (1995) has posited that mathematics can be interpreted as a technology of distance; that is, by assuming the face of objectivity, information conveyed by numbers is able to gain more trust and credibility from a greater number of people. Skovsmose and Yasukawa (2009) have focussed on mathematical models and the propensity for people to use models to reflect what exists, but also to reshape what exists and in some cases bring into existence new realities. The focus of these studies has been the ways in which the interests of the people who construct and appropriate the models are enacted through the operation of these models. Mathematics itself is not a technology in a material sense; it is not a physical artefact in the sense that a hammer or a sewing machine or an aircraft is. It is an invisible technology that can shape people's way of seeing and making sense of the world. It configures ways of acting in the world. An analysis of mathematics as a technology, therefore, also turns into an analysis of the way particular kinds of rationality are constructed, cultivated, materialised, and brought into operation.

Challenging a classic way of thinking about mathematics, that mathematics itself is value-free and neutral, the notion of "mathematics in action" (Skovsmose, 2005) is used to describe how mathematics can be interpreted as interacting with other actors in society as a "formatting power" (Skovsmose, 1994; Skovsmose and Yasukawa 2009). In the field of STS, there have been theoretical developments, debates, and variations around the relationship between technology and society leading to the thesis of the Social Construction of Technology (SCOT) (see, for example, Bijker, Hughes, & Pinch, 1987; Hacking, 1999; Hughes, 1989, 2004; Winner, 1986), and actor-network theory (ANT) (Callon, 1987; Latour, 1987; Law, 1987). In this chapter, we seek to reflect on, re-examine, and find a distinctive location of, the concept of mathematics in action in relation to the theoretical constructs from the STS literature. In this way, we hope to further develop conceptual tools for understanding the relationship between mathematics and society. In particular, we want to show how the continued development of conceptual tools for the examination of mathematics in society is needed in contemporary social theorizing to create avenues and possibilities for some radical reflections about some of the taken-for-granted assumptions about the role of mathematics in society.

In the first section of this chapter, the notion of mathematics in action will be presented to argue that mathematics is an active constituent in technologies that shape and are shaped by, and in, the wider social milieu. In the second section, we examine the notion of Thévenot (1984) of codification as a "form-giving activity", and a further elaboration of that in a framework developed by Boltanksi and Thévenot (1999) on "modes of justification" in different social worlds that are used to manage disagreements. These ideas are used to examine mathematics as a technology of rationality. In the third section, we examine mathematics as a technology of rationality with reference to theoretical developments in STS to see what explanatory benefits can be afforded by these theories. In the fourth section, presenting our conclusions, we point out implications of the notion of mathematics as a technology of rationality for social theorising more broadly.

MATHEMATICS IN ACTION

We can approach the study of the relationship between mathematics and society from two directions. On the one hand, we can start by considering how society is influencing mathematics as a body of knowledge, for instance, how social forces such as cultural norms, military interests, and economic priorities are influencing what is researched in mathematics. Thus, one can seek to examine how social concerns are reflected in the scholarship of mathematics. On the other hand, one can start by considering what can be influenced by mathematics. In this latter instance, one is concerned about how mathematical techniques and ways of approaching a problem are reflected in the social sphere, for instance, how metrics – for instance in terms of productivity, wage rates, school performances on standardised tests – are imposed on human activities to value their worth. Naturally, these two directions of influence are closely connected, and when talking about mathematics in action we particularly emphasise and address how mathematics has an interactional relationship with the social world in the way it influences, and is influenced by, social practices and social change. We find that this emphasis is an important contribution to social theorising; when we seek to understand and explain, and perhaps even create or resist, social change, we cannot ignore the part played by mathematics in action.

An earlier notion developed by us on the relation between mathematics and society was the "formatting power of mathematics" (Skovsmose, 1994; Skovsmose & Yasukawa, 2009); this notion expresses how mathematical thinking can form the way people view, imagine, and negotiate the world around them. Using this notion, it was argued that mathematics could influence social and technological environments. The problem, however, with this notion is the ease with which we slip into a view that it is mathematics as such which exercises power in some autonomous way. We reject the idea that mathematics can exist or have influence on the social sphere independent of any social agency. The notion of mathematics in action encompasses a broader set of ideas that emphasise how mathematics can be an active and significant constituent of actions in, on, and with society. In that sense one could also talk about mathematics-based human actions. Through such actions mathematics can exercise a formatting power.

Mathematics in action can be understood in terms of three aspects of the manifestation of mathematics in social affairs: (1) how mathematics plays a role in generating *technological imaginations*; (2) how mathematics may facilitate *hypothetical reasoning*; and (3) how mathematics is realised socially as a *formatting power*, for instance as algorithms or principles of a functioning technology.

Technological imaginations. The first aspect of mathematics in action is technological imaginations. Thus, mathematics provides a tool for expressing one's model of a not-yet-realised technology. The technology may be a material artifact such as an electric motor or a temperature gauge, or one that is conceptual, such as a formula for weight loss or an algorithm for calculating pay. Mathematics, in the form of a quantitative model, describes to others (in a particular way) what one person has in mind about a new piece of technology. In this way, mathematics facilitates discussion about technological possibilities that otherwise only exist as an imagination in a person's mind. Mathematics can help to describe what a technological artefact can look like, in terms of its physical dimensions or symbolic appearance (e.g. tables, charts, and maps) and it can also help to describe how a technological artefact will function in terms of its dynamic processes. Thus, this descriptive action that mathematics performs not only enables the production of a more concrete representation of what is in one's imagination, but it also enables communication to take place among groups of people about different aspects of what had only been in one's imagination.

Hypothetical reasoning. The second aspect of mathematics in action can be understood as that of identifying actions following the technological imagination. Mathematics can be used to predict what can happen under certain conditions, sometimes with confidence, sometimes with much less confidence, depending, for instance, on the size of samples that were available in statistical modelling of a process. Once there is a mathematical representation of an imagined technological system, one can investigate what would happen *if* a value of a certain dimension was doubled on the device, or *if* this new factor were introduced into the management system, or *if* the person needing dietary advice were 30 years of age and weighed 50 kg rather than 40 years of age and weighed 48 kg. These hypothetical "experiments" are critical tools for making decisions in technological design and social policy making. For example, the mathematical models may help to determine the increased cost of a device if the size of a device were changed, or the amount by which an organisational unit would need to increase its productivity in order to meet a likely budget cut, or if a particular dietary regime could be recommended to everyone. At the same time the tremendous risk that is associated with technological development is precisely linked to the mathematical format of the hypothetical reasoning.

Formatting power. The third aspect of mathematics in action relates to the ways in which mathematics is an active constituent in a technological system and is used to exercise power – that is, facilitate control or ordering of what can and cannot happen. Often, the mathematical component within the system is invisible to the user or those who are affected by the system. An airline booking system operates on the basis of a mathematically-based algorithm that seeks to ensure an optimal number of passengers can be transported per flight, taking into account probabilities of late cancellations, last minute reservations, and changes to bookings. A biometric security system that relies on the unique features of the human iris or fingerprint is also based on a mathematically based pattern recognition algorithm that "decides" whether a person is the person they claim to be or not. These functions of mathematics shape and influence ways in which society is organised at a number of different levels; indeed it can be argued that they have a way of prescribing the ways people and groups of people organise and behave in society by providing a reason or a justification for the decisions that are made.

Davis and Hersh (1988) have used the terms *descriptive*, *predictive,* and *prescriptive* to characterise the functions of mathematics in the world. While Davis and Hersh do not pursue a social theorising of mathematics, their way of examining the prescriptive functions of mathematics, in particular, resonates with aspects of mathematics in action. For example, one can use mathematics to describe the expenditure of an organisation that goes to wages. This description can then be used by the same organisation to predict the budgetary implications of retaining the same staffing levels for the following year. The prediction may then lead the management of the organisation to prescribe certain actions, such as changes to the staffing levels, or expansion or contraction of the work that the organisation does: it is through mathematical calculations that changes are enacted on the structure and size of the workplace.

The thesis of mathematics in action reflects the idea that one can make descriptive, predictive, and prescriptive uses of mathematics. In all such forms of action, we find that mathematics can both limit and open up possibilities. In the previous example, depending on what the budget predictions look like, mathematics in action in the organisation can limit the tenure of employment of a staff member or open up possibilities for prospective employees to work there; it can lead to the organisation contracting its business or expanding into new services or markets. And mathematics in action operates not as an isolated source of power, but through interactions with other resources, including other forms of technology.

We have previously sought to investigate the formatting power of mathematics by examining the role of mathematics in a software encryption package called Pretty Good Privacy, or PGP as it is commonly known (Skovsmose & Yasukawa, 2009). In this study, it became apparent that a piece of mathematics (for example, theorems and properties of numbers), that may not have attracted much attention beyond the circle of academic mathematicians, can assume new and unexpected significance when it is brought into action in a new period in history and in a new social and technological environment. This may not be such a radical view when considering mathematical concepts such as the normal distribution, and how the meaning of this distribution carries different social significance, say in the parallel developments of statistics and studies of criminal behaviour and deviance, genetics, evolutionary biology, and insurance from the late eighteenth century onwards (see for example, Gigerenzer, Swijtink, Porter, Daston, Beatty, & Krüger, 1989; Hacking 1990). However, there are some mathematical results in pure mathematics that are often deemed to be insignificant outside the internal world of mathematics. The PGP case study showed how a new meaning and expression of the fundamental notion in social relationships – trust – is afforded through computer algorithms that are based on the Fundamental Theorem of Arithmetic (which states that any integer greater than 1 can be written as a product of prime numbers in one and only one way). Indeed, the mathematician G. H. Hardy emphasised the following in his *A Mathematician's Apology*:

> If useful knowledge is ... knowledge which is likely, now or in the comparatively near future, to contribute to the material comfort of mankind, so that mere intellectual satisfaction is irrelevant, then the great bulk of higher mathematics is useless. (Hardy, 1967, p. 135)

In the concluding pages of *Apology*, Hardy made the following statement concerning his own mathematical work:

> I have never done anything "useful". No discovery of mine has made, or is likely to make, directly or indirectly, for good or ill, the least difference to the amenity of the world. (Hardy, 1967, p. 150)

Hardy's counter-thesis to the concept of mathematics in action is of course extreme, and, arguably, it is not uncommon for there to be little prospect of practical applications of many new mathematical results immediately following their discovery. It nevertheless suggests that a strong tradition exists within the academic world of mathematics that emphasises and values the distance between mathematics and society. What is of greatest concern from a sociological perspective is that mathematics, as an active constituent of social and technological systems, is actually playing a significant role in shaping the social world, although often invisibly and therefore unnoticed. It is for this reason that attention to mathematics in action cannot be ignored; without this attention, it will be difficult to encourage reflections on the role of mathematics in social affairs.

Another case study of mathematics in action, though not framed in those terms, that warranted a closer and critical examination of how mathematics was actively shaping human activities is the example of wage rates for casually employed academics in an Australian university (Yasukawa & Brown, Chapter 11, this volume). Casual academics in Australia are employed for the length of an academic semester for their work as a lecturer, tutor, or laboratory demonstrator and are paid at an hourly rate. The hourly rate is worked out as

> (relevant full-time salary/52)/37.5 + x% (relevant loading)
> (Australian Industrial Relations Commission (AIRC), 2002, p. 7)

And the number of hours to be paid for each hour of delivery is based on the following:

> A casual academic required to deliver a lecture (or equivalent delivery through face to face teaching mode) of a specified duration and relatedly provide direct associated non-contact duties in the nature of preparation, reasonably contemporaneous marking and student consultation will be paid for at a rate for each hour of lecture delivered ...
> Basic lecture (1 hour of delivery and 2 hours of associated working time) (AIRC, 2002, p. 8)

and is based on an industrial award that had been negotiated between the higher education industry trade union and the organisation representing the employers in the university sector. While there is nothing particularly secretive about the way that the pay is calculated, the casual academics' experience was the formula "in action", not "in the making". And their experience was the formatting effect, that is, the amount of pay in their bank accounts, which did not seem to match the way they would themselves value their work (Yasukawa and Brown, Chapter 11, this volume). But so long as the formula that was enshrined in an industrial award could not be challenged, it left little leeway for the casual academics to argue for pay that they felt was more commensurate with the work they did. As an employer, the university could rationalise the small pay because there was a formula that justified the structure and amount of pay that the casuals received.

MATHEMATICS AND MODES OF JUSTIFICATION

Of particular interest is how mathematics acts as a technology of rationality, that is, as a tool for reasoning and making decisions. Thévenot (1984) studies the phenomenon of organisations' "investment in form-giving activities" (p. 2). He describes how organisations invest in codifying the relations between management and their labour force by classifying occupational tasks using methods such as statistical coding. Frederick Taylor's mechanism of scientific management is used to illustrate this idea.[2] By codifying occupational characteristics sufficiently, it becomes possible to evaluate equivalences between different individuals who are coded in a similar fashion. Recruitment and the deployment of the workforce become simpler because – the argument goes – the employer simply equates the task that the employee will perform, their qualifications or other codified requisites, and the wages set for that occupational task, and does not need to pay attention to individual characteristics that are not part of the codification structures.

While not located within an organisational context, the example of mathematics in action in encryption systems, as referred to earlier, can be understood as a form-giving activity on two different levels. Computer encryption systems rely on the use of large prime numbers to generate encryption keys. The larger the prime number that forms part of the key, the more difficult it is to break the code, or in other words, guess the secret key. Classifying encryption systems based on the number of digits of the encryption key is a way of giving "form" to establishing trust between two parties; a secure channel of communication is one that is protected by a strong encryption system. But mathematics in an encryption system also gives form in another way. It gives form to the level of trust that the people exchanging confidential information over the Internet can afford each other. Trust – a very human dimension of social relationships – is given form in a way that can be evaluated, based on the number of digits or bits that are used in the encryption system that is used to communicate between the parties in the virtual world. When a key is deemed large enough, a person can rationalise the trust that they invest in the person sending them information or to whom they are sending information.

Form-giving activities are compared by Star and Lampland (2009) with the "human use, creation, and disuse of standards" (p. 4). They argue that there is an increasing propensity in modern society to standardise humans, objects, and activities in a wide range of socio-cultural endeavours. They identify five features that are characteristic of standards: 1) standards are "nested" within other standards – the need for, or the approach to, standardisation of one process is often influenced by the way another related process has already been standardised; 2) the enforcement of standards is uneven in the larger socio-cultural landscape – standards are political; 3) the acceptability and suitability of standards are culturally and socially contingent – one society's standards may have little meaning or benefit in another; 4) standards increasingly are integrated within standards operating within and across larger networks of organisations and systems; and 5) standards "codify, embody or prescribe ethics and values, often with great consequences for individuals" (p. 5). We can add that standardisation is a principal example of mathematics in action. Using a formalised instrument as a resource for making decisions can have consequences that are not only felt by the individual creators of the instrument, but that are much more far-reaching and widespread.

Giving form to "trust" and committing crime against the state

Computer encryption system can be examined as a standard using Star and Lampland's characterisations. The algorithm used within PGP operates within a larger set of standards about how data sets are represented on digital computers. The standards used to represent numbers within a computer influence the speed at which the encryption algorithm can operate, and at the same time, the speed at which a hacker trying to break the code might successfully hack the system. Access to encryption technologies is uneven. As documented by the creator of the PGP encryption system (Zimmermann, 1991, updated 1999), there is a history of the United States Government attempting to restrict the distribution of the PGP encryption systems outside its own borders, and in that way making the distribution of encryption system a political process. Encryption systems are deemed necessary in some transactions, such as commercial transactions over the Internet, or other exchange of information where it is desired that some element of what is exchanged remains confidential to the parties involved. The process of encryption, in the way computer encryption systems operate, may be deemed unnecessary and indeed completely out of place if applied to a simple e-mail message to a friend. On the other hand, some encryption systems that may be adequate for encrypting one's password to an intranet site within a community-based organisation may be deemed inadequate for the purposes of highly sensitive commercial or military interactions. The last two characteristics that Star and Lampland attribute to standards are interesting in drawing the analogy between the mathematics in action in an encryption package and standards.

What ethics and values can be codified, embodied, or prescribed by mathematics in action within encryption systems? Encrypting a message or a set of data, by implication, suggests that that message or set of data is confidential – not open for public consumption. Thus, by virtue of investing in the creation and use of encryption systems (and disuse of hacked encryption systems) an assumption is that the information being transmitted with the use of this system is private. While there may be consensus that a person's password or pin number of their personal bank account ought to be able to be kept secret, there are other pieces of information for which consensus may not be easily reached. The controversy over Wikileaks, for example, raises, among many questions, the question of whether the information should have been kept secret or that releasing it was a heroic act for democracy. Surrounding the creation of PGP itself was met with conflicting political views. The creator of PGP wrote:

> It's personal. It's private. And it's no one's business but yours. You may be planning a political campaign, discussing your taxes, or having a secret romance. Or you may be communicating with a political dissident in a repressive country. Whatever it is, you don't want your private electronic mail (email) or confidential documents read by anyone else. There's nothing wrong with asserting your privacy. Privacy is as apple-pie as the Constitution. (Zimmermann, 1991, updated 1999, paragraph 2)

He felt that everyone had the right to encryption systems because everyone had the right to privacy of communication. The United States Government, on the other hand, had a different view. They wanted to have the right to control the use of encryption systems. In his Senate Testimony where he argued for changes to the restrictive export policy of encryption software, Zimmermann explained:

> In 1991, Senate Bill 266 included a non-binding resolution, which if it had become real law, would have forced manufacturers of secure communications equipment to insert special "trap doors" in their products, so that the government could read anyone's encrypted messages. Before that measure was defeated, I wrote and released Pretty Good Privacy. I did it because I wanted cryptography to be made available to the American public before it became illegal to use it. I gave it away for free so that it would achieve wide dispersal, to inoculate the body politic. (Zimmermann, 1996, paragraph 14).

The creator and disseminator of PGP, Phil Zimmermann, saw his invention as an ethically defensible act, whereas the United States Government saw it as illegal and a threat to national security. Thus, the packaging of mathematical results into a software program to enable the private exchange of information as a basic human right became entangled with political and diplomatic interests and systems of law enforcement of the most powerful nation in the world to the extent that:

> The government responded, in turn, by claiming patent infringement and later by classifying the RSA algorithm [used within PGP] under the International Tariff on Arms Restrictions (ITAR). In so doing, for the first time, three lines of computer code become officially classified as a munition. Exporting it without official sanction from the US State Department was no small crime. If convicted of exporting RSA, PGP, or any other encryption program, one could face a $1 million fine and 10 years in jail for each export violation. (Thomas, 2005, p. 656)

Such were the consequences for Zimmermann who for three years was the target of a criminal investigation by the US Customs Service, when the released PGP started to spread outside the United States (Zimmermann, 1996).

A feature of the encryption system such as PGP is that the algorithm is standard – it works on the same logic to encrypt any message, but the same message would be encrypted differently depending on the key that is used. By creating a standard algorithm, the algorithm could be "packaged" and disseminated widely across the World Wide Web, and this was in fact the source of the threat felt by those who tried to stop its distribution and who wanted to control who could and who could not communicate in private. Had the PGP encryption system been a one-off, custom-tailored system that did not rely on the same algorithm as any another, the power of the system would have been severely limited. Creating PGP as a standard encryption algorithm could therefore be regarded as an "investment in forms".

Justifying with numbers and mobilising through collective interest

In the study of mobilising a collective response of casual employees in a higher education institution, Yasukawa and Brown (Chapter 11, this volume) explained how these workers ultimately challenged the way their work was being formulated by a coding or standardising of a casual worker's work within an Enterprise Agreement. As Star and Lampland (2009) would say, the standardisation of their pay structure was nested – they had signed a standard contract used by their department for paying casual academics; the form of the contract was the standard contract used for casual academics within the institution; the contract was designed to meet the industrial standards set for the institution through the process of negotiations between the university management and the trade union; what was ultimately agreed in these negotiations had to satisfy standards set by the employer peak body of the higher education system as well as those set by the national office of the trade union; these standards in turn had to meet the federal industrial legislation standards. Thus, the statement "the rate you're being paid is the standard rate" is not just about the formula that was applied to calculate the casual academics' pay, but is a many-layered, historically imbued statement that is difficult for an individual to challenge.

Rates of pay, like the codes that are used in the types of forms Thévenot (1984) describes, reduce the complexity of characteristic of work to something that can be easily recorded and compared. Boltanski and Thévenot (1999), in their paper "The sociology of critical capacity" explain the importance that people in different spheres of life place on being able to make comparisons and draw equivalences as part of their decision-making and justification processes. They explain that there are different regimes of justification depending on the nature of the situations, and that different orders of worth exist in each regime.

The statistical coding of occupational tasks that Thévenot (1984) examined offers a tool for justification in what Boltanski and Thévenot (1999) call the industrial world. In this world, the order of worth is based on measures of efficiency and productivity. It is this world that was the focus of Thévenot's work on "investment in forms" – the statistical coding of occupational tasks such as in the Taylorist factories. In this world, the standardised codes, such as the time it should take to complete twelve identical tasks, or the wage rate appropriate to a casual lecturer provide a mechanism for minimising disputes in the workplace relationship. By establishing a "form" or a "standard" for what work is done and how within any occupational group, the task can be objectified away from the person, and determination of the worth of the task can be evaluated in relation to what is deemed worthy within the industrial world – efficiency and productivity, for example.

What was shown by Yasukawa and Brown (Chapter 11, this volume) was that rather than trying to resolve the dispute wholly within the regime of justification of the industrial world, the workers and their union delegates simultaneously worked in what Boltanski and Thévenot call the civic world, where the "order of worth" is not based on measures of efficiency or productivity, but rather on the collective interest whose value is recognised by the level of solidarity or mobilisation that it can facilitate. They were able to show collectively, rather than just as individuals, that the common work they were each doing took much longer than the standard formula said it would take. They could demonstrate an equivalence amongst this group of the time they each took to mark twenty essays, for example, but they could not establish equivalence of this effort with what the formula said. The formula could no longer be accepted as equivalent to the way they conducted their work, and it could no longer rationalise the kind of pay that the workers were receiving.

EXPLAINING MATHEMATICS AS A TECHNOLOGY OF RATIONALITY

Classical results from mathematics provide the justification for the PGP designers to claim that if one chooses a key large enough, the user could feel confident about the security of their encrypted message. The mathematics in PGP is invisible to the user, yet, without it, the encryption technology would not work – or, for that matter, exist. Mathematics is an invisible, but active, constituent in the confidential communication that takes place over the Internet. Wage rate formulae as established for the casual academics constitute another mathematically based technology that facilitates the justification of paying the workers for a particular amount of time at a particular rate. In the study cited above, the rationality imposed by the formula was challenged, not by a superior mathematical formula, but rather by a circuitous approach drawing on other regimes of justification based on the collective interests and solidarity of the workers.

For neither of these examples does a simple determinist theory of technology suffice as an explanation of the relationship between the "mathematics in action" and the social groups that were affected. It is true that, unlike what Hardy predicted and wished, theoretical results in mathematics such as the Fundamental Theorem of Arithmetic have had social and economic effects. Operating invisibly in a software package, the mathematical result has enabled secure communication among human rights activists (see, for example, testimonials on Zimmermann, no date), as well as business and military personnel, to do their business in ways that would otherwise be difficult. It would be possible, at least momentarily, to accept the thesis that once created, technology takes on a life of its own: it's unstoppable. Indeed PGP's history shows that once invented, even the United States Federal Government struggled to stop its diffusion across the World Wide Web. However, as MacKenzie and Wajcman (1999) would argue, caution is needed in accepting that somehow, the technology – in this case, the Fundamental Theorem of Arithmetic – is in itself the cause of society's increased reliance on encryption systems such as the PGP in commercial and other transactions. The mathematics had an effect. So did computer technologies, including the World Wide Web, which, while also relying on mathematics in their development, relied on other scientific and technological developments to continue to develop and become more powerful and ubiquitous. So did transport technologies that allow groups of people to be much more mobile than they used to be, so for example, allowing social activists as well as business executives from one country to organise their activities in another country. But there are factors that are not strictly technological that have shaped the emergence of technologies like the PGP. The growing global political, economic, and military influence of the United States compared to when Hardy was writing his *Apology* cannot be ignored as a factor in trying to stop Zimmermann from providing ordinary people with tools that might help them to communicate information that the United States Government cannot easily monitor. The way that mathematics through PGP has come to be an active constituent in society requires a more complex explanation.

Social Construction of Technology (SCOT) as a tool for understanding mathematics in action

Two related theoretical developments in STS provide ways of theorising the relationship between technology and society in more complex and nuanced ways than the determinist theses that look for a causal relationship, be it that technology causes social/economic or political change or that social change drives technological change. One is the Social Construction Of Technology (SCOT) thesis that emerged from the work of Pinch and Bijker (1987). A key contribution from this work is the foregrounding of the symmetrical relationship between technical artefacts and social groups. They introduced the concept of symmetry to explain that the ways in which the value of a new technology is interpreted are flexible, that is contingent on who is viewing the development of this technology and how they perceive the way the technology works, and that what works for one group may not be workable or even need to be workable for another group (Pinch & Bijker, 1987, p. 40). Technologies cannot be evaluated purely on the basis of meeting the original technical specifications that led to their design; rather they are evaluated by a number of different people holding different perspectives and interests in how they want the technology to work. This interpretive flexibility allows technical artefacts to be refined to make them more viable for certain social groups, particularly those groups that have the power to dominate, and one could trace the developmental paths of an artefact by tracing the ways in which variations of a technical artefact emerge in response to what different groups want to see. The developmental paths of bicycles, for example, are illustrated as a case study of interpretative flexibility by Bijker (1995).

In the example of the mathematics underpinning PGP technology, the technological paths to becoming an active element within the PGP encryption algorithm took some two millennia since Euclid first proved the Fundamental Theorem of Arithmetic. The trajectory of the Fundamental Theorem from Euclid's time to the last three decades when its inclusion into encryption packages such as the PGP occurred can be examined in more detail to show how the theorem helped to give rise to other results in number theory, some that are also embedded in encryption systems (see for example, Koblitz, 1994). Each of these mathematical developments is valued differently by different interest groups. At one extreme, Hardy valued each development for the advancement of mathematical knowledge with no regard for its worth outside this realm. By contrast, others, such as cryptologists, could see the value of some of these results for their application in developing a resource with practical use. While Hardy and other purist mathematicians might have liked to think that these developments were motivated by an interest to extend their intellectual pursuits about the abstract structures and properties of numbers, as SCOT would tell us, the ways in which the results are valued differ depending on whose use is influencing the evaluation. Furthermore, the different groups interested in the further development of number theory – for its own sake or for improving cryptographic systems – may each play a part in assisting the development. For example, research grants in number theory may be more forthcoming from some funding sources if the outcome of the research could be justified in terms of its contribution to national security in the form of stronger encryption systems (or ability to more quickly break other people's codes to monitor electronic communication).

SCOT could also provide a reason to examine the developmental paths of the wage rate formula for casual employees. As explained earlier, the formula has a history going back to a 2002 industrial award that was being used as a basis for the casual employees' pay in 2004 when the research cited was taking place. Closer examination of this award reveals that it also has a history of a previous 1987 award (AIRC, 2002), and that both of these awards were negotiated between the representatives of the university employees and the employers. The actual rates at each university are a product of enterprise-based negotiations that take place at each of the universities between the local university management and the union representatives. Like all negotiations, there are wins and losses – trade-offs – on both sides of the negotiating table. The final pay rate that emerges at the conclusion of the negotiations is a product of these negotiations, and part of a larger package of working conditions and remuneration. The package is what the ordinary employee sees, not the history of the package. Thus, in this case, even though the mathematics of the pay is visible, unlike the mathematics within PGP, the history of the mathematical formula is opaque. The formula is not amenable for re-formulation. Another factor that makes the formula act in ways that seem contrary to what it has supposedly codified – the pattern of work of casual academics – is that the negotiations over how the work should be codified has typically been done by full-time employee representatives of the union, not the casual employees. Whether their voice is heard at all has often been dependent upon the initiative (or lack thereof) of the full-time employee representatives to bring them closer to the negotiating table. Thus, the "interpretative flexibility" that the SCOT theorists have identified in technological development must also be subjected to a political analysis: whose views and voices about the design of a technology are heard, and whose are left out?

Actor-network theory (ANT) and the black-boxing of mathematical rationality

SCOT refers to the social construction of technology. This is a very important idea for any broader understanding of technology. However, it is also possible to consider the technological construction of society, or the dialectical process of society shaping technology that in turn shapes aspects of the social world. In our interpretation of mathematics, we emphasise that mathematics plays an important role in the dialectical process, both by being constructed and being constructive. One of the ways in which this dialectical process has been examined in STS is Actor Network Theory (Callon, 1987; Latour, 1987; Law, 1987). Instead of assuming a clear separation between human actors and material things in the evolution of new technological systems, ANT removes this separation and assumes that technologies form part of the evolution of a heterogenous network, that is, connections among both human and non-human "actors". Their focus of attention is less on the effects or impact of a technology than it is on how they come to be through interactions and negotiation of goals and interests among the different actors. While we do not subscribe to the idea that mathematics has agency in the same sense of a person making conscious decisions on how to act in any given situation, we find the idea of human and non-human (mathematical) actors connected in a heterogenous network useful. It is useful as a metaphor for thinking about situations in which an individual or society makes or justifies a decision as "common sense" or "logical" without noticing that their common sense and logic reflects the mathematical rationality that they have subscribed to. Measuring the worth of a person's work according to a pay rate formula, or trusting the confidentiality of a piece of online communication based on the size of the numerical encryption key, are examples.

The propensity for people to rely so readily on mathematics as a technology of rationality in different situations without trying to uncover the interests and the history of the particular mathematical reasoning can be explained using another central concept in ANT. This is "translation", the transformation of interests and goals of the actors struggling to form a stable network – that is, a group of actors with shared, or at least mutually compatible, interests (Latour, 1987). The chain of connections and negotiations between the different human and non-human actors can be long and complex. Where a new technological system does emerge and is accepted and the network is said to have stabilized around it, the technological "product" becomes a black box whose history and its constitutive interests become opaque.

The idea of technologies being black-boxed is one that draws resonance with an encryption algorithm being "packaged" in a computer software package such as PGP. Because the user acquires the package, not its constituent components, the fact that a classic mathematical theorem dating back to Euclid forms part of the package would be elusive to the user. Similarly, in the case of the casual employees' pay rate formulae, the negotiations that led to the agreement over the formulae are history that is invisible to the employee. The employee interacts with the formulae in a purely ahistorical manner.

Thus, what we are interested in, from the perspective of ANT, is the way in which these black boxes, packages, or formulae come into being. We are interested in exposing, rather than censoring or discarding, the history. This uncovering and exposing is an important theoretical resource in the study of mathematics in action for the same reason that it has aided the goals of STS quoted earlier, that is to enable us to see social change as "a process that is actively, and democratically shaped" (Mackenzie and Wajcman, 1999, p. 5). Mathematics plays an active role in human decision making – in determining the pay of an employee, deciding that communication is secure with another party on the Internet. It is a technology of rationality, and operates in the industrial world described by Boltanski and Thévenot (1999) as well as in the virtual world of the Internet. It gives form to the decision-making process – the formulae say you get paid this much, or the package says you are safe if you use a key size of so many digits. These forms are difficult to de-form – in the one case because its origins are opaque to all but the few, and in the other because it is written and buried into the package in humanly indecipherable ways; the forms acquire power as a result of the intractability of their constituent elements. They give the illusion that the technology operates as a neutral tool.

CONCLUSIONS

In studying what we have called mathematics in action, attention turned to theories that help us to try to understand, and perhaps ultimately shape or resist, the powers manifested through mathematically based actions. We have established a strong connection between the concepts from ANT and SCOT from the STS literature and the concept of mathematics in action in the social theorizing of mathematics. Focusing on the concepts of science and technology alone does not enable us to engage in a radical reflection and critique of society. As is commonly said, mathematics forms a "foundation" for science and technology; this suggests that in order to get to the core of a critical reflection on the nexus between technology and society, we must examine mathematics itself. However, this examination takes some effort because foundations often become taken for granted and invisible as other elements are built and layers are laid on top of them. In particular, mathematics, like other forms of technology, can be understood to be in a dialectical relationship with society; mathematics is both created and shaped by society, and under the guise of mathematical reasoning, society creates and shapes social values and norms. This connection is not just a parenthesis in our intellectual endeavour. Social theorizing through the concept of mathematics in action points to a critical component for understanding social change today, namely mathematics as a technology of rationality.

Thus, we have also shown that mathematics as an active constituent in society is often invisible, yet plays an important role in providing and limiting the lens that we use to see, understand, and act on the world, particularly when we reflect on the reasoning behind many management and other technologies for social engineering. If we do not pay special attention to mathematics, then we risk overlooking and therefore taking for granted the blinkered understanding of what is, what could be, and what to avoid in the social world.

We have also made connections between mathematics in action and particular types of technologies of rationality, or, as Boltanski and Thévenot (1999) call them, modes of justification. In particular, we explored how mathematics plays a critical role in the industrial world where there is a propensity to standardise and codify occupational tasks. This was seen as an example of a broader trend in contemporary society towards standardisation, as theorised by Star and Lampland (2009).

There is much that the study of mathematics as an active constituent of social change can gain from the theoretical developments in Science and Technology

Studies. Equally, we leave this paper with the proposition that much can be gained in the further development of STS as an endeavour to provide radical reflection and critique of society, by focussing on the active role of mathematics in action in the creation of technological systems, both large and small.

The thesis of SCOT, the Social Construction of Technology, and its variants grew out of challenges to the thesis of technological determinism – that is, that technologies, once developed, start to take a life of its own, and in particular, become an autonomous agent of technological and social change. Our position is that mathematics is also socially, culturally, and historically contingent. In providing case studies of mathematics in action, we illustrated that once mathematical reasoning gets "black-boxed" or encapsulated, for example in a computer algorithm or formula, the social history of the mathematics-based reasoning afforded by the algorithm or formula becomes invisible to the user or consumer of this resource. Mathematics as a technology of rationality, in contrast to other material technologies, can be brought into action much more invisibly than material artefacts. Thus, while we are in general agreement about the SCOT thesis as an explanation of mathematics as a technology, we emphasise the need to remain vigilant towards the invisible ways in which mathematical reasoning is being used to influence social change. It is not an autonomous force, but the consequences of mathematical reasoning can be felt like an inevitable consequence if society is blind to the power of mathematics as a resource or technology of justification.

NOTES

1. There have, howevere, been some exceptions. See, for example, Restivo (1992), Restivo, Van Bendegem, & Fischer (1993), MacKenzie (2001), and Greer & Mukhopadhyay (Chapter 10, this volume).
2. For a discussion of Taylor and scientific management, see also Skovsmose (2009).

REFERENCES

Ascher, M. (2002). *Mathematics elsewhere: An exploration of ideas across cultures*. Princeton, NJ: Princeton University Press.

Australian Industrial Relations Commission (2002). *Higher Education Academic Salaries Award 2002*, AW820200 PR925551. Retrieved from http://www.airc.gov.au/alldocuments/PR925551.htm

Bijker, W. E. (1995). *Of bicycles, bakelite, and bulbs: Toward a theory of sociotechnical change*. Cambridge, MA: MIT Press.

Bijker, W. E., Hughes, T. P., & Pinch, T. (Eds.). (1987). *The social construction of technological systems: New directions in the sociology and history of technology*. Cambridge, MA: MIT Press.

Bloomfield, B. P., Coombs, R., Knights, D.,& Littler, D. (Eds.). (1997). *Information technology and organizations: Strategies, networks and integration*. Oxford: Oxford University Press.

Boltanski, L., & Thévenot, L. (1999)..The sociology of critical capacity. *European Journal of Social Theory*, *2*(3), 359–377.

Callon, M. (1987) Society in the naking: The study of technology as a tool for sociological analysis. In W. E. Bijker, T. P. Hughes & T. J. Pinch (Eds), *The social construction of technological systems: New directions in the sociology and history of technology* (pp. 83–103). Cambridge, MA: MIT Press.

Cowan, R. S. (1983). *More work for Mother: The ironies of household technology from the open hearth to the microwave*. New York: Basic Books.

D'Ambrosio, U. (1985). Ethnomathematics and its place in the history and pedagogy of mathematics. *For the Learning of Mathematics*, *5*(1), 44–48.

Davis, P. J. and Hersh, R. (1988). *Descartes' dream: The world according to mathematics.* London: Penguin.

Gebre, A. H., Rogers, A., Street. B., & Openjuru, G. (2009). *Everyday literacies in Africa.* Kampala: Fountain Publishers.

Gigerenzer, G., Swijtink, Z., Porter, T., Daston, L., Beatty, J. & Krüger, L. (1989). *The empire of chance: How probability changed science and everyday life.* Cambridge: Cambridge University Press.

Grint, K., & Woolgar, S. (1997). *The machine at work: Technology, work, and organisation.* Cambridge: Polity.

Hacking, I. (1999). *The social construction of what?* Cambridge, MA: Harvard University Press.

Hughes, T. P. (1998). *Rescuing Prometheus.* New York: Vintage Books.

Hughes, T. P. (2004). *Human-built world: How to think about technology and culture.* Chicago: University of Chicago Press.

Joseph, G. G. (2000). *The crest of the peacock: Non-European roots of mathematics (New edition).* London: Penguin.

Koblitz, N. (1994). *A course in number theory and cryptography.* New York: Springer-Verlag.

Latour, B. (1987). *Science in action: How to follow scientists and engineers in action.* Cambridge, MA: Harvard University Press.

Lave, J. (1988). *Cognition in practice: Mind, mathematics, and culture in everyday life.* Cambridge: Cambridge University Press.

Law, J. (1987). Technology and heterogeneous engineering: The case of the Portuguese expansion. In W. E. Bijker, T. P. Hughes & T. J. Pinch (Eds), *The social construction of technological systems: New directions in the sociology and history of technology* (pp. 111–133). Cambridge, MA: MIT Press.

MacKenzie, D. (2001). *Mechanizing proof: Computing, risk, and trust.* Cambridge, MA: MIT Press.

MacKenzie, D., & Wajcman, J. (1999). Introductory essay: The social shaping of technology. In D. MacKenzie & J. Wajcman, J. (Eds.), *The social shaping of technology (Second Edition)* (pp. 3–27). Milton Keynes, UK: Open University Press.

Nunes, T., Schliemann, A. D., & Carraher, D. W. (1993). *Street mathematics and school mathematics.* Cambridge: Cambridge University Press.

Pinch, T., & Bijker, W. B. (1987). The social construction of facts and artifacts: Or how the sociology of science and the sociology of technology might benefit each other. In W. B. Bijker, T. P. Hughes & T. Pinch, T. (Eds.), *The social construction of technological systems: New directions in the sociology and history of technology* (pp. 17–49). Cambridge, MA: MIT Press.

Porter, T. (1995). *Trust in numbers.* Princeton, NJ: Princeton University Press.

Restivo, S. (1992). *Mathematics in society and history: Sociological inquiries.* Dordrecht, The Netherlands: Kluwer.

Restivo, S., Van Bendegem, J. P., & Fischer, R. (1993). *Math worlds: Philosophical and social studies of mathematics and mathematics education.* Albany, NY: SUNY Press.

Skovsmose, O. (1994). *Towards a philosophy of critical mathematics education.* Dordrecht, The Netherlands: Kluwer.

Skovsmose, O. (2005). *Travelling through education.* Rotterdam: Sense Publishers.

Skovsmose, O. (2009). *In doubt: About language, mathematics, knowledge and life-worlds.* Rotterdam: Sense Publishers.

Skovsmose, O and Yasukawa, K. (2009). Formatting power of "mathematics in a package": A Challenge for Social Theorising? In P. Ernest, B. Greer and B. Sriraman (Eds.), *Critical Issues in Mathematics Education* (255–281). Charlotte, NC, USA: Information Age Publishing.

Star, S. L., & Lampland, M. (2009). Reckoning with standards. In M. Lampland & S. L. Star (Eds.), *Standards and their stories: How quantifying, classifying, and formalizing practices shape everyday life* (pp. 3–24). Ithaca, NY: Cornell University Press.

Street, B., Baker, D., & Tomlin, A. (2008). *Navigating numeracies: Home/school numeracy practices*. New York: Springer.

Thévenot, L. (1984). Rules and implements: Investment in forms. *Social Science Information, 23*(1), 1–45.

Thomas, D. (2005). Hacking the body: Code, performance and corporeality. *New Media and Society, 7*, 647–662.

Winner, L. (1986). *The whale and the reactor: A search for limits in an age of high technology*. Chicago: University of Chicago Press.

Yasukawa, K. (2003). Towards a social studies of mathematics: Numeracy and actor-network theory. In S. Kelly, B. Johnston & K. Yasukawa (Eds.), *The adult numeracy handbook: Reframing adult numeracy in Australia* (pp. 26–34). Sydney and Melbourne: New South Wales Adult Literacy and Numeracy Australian Research Consortium.

Zaslavsky, C. (1999). *Africa counts: Number and pattern in African culture (Third edition)*. Chicago: Lawrence Hill Books.

Zimmermann, P. (1991, updated 1999). *Why I wrote PGP*. Retrieved from http://www.philzimmermann.com/EN/essays/index.html

Zimmermann, P. (1996). Testimony of Philip R. Zimmermann to the Subcommittee on Science, Technology, and Space of the US Senate Committee on Commerce, Science, and Transportation. Retrieved from http://www.philzimmermann.com/EN/testimony/index.html

Zimmernann, P. (no date). Letters to Phil Zimmermann from human rights groups. Retrieved from http://www.philzimmermann.com/EN/letters/index.html

Keiko Yasukawa
University of Technology, Sydney

Ole Skovsmose
Aalborg University, Denmark
State University of São Paulo, Brazil

Ole Ravn
Aalborg University, Denmark

PART IV

SEARCHING FOR POSSIBILITIES

CHAPTER 13

POTENTIALS, PITFALLS, AND DISCRIMINATIONS: CURRICULUM CONCEPTIONS REVISITED

EVA JABLONKA AND UWE GELLERT

INTRODUCTION

Access to the forms of mathematics valued in institutions of tertiary education is an important goal of secondary mathematics education, and algebra often acts as a gatekeeper for more advanced (not compulsory) mathematics courses. The chances for advanced mathematics enrolment have been shown to be lower for students of lower socio-economic status, female students, and minorities (see, for example, Oakes, 1990; Valades, 2002). In tertiary education, formal theoretical parts of written mathematical discourse are valued more than informal and practical mathematics, an issue that caused a lot of discussion about a curriculum mismatch between secondary and tertiary mathematics (see, for example, Kajander & Lovric, 2005). In some professional contexts, highly specialised vocation-specific mathematics is used and further developed through mathematical modelling, whereas in others it tends to be incorporated into black-boxes and computerised tools, often with graphical interfaces (see Clayton, 1999). However, preparing students for a career that relies on the cultural and symbolic capital associated with the completion of advanced or specialised mathematics courses is not the only goal of compulsory mathematics education. Rationales for teaching mathematics include the enculturation of the students into the cultural heritage of mathematics with its intrinsic delights, effective functioning in everyday life (mostly in financial transactions), enhancement of the students' general reasoning capacity and critical thinking. As school curriculum conceptions for mathematics education are the product of a social process, including ideological struggles between stakeholders pursuing diverse economic and political goals, the result represents a compromise between different or differently nuanced social positions and agendas.

With this contribution, we intend to initiate a discussion of alternative curriculum conceptions in terms of how these might facilitate or restrict access to forms of mathematical practices and discourses valued in tertiary education, in specialised professions, or more recently, valued by an educational policy discourse asseverating an increasing need for mathematical modelling skills in the face of technological development and economic challenges. The consequences of mathematics curricula for different student groups in terms of their access to those valued forms of mathematical discourse, their formation of mathematical identities,

O. Skovsmose and B. Greer (eds.), Opening the Cage: Critique and Politics of Mathematics Education, 287–308.

and their positioning in the "knowledge society" are rarely directly visible, as curriculum conceptions often represent ideological hybrids. However, these consequences are not simply more or less accepted side effects of the practice of schooling.

Before addressing the potentials, pitfalls, and discriminations that curriculum conceptions might provoke for different social groups, we expose a conceptual framework for our description and analysis. By means of this framework we intend to overcome a characterisation of curriculum conceptions by a simple dichotomy of mainstream and alternative conceptions. Further, we describe conceptions of school mathematics as realisations of a process of dual recontextualisation that draw in different ways on the practices of professional mathematicians on the one hand, and on vocational, domestic, or leisure time activities on the other hand. The conclusion of the chapter focuses on the issue of how access to privileged discourses can best be provided by mathematics education practices.

MAINSTREAM CURRICULUM AND POSITIONS OF RESISTANCE

To the extent to which curriculum documents are results of compromises, they leave more or less space for alternative readings by teachers and students. Identifying these spaces requires a separate analysis. As curriculum conceptions construct their ideal readers, with distinct dispositions for mastering their explicit and implicit demands, differences in "orientations to meanings" (Bernstein, 1990) generate patterns of achievement in line with social differences (such as gender, ethnicity, social class).

The students are the "consumers" of the privileged meanings established in the curriculum, and, if they successfully acquire the intended interpretations, the resulting certificate, and/or the mathematical knowledge, is of symbolic value and eases access to further education. Alternative curriculum conceptions aim at redistributing access to privileged discourses. This can be achieved at different levels. Protagonists might be concerned with expanding the repertoire of individual students with a focus on marginalised groups and their orientations to meanings, without challenging the available reservoir of cultural meanings (such as "mathematics as thinking and problem solving" or "mathematics as a universally applicable technology"). On the other hand, more radical alternatives challenge the available reservoir of meanings. The first option might be described as a form of tactical resistance, whereas the second is aiming at deconstruction of culturally inherited meanings.

In interpreting mathematics curriculum conceptions as texts in a social context that position their readers (the students and the teachers alike), one could attempt to classify alternative conceptions according to their position towards mainstream conceptions in a similar way as Martin (1993) differentiates resistant reading positions in the context of research on literacy: as tactical resistance versus deconstructive resistance and oppositional versus subversive deconstructions (see figure 1). In a given context, conceptions that are alternative to the curricular mainstream might then be classified according to this scheme.

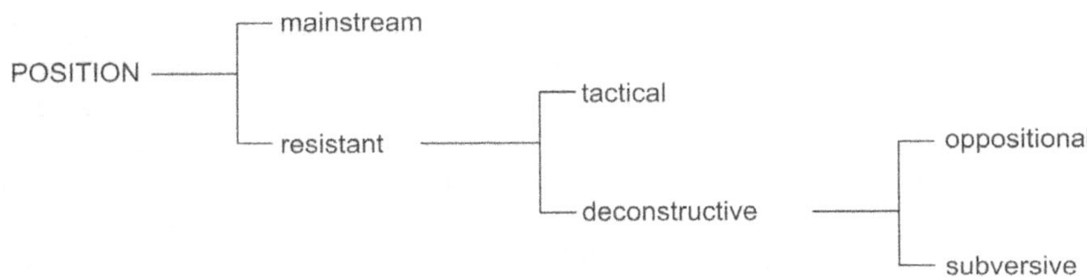

Figure 1. Dimensions of position (modified from Martin, 1993, p. 159)

However, what counts as the curriculum mainstream is different in different social and political settings. What currently is mainstream in one place might resemble, for instance, a tactical resistance position in other places, or a short-lived reform that has been followed by a counter-reform. The world of school mathematics curricula is not (yet) fully uniform.

Moreover, tactically resistant positions tend to aim at becoming the mainstream, hence the label "tactical". As a consequence, it is often difficult to identify what the mainstream position exactly consists of, even in a rather local setting. Practices of mathematics instruction are constantly (even if only slightly) changing, integrating aspects of tactical resistant positions into the mainstream. Bernstein (1996, p. 48) distinguishes between an "*official recontextualizing field* (ORF) created and dominated by the state and its selected agents and ministries, and a *pedagogic recontextualizing field* (PRF)". The PRF consists of teachers, teacher educators, researchers, private research foundations etc. What is considered as mainstream might be different in these recontextualising fields and amounts to conflicts.

It can be argued that opposition and dissent to the mathematical discourses, and to positions distributed through mainstream curriculum conceptions, require insight into the discourses that are the focus of critique. Apparently, there is a tension between a pedagogy of access and a pedagogy of dissent. Is access to valued forms of mathematical knowledge a precondition for a critique of social mathematical practices and their constituting discourses, or is access to valued forms of mathematical knowledge possible by critiquing mainstream discourses? How can these two poles be balanced? The tension is linked to the ideological split between agendas that emphasize all individuals' rights to improve their living conditions and to participate in public life, and agendas that stress the relationship between education and economy in terms of developing a stock of human capital for employment in science and technology.

MATHEMATICS CURRICULA AS A PRODUCT OF DUAL RECONTEXTUALISATION

Curriculum conceptions for mathematics education can be described as the specific product of a dual recontextualisation. On the one hand, school mathematics can be seen as the result of a subordination of the practices of generating new mathematical knowledge (exploration, systematisation, proof) to the pedagogic and didactic principles of the transmission of knowledge. On the other hand, school mathematics recontextualises vocational, domestic, and leisure time activities by subordinating them to a mathematical gaze. There is a variety of ways in which this dual recontextualisation can be realised in the mathematics curriculum. Some common versions of the mathematics curriculum in place, in which this dual recontextualisation constructs a hybrid between domestic and mathematical knowledge, often manifest in word problems, have been shown to be socially biased and largely self-referential.

Different alternative ways (focus on investigations and problem solving, ethnomathematics, mathematical modelling, critical mathematics literacy) implicate different potentials, pitfalls, (dis-)advantages and discriminations for different social groups. They differ in what knowledge is accessed in classrooms and in how this knowledge is made accessible. In an elaboration of Bernstein's sociology of education (Bernstein, 1996), the underlying principles of these differences can be termed *classification* and *framing*:

> I will now proceed to define two concepts, one for the translation of power, of power relations, and the other for the translation of control relations, which I hope will provide the means of understanding the process of symbolic control regulated by different modalities of pedagogic discourse. [...] I shall start first with power. We have said that dominant power relations establish boundaries, that is, relationships between boundaries, relationships between categories. The concept to translate power at the level of the individual must deal with relationships between boundaries and the category representations of these boundaries. I am going to use the concept of *classification* to examine relations between categories, whether these categories are between agencies, between agents, between discourses, between practices. (Bernstein, 1996, pp. 19–20)

In the context of mathematics education, classification refers to categorizing areas of knowledge within the mathematics curriculum. Strong internal classification means that clear boundaries between mathematical sub-areas are maintained. Strong external classification indicates that few connections are made to other disciplines or everyday practice.

Framing draws on the nature of the control over the selection of the communication, its sequencing, its pacing, the evaluation criteria, and the hierarchical rules as the social base which makes access to knowledge possible (p. 27):

> I am going to look at the form of control which regulates and legitimizes communication in pedagogic relations: the nature of the talk and the kinds of spaces constructed. I shall use the concept of framing to analyse the different forms of legitimate communication realized in any pedagogic practice. (p. 26)

The concepts of classification and framing are useful to describe the kind of knowledge emphasised in alternative curriculum conceptions as well as the way in which this knowledge is assessed.

As we will argue, different alternative ways of recontextualising practices of professional mathematicians as well as everyday practices, implicate different potentials, pitfalls, (dis-)advantages and discriminations for different social groups. We will attempt to link the discussion to the political bases of the alternatives we have chosen to discuss. In order to facilitate comparing and contrasting, we describe variations of curriculum conceptions in terms of common shared principles rather than as multifaceted or mixed. The following selection of alternative curriculum conceptions is made on the grounds that some of these conceptions were positioned as non-mainstream when they emerged, even though they might in the meantime have become mainstream in some places, while others still represent resistant construals of mathematical meaning. Examples and references are exemplary and not representative of the conceptions.

INQUIRY-BASED MATHEMATICS EDUCATION

Inquiry-based mathematics education starts from the assumption that young learners can be regarded as miniature scholar-specialists whose mathematical activity is not qualitatively different from that of a mathematician. Academic mathematics, often described as "the science of patterns", is mirrored in the mathematics classroom where students are engaged in discovering and exploring regularities, identifying relationships, and applying their mathematical knowledge in new mathematical situations. The general idea behind has been summarised by Bruner (1960, p. 14): "Intellectual activity anywhere is the same, whether at the frontier of knowledge or in a third-grade classroom." More provocatively: "In teaching from kindergarten to graduate school, I have been amazed at the intellectual similarity of human beings at all ages, although children are perhaps more spontaneous, creative, and energetic than adults" (p. 40). This view has been criticized as "romantic" (Tanner & Tanner, 1980, p. 535), as it neglects the fundamental differences between the production of knowledge and its reproduction in schools, as witnessed in the following quote:

> The pedagogical tradition calls for transmittal of the "given". It is a tradition of the transmittal of certainty, not of doubt. But doubt is precisely the quality of the scholar. The scholar, taken as an intellectual, is one "who makes the given problematic". Our pedagogical tradition does not deal with problematic material. If we obey our tradition, we take what is problematic and make it into sets of certainties, which we then call upon the students to "master". In too many instances, our sets of certainties come dissociated from the fields of knowledge out of which they originally grew. In some cases, the contrast between the school subject and its underlying field of knowledge is ludicrous. (Foshay, 1961, p. 32–33)

An inquiry-based mathematics curriculum can be understood as an attempt to overcome this pedagogical tradition by reconciling content and method: to find material that can be made problematic in order to develop knowledge both about how material is to be made problematic, and about the mathematical generalizations. Inquiry-based mathematics education is thus working in a combination of the inductive and the deductive mode.

The inductive part of inquiry-based mathematics education has been characterized by Dowling (2009) as involving skills and, moreover, tricks. In discussing a typical example of school mathematical investigations he shows how the "investigative" approach to school mathematics is actually introducing new areas of weakly classified strategies – skills, tricks – in a discipline that is apparently strongly classified. This might be misleading for some students, as the latter is generally preferred in mathematics. What makes a skill or a trick mathematically meaningful can tacitly be decided on the grounds of previously acquired mathematical knowledge. In most cases, however, this decision is made through the mathematical authority of the teacher in the face of the standards of mathematical knowledge to be acquired – the above-mentioned sets of certainties.

The construction of mathematical meaning through generalization of weakly classified activity and idiosyncratic notation of findings is a crucial component of inquiry-based mathematics instruction. For establishing generalized mathematical meanings when students are engaged in such activity in the mathematics classroom, two conditions (at least) have to be fulfilled. First, there have to be students who have already acquired the sufficient mathematical skills and tacit knowledge about what to look for and what to strive for when confronted with an open investigative mathematics problem; otherwise no valued generalization can be made at all. Second, only highly qualified teachers will be able to develop mathematical generalizations from the students' idiosyncratic and often not fully developed problem solutions. In many places of the world, these conditions are only partly met, and the inquiry-based curricular approach to mathematics education appears as a rather elitist option.

Inquiry-based mathematics education is problem-centred and characterized by strong external classification – although the concept of "problem" appears weakly classified and ambivalent in many descriptions. Inquiry-based mathematics education has been legitimised as a contrast to the conception of the "core curriculum":

> In the past, we saw a reality that the problems of life do not come in 'disciplined' packages. For example, a good many of the public problems we must deal with – housing, crime, transportation, and the like – go beyond the boundaries of any one discipline and must be studied on a multi-disciplinary basis. The most notable of the curriculum reforms intended to deal with this reality was the core curriculum, a problem-centered approach to learning, in which the mode of inquiry was to be dictated by the nature of the problem itself. We don't want our students ill-prepared for the practical problems of life, but there is another reality which we have tended to overlook. This second reality is that each of the disciplines, as they are organized, contains within its domain and methodology the best thought about reality in its own field. For example, one who knows how a chemist thinks can see more deeply into what is 'chemical' about an industrial problem than one who does not know how a chemist thinks (Foshay, 1961, pp. 33–34).

If argued like this, the conception resembles a tactical resistance position as it tries to point out why a conception with a focus on mathematical modes of inquiry is better suited for engaging with the same public reality as a curriculum stressing factual knowledge, that is, pursuing the same educational ends by different means. The details of this public reality given in the quote above – e.g. industry, economy, crime – and the claims suggest a prospective neo-conservative ideology.

In the course of the reforms and counter-reforms of the mathematics curriculum in Victoria, Australia, the inquiry-based curriculum seemed to have a different ideological base, the main focus being on offering access for all students through a conception that overcomes the levelled hierarchical nature of the traditional mathematics curriculum: "It had the potential to generalize the social reach of mathematics and to place school curricula on a new basis" (Teese, 2000, p. 169). However, the results on the "investigative project 1992" turned out to be disastrous for working class girls: 43% percent received the lowest possible grade or could not even master the minimum criteria for getting a grade (Teese, 2000, p. 171). But it was not the concern for exclusion of disadvantaged groups but the judgement by academic mathematicians that students would not learn enough, and that the most talented students would be "punished", that marked the end of these reform efforts.

ETHNOMATHEMATICS

Ethnomathematics as a programme emerged in opposition to mainstream discourse in mathematics education. A Eurocentric bias of mathematics education is most salient in curricula and textbooks developed in industrialised states and imported into former colonies. Vithal and Skovsmose (1997) interpret the emergence of ethnomathematics as a reaction to naïve modernisation theory and the cultural imperialism implied by it. By uncovering the cultural bias in historical accounts of mathematics and by documenting and analysing local mathematical practices, ethnomathematics set out to deconstruct mainstream discourse and offer new views on what counts as mathematics. Earlier work was often carried out from the perspective of cognitive anthropology, as witnessed in the reference list "Ethnomathematics: A Preliminary Bibliography" provided by Scott (1985) in the first Newsletter of the International Study Group on Ethnomathematics. The term "ethnomathematics" suggests a broad interpretation of both mathematics and "ethno". The latter encompasses "identifiable cultural groups, such as national-tribal societies, labor groups, children of certain age brackets, professional classes, and so on". (D'Ambrosio, 1985, p. 45)

In line with this agenda, a base for the development of an ethnomathematical curriculum consists of uncovering and describing the mathematical concepts and procedures that are more or less implicit in practices of sub-ordinated and oppressed people and marginalised groups. This type of research can be described as ethnographic. Ethnographic work can be done from different positions, as for example from a dominant position of racial classification, such as the "White-on-black" research, witnessed in several studies carried out in South Africa (see Khuzwayo, 2005). Ethnographic ethnomathematical research finds itself in a difficult position because there remains the issue in whose terms the ethnomathematical practices are to be described. When incorporated into the curriculum, there is a related problem. For local practices that might be of interest to the students and are identified to contain some mathematics, there is a risk that incorporation into classroom discourse amounts to a recontextualisation for the purpose of exploitation in terms of traditional school mathematical topics. Fantinato (2008) points to the difficulties that might be faced at the level of classroom interaction:

> However, it is important to keep in mind that the mathematics teacher stands for the official mathematics image in the classroom. This person holds a knowledge considered superior to students' daily knowledge due to its privileged social position in our society. This uneven status position interferes in the relations among different types of knowledges, which take part in the classroom cultural dynamics. When voicing students' knowledges, the dialogic attitude of the teacher entails an awareness of the mythical status of his math and the depreciation of other math as an effort to reverse this difference. (pp. 2–3).

Curriculum alternatives more closely linked to the original conception of ethnomathematics include the use of (historical) examples of culturally relevant practices as a springboard for developing mathematical notions (e.g. Jama Musse, 1999) or mathematical analysis of traditional artefacts, as for example decorative pattern designs (e.g. Gerdes, 1990).

The first alternative might assist in overcoming cultural alienation, but faces the same problématique as developing school mathematics on the basis of recontextualised domestic practices. The recontextualisation of everyday domestic practices, which amounts to a collection of their traces in the form of contextualised tasks, generally has a tendency to amount to an implicit pedagogy with weakly classified content that disadvantages marginalised groups (e.g. Chouliaraki, 1996; Cooper & Dunne, 2000; Gellert & Jablonka, 2009; Hasan, 2001; Lubienski, 2000; Morais & Miranda, 1996). A similar pitfall is inherent in some versions of a mathematical modelling conception (see below).

The following task (see figure 2) provides an example of the second alternative, used in a teacher education course (Gerdes, 1998). How many possible band patterns of the *sipatsi* type of given dimensions p and d exist, where p denotes the period of the respective decorative motif and d its diagonal height? Figure 2 shows the possible patterns of dimensions 2x4. The images on the left side display the generating motifs.

As to the classificatory principle, this task (if posed without an initial introduction into the mathematical description of the pattern dimension) resembles an inquiry mathematics task. If already mathematised, the criteria for the inquiry become more explicit. It is then a mathematical task on a comparatively advanced level. The prospect of producing computer-generated imitations of pattern designs, based on such mathematical explorations, might for some amount to a disenchantment of the wisdom and skills of traditional crafts. Kaplan (2003), for example, presents a process for creating computer-generated Islamic Star Patterns on a web page on which one can play around with a Taprats Applet. If the complexity of the mathematical algorithm provides an argument for the complexity of the skills involved in traditional crafts, then this value judgment privileges Western mathematics. The incorporation of local practices through their (school) mathematical recontextualisation in order to ease access represents a tactical position.

Figure 2. Sipatsi patterns and generating motifs (Gerdes, 1998, p. 44)

D'Ambrosio (2007) locates ethnomathematics within a wider project of social change that points to the responsibility of mathematicians and mathematics educators in offering venues for Peace (p. 26). He proposes a curriculum that is conceptualised as a modern trivium, including Literacy, Matheracy and Technoracy, that aims at providing "in a critical way, the communicative, analytical and technological instruments necessary for life in the twenty-first century" (p. 28). Matheracy is connected to the capability of inferring, proposing hypotheses, and drawing conclusions, that is, to classical academic virtues associated with mathematical thinking, access to which has been restricted to an elite. This conception indicates a critical stance towards teaching mathematical modelling and applied mathematics and also a departure from earlier envisaged forms of ethnomathematics. He also stresses that teaching "ethnomathematics of other cultures, for example, the mathematics of ancient Egypt, the mathematics of the Mayas, the mathematics of basket weavers of Mozambique, the mathematics of Jequitinhona ceramists, in Minas Gerais, Brazil, and so and so, it is not because it is important for children to learn any of these ethnomathematics" (p. 33). The main reasons for doing so include to "de-mystify a form of knowledge [mathematics] as being final, permanent, absolute, unique". This is to overcome the damaging misperception "that those who perform well in mathematics are more intelligent, indeed 'superior' to others, and to illustrate intellectual achievement of various civilizations, cultures, peoples, professions, gender" (pp. 33–34). If such a conception aims at challenging views conveyed through mainstream curriculum, it resembles a deconstructive resistant position.

Knijnik (2000) provides an example from her work with settlers of the Landless People's Movement (MST) in Brazil where the practices of production and sale of melon crops were "naturally" changed through the process of confrontation and translation of different forms of knowledge. She argues that if the pedagogical process were limited to the recovery of the native knowledge, this would restrict access to useful knowledge and as a consequence reinforce social inequalities. Such a restriction is sometimes also recognised by the students themselves, and it can be difficult to motivate students by relating mathematically to their "real life", because "their 'real life' is precisely what they want to overcome!" (Fioriti & Gorgorió, 2006, p. 3). Identifying practices that could profitably be transformed by a mathematical recontextualisation remains a major and continuous task for overcoming problems of discontinuity and disjuncture between different mathematical practices and school mathematics. Which out-of-school practices are to be selected as representative of the students' cultures remains a political issue.

MATHEMATICAL MODELLING

"Mathematical modelling" is a rather vaguely defined term for a curriculum conception that comprises many different classroom practices. Modelling conceptions can be distinguished by the strength of the internal and external classification of the respective knowledge domains as well as by the value attributed to the different knowledge domains in classroom modelling practice. A version of school mathematical modelling that stresses that the external classification remains as strong as in mainstream curriculum conceptions, is provided by Zbiek and Conner (2006):

> The primary goal of including mathematical modeling activities in students' mathematics experiences within our schools typically is to provide an alternative – and supposedly engaging – setting in which students learn mathematics without the primary goal of becoming proficient modelers. We refer to the mathematics to be learned in these classrooms as "curricular mathematics" to emphasize that this mathematics is the mathematics valued in these schools and does not include mathematical modeling as an explicit area of study … we recognize that extensive student engagement in classroom modeling activities is essential in mathematics instruction only if modeling provides our students with significant opportunities to develop deeper and stronger understanding of curricular mathematics. (pp. 89–90)

Such a version is reflected in the approach of *Realistic Mathematics Education* (RME) where models are seen as vehicles to support "progressive mathematization" (Treffers & Goffree, 1985), as van den Heuvel-Panhuizen (2003) points to:

> Within RME, models are seen as representations of problem situations, which necessarily reflect essential aspects of mathematical concepts and structures that are relevant for the problem situation, but that can have various manifestations. (p. 13)

A version of school mathematical modelling in which the external classification is weakened considerably constructs modelling as new (but vague) content. This version is sometimes referred to as *emergent modelling*:

> This second perspective [RME is the first one], which we favour, does not view applications and modelling primarily as a means of achieving some other mathematical learning end, although at times this is valuable additional benefit. Rather this view is motivated by the desire to develop skills appropriate to obtaining a mathematically productive outcome for a problem with genuine real-world connections ... Here the solution to a problem must take seriously the context outside the mathematics classroom, within which the problem is located, in evaluating its appropriateness and value … While the above approaches differ in the emphases they afford modelling in terms of its contribution to student learning, they generally agree that modelling involves some total process that encompasses formulation, solution, interpretation, and evaluation as essential components. (Galbraith, Stillman & Brown, 2006, p. 237)

Given the diversity of agendas and examples, the unifying principle of the modelling discourse in mathematics education can be seen in the differences constructed in relation to mainstream school mathematics without applications or in the differences to other forms of insertions of non-mathematical practices (such as word problems). There are some characteristic knowledge claims reflected in mathematical modelling: an ontological realism that acknowledges an independently existing reality that is the object of knowledge and the properties of which provide objective limits to how we can know it. However, these are seen as open to revision: a fallibility principle is acknowledged. This is a difference in comparison to school mathematics with a focus on both procedures and algorithms as well as on mathematical relationships and proof.

Julie (2002) summarises the differences as follows: a change of criteria towards acceptance of different non-equivalent answers, unrestricted time, acceptance of the provisional status of the outcome, and presentation in a format chosen by the students. The social base changes from individualistic to working in collaborative teams. Texts are not objects to be mastered, but used as resources. In a classroom such a shift would indicate a shift in the authority relationship between teachers, texts, and students that indicates a weakening of framing. Underpinned by learning theories that stress the agency of the learners, school mathematical modelling activities are also intended to encourage students to communicate their own ideas and to scrutinise the ideas of others (English, 2006).

The situation chosen as a starting point for modelling might be selected because of mathematical reasons or because of social reasons (Julie, 2002). In the first case the context is arbitrary and the mathematical concepts, procedures etc. are those specified in the curriculum; in the latter, the context is given (or selected by the students) and the mathematics is arbitrary. But any mathematics curriculum in the end prescribes a set of forms of mathematical knowledge rather than a set of situations to be dealt with mathematically. Only a critical mathematics literacy curriculum (see below) explicitly specifies the contexts in which mathematical knowledge is to be developed or applied. Mainstream curricula do so only implicitly (Dowling, 1998).

Different versions of mathematical modelling in the classroom imply variations of classification. If the situation chosen to be modelled is selected because of mathematical reasons, the external classification might still be strong whereas the internal classification becomes weaker as a mix of different mathematical topics and procedures is legitimate. If, in contrast, the situation chosen for a modelling activity is selected because of social reasons, then the external (as well as the internal) classification might be rather weak.

It can be argued that the two modes of school mathematical modelling, which are different in terms of the relationships between the knowledge domains involved, relate to differential access to mathematical knowledge (see Jablonka & Gellert, 2011). If modelling is not subordinated to the principles of school mathematics, then the question arises to the principles of which discourse it refers. As mathematical modelling is not a uniform practice, but a set of interrelated activities in different domains, there is no set of uniform criteria for performing mathematical modelling. Consequently, the discourse of school mathematical modelling, if it is not subordinated to accessing mathematical knowledge, leaves an open space for promoting different agendas, such as developing human capital by channelling students into an engineering career pipeline, expressing and rethinking cultural identity, educating critical consumers or promoting social change (see the section below on critical mathematics literacy).

When mathematical modelling is seen as a way to promote "curricular mathematics" (cf., Zbiek & Conner, 2006), then it hardly can be regarded as a resistant position towards the mainstream. In fact, the implementation of conceptions like RME demonstrates how well established the focus on mathematical models and progressive mathematisation already is. In contrast, the focus on skills appropriate to obtaining a mathematically productive outcome for a problem with genuine real-world connections defines a resistant position to a curriculum structured by mathematical domains. This resistance is tactical when it (1) aims to complement rather than to overcome mainstream mathematical education practice and, simultaneously, (2) does not question the mathematical structure of the mathematics curriculum by imposing an order that takes the out-of-school problems to be modelled into account. There is no serious intention to deconstruct or subvert the mainstream mathematics curriculum.

The conceptualisation of modelling as a set of generic competencies that could be provided by mathematics education seemingly transcends the difficulties arising from cultural differences and economic inequalities because the activity of constructing mathematical models, through which these competencies are to be developed, is not seen as culture-bound and value-driven. Such a conception masks the fact that the construction of mathematical models depends on the perception of what the problem to be solved with the help of mathematics consists of and what counts as a solution. But depending on the subject position of the "modeller" in a practice, there are different models of the same problematic situation:

> For example, if the problem of a bank employee, who has to advise a client (aided by a software package), is the comparison of financing offers for a mortgage, for the manager of the bank this is a problem of profitability, and for the customer it is one of planning her personal finances. (Jablonka, 2007, p. 193)

This is not to suggest that mathematical models should be scrutinised exclusively in terms of the values connected with the underlying interests. But the discourse of mathematical modelling as providing individuals with generic competencies that enable them to become adaptive to the conditions of technological development, to overcome the limitations of specialised knowledge, to gain competitive advantage on the labour market, and become critical consumers and democratic citizens, is mythologizing mathematical modelling because the causality between participating in mathematical modelling activities and the diverse educational potentials attributed to this experience is mythical. The myth embodies the claim of the ethical neutrality of mathematical modelling practices.

The popularity of modelling can be explained by the fact that it achieves a fictitious marriage between two strands of critique of a strongly classified mathematics curriculum. Such critique is on the one hand an outcome of an attack on a neo-conservative defence of canons of disciplinary specialised knowledge, which (at least historically) comes together with the reproduction of inequality of access to such specialised knowledge. On the other hand, the critique of strongly classified curricular knowledge comes from the side of those called "technical instrumentalists" by Moore and Young (2001) who advertise economic goals. Preparation for the "knowledge-based economy" is a major concern. Moore and Young observe that the scope of instrumentalism has extended from vocational training to general education under the guise of promoting the employability of all students. There is a danger that the myth of the neutrality of generic modelling skills discards the tension between neo-liberal ideology with a focus on human capital preparation and a conception of education for social change.

CRITICAL MATHEMATICS LITERACY

Critical mathematics literacy is here used as an umbrella term that includes conceptions that aim at identifying and analysing critical features of social realities and at contributing to the development of social justice. One strategy of pursuing these goals is sensitising students to social problems and helping them to articulate their interests as citizens. These social problems include the particular hidden injustice students face because of their race, social class, cultural origin etc. Mathematics could then be one of their resources for political action. A second strategy is directed towards the analysis of mathematics itself because of its function as part of technology, including social technology. Critical mathematics literacy is also concerned with the discriminative practice of mathematics education itself: How does mathematics education reproduce or reinforce social inequalities? (For a discussion of different strategies see Jablonka, 2003; Skovsmose & Nielsen, 1996).

Published experience with critical mathematics literacy in (most often) secondary schooling has mainly focused on two features. On the one hand, critical mathematics literacy is strongly connected to the construction and use of data and statistical diagrams. Examples include a discussion of a "race & recess chart" (Powell & Brantlinger, 2008) and "supposedly random traffic stops" (Gutstein, 2008). On the other hand, critical mathematics literacy is directed at the official use and interpretation of socially relevant data in form of quantitative arguments. Examples include analysing the "discounting of Iraqi deaths" (Greer, 2008) and the ways numerical information can be presented in order to augment or reduce its comprehensibility (Frankenstein, 2008).

The subversive rather than oppositional deconstructive resistant position of critical mathematics literacy is apparent as critical mathematics literacy explicitly aims at demolishing the correlation between social class, race, and academic achievement by demystifying the "naturalness" of this relation (Martin, 2010). It is subversive because it aims at eroding and undermining hidden principles of school mathematics instruction and social stratification. These principles serve to perpetuate the hierarchical structure of society and societies. Critical mathematics literacy scrutinizes the mechanisms by which race and social class structures are reinforced.

The examples presented above point to a common problématique: Critical mathematics literacy intends to be simultaneously a pedagogy of access and a pedagogy of dissent (McLaren, 1997; Morrell, 2007). This includes access to higher education, to rewarding professional employment, and to civic life, particularly for marginalized populations, though access might also be understood in terms of personal and social emancipation. However, advanced mathematics literacy does not automatically translate into power, and it does not translate into power equally for everyone who possesses it. In a pedagogy of dissent students develop a language of critique of systems of social reproduction and of inequitable power relations in society. They critically analyse the role that mathematics and mathematics education play in legitimating and perpetuating these conditions aiming at spaces for transformative action. Is this *simultaneously* possible?

Eric Gutstein has worked in a setting characterised by a separation of pedagogies of access and dissent:

> The class I refer to here has intermittently completed social justice mathematics projects since the week they started school. Although we have only spent perhaps 15% of our total time, on three or four projects a year, they have been evidently been sufficient meaningful and memorable to students that none reported it as unusual to hear that particular framing of mathematics. (Gutstein, 2008, p.15)

Though the intention is to reverse the proportion of "standards-based mathematics" and "social justice mathematics" (p. 18), what students learn and develop in both cases is structurally different. The discourse of a mathematics pedagogy of dissent is necessarily weakly classified. Within the Freirean students' generative themes, which are the focus within a pedagogy of dissent, the mathematics might play a crucial role yet not in a very simple and visible way (Jablonka & Gellert, 2007). For the prominent case of "random traffic stops" (Gutstein, 2008), Dowling (2010, p. 4) argues:

> One might suppose that police are often not able to estimate the ethnicity of a driver until after they have made the stop. This would seem to suggest that, if there is a correlation between ethnicity and the probability of being stopped, then we might look for the presence of intervening variables for an explanation; a correlation between ethnicity and relative poverty and the association of the latter with the use of elderly and poorly maintained vehicles having visible defects, for example.

The traffic stop problem is apparently much more difficult and not easy to grasp mathematically, as most real life issues are not governed by single causes and single effects. Simplifying reality is a strategy to allow for mathematical application, but it is crucial to recognize the risks inherent in "dangerously naïve" problems (de Freitas, 2008, p. 87) if critical mathematics literacy is at stake. As a consequence of the simplification, the relation between the weakly classified social justice issue and the strongly classified relevant mathematical knowledge is obscured. In fact, students may get used to handling a piece of legitimate school mathematics – expected value in random experiments – but this at the risk of misjudging the relationship between mathematics, social structure, and social technology. In terms of a pedagogy of dissent, the dissent is only constructed towards a critique of societal power relations, but not towards the role mathematics plays in formatting these power relations. In terms of access, access is given to applications of the mathematical concept of expected value, though this in a context only marginally relevant for professional promotion or academic success.

Examples of a strategy that is directed towards the analysis of social technologies that involve more or less visible mathematical models are provided by Skovsmose (1994). He reports from project work that aimed at developing awareness of economic relationships in the world of a child through discussing children's financial transactions and through analysing and critiquing social policies that distribute money to families based on their size. A critique of mathematical practices that affect political decisions and economic living conditions is, in general, relying on sophisticated methodological considerations and on specialised domain-specific knowledge, which students and their mathematics teachers might not possess. This seriously draws into attention the tension between access and critique, as it might not be possible to develop such knowledge in the context of a school classroom.

Turner (2003) reports from a mathematics project, in which urban middle school students expanded their understanding of some mathematics related to measurement and ratios when trying to argue that their school was significantly more crowded than a comparable middle school in the same district. Based on some of the students' suggestions, she states that mathematics might not always provide the most powerful tools for arguing:

> … and Joel's idea of creating a video, which would have drawn on visual images rather than quantitative data to convince the district of the space problems at the school, would have been more powerful. Perhaps photographs of students struggling to pass through hallways, or video of students running into the poles in the gym during a basketball game would have better communicated the urgency of the problem. (Turner, 2003, p. 235)

If critical mathematics literacy were only a pedagogy of dissent, then the students' viewing the world with a critical mind set could also challenge their participation in the mathematical activities during such a project.

TOWARDS A "RADICAL CONSERVATIVE PEDAGOGY" IN MATHEMATICS EDUCATION?

According to Bernstein's (1990) characterisation of a conservative pedagogy, a traditional strongly classified mathematics curriculum that establishes an explicit hierarchical relationship between teacher and students and includes explicit sequencing rules, as well as explicit specific criteria, is an example of a realisation of such pedagogy. It is underpinned by a theory of instruction that focuses on intra-individual changes in terms of individual's competences or performances rather than on changes in the relation between social groups. Consequently, it highlights neither shared competencies nor the sharing of experiences.

Inquiry-based mathematics and mathematical modelling do not solve the problématique of providing alternative conceptions that relate to social justice for a socially marginalized student population. As Bernstein argues, this is particularly due to focussing on the mathematical capability and development of the individual student. He also deconstructs "progressive" pedagogy because of its differential effect stemming from the implicitness of the recontextualisation principle, which makes invisible the classificatory principle of the knowledge to be acquired.

Ethnomathematics and critical mathematics literacy explicitly focus on groups of marginalized students and look for the empowerment of social groups. However, the tension between, on the one hand, the students' generative themes or cultural heritage, and, on the other hand, an institutionally valorised mathematics can only partly be mitigated by these curriculum conceptions. However, the strategy of establishing engagement for these groups of students through starting with their generative themes might indeed help to change their views of the value of participating in a school mathematics classroom.

A further position of subversive resistance has been theoretically argued by Martin (1993) and outlined by Bourne (2004). Both draw on Bernstein (1990) who sketches "an apparently conservative pedagogy yet to be realized" (p. 214). In a radical realisation of a conservative pedagogy the emphasis is on "the *explicit* effective ordering of the discourse to be acquired" (p. 214). As

Bourne (2004) demonstrates in a case of literacy teaching, by establishing an overtly highly regulated discourse the teacher successfully inducts students to valued and powerful new discursive opportunities and, at the same time, coordinates the everyday discourse that students are familiar with. By managing changes in place, pace, and deportment, the teacher makes the strong classification of school and community knowledge visible. As Bourne (p. 65) remarks: "Visible pedagogy is explicit in acknowledging responsibility for taking up a position of authority; invisible pedagogy (whether progressive or 'emancipatory') simply masks the inescapable authority of the teacher." Bourne's literacy teacher is practising a strictly authoritative teaching method that helps students to realise and internalise the rationales underlying classroom rules and to operate within these rules on their initiative (Hughes, 2002; McCaslin & Good, 1992).

The example from Turner's (2003) study (above) indicates that a focus on mathematical investigations (instead of producing a video, for example) only can be achieved through the teacher's strict ordering of the discourse. The teacher's intention to show the power of mathematics as a resource must be explicit. If any outcome in terms of dealing with certain mathematical domains has to be achieved, the space for the students necessarily needs to be limited. Otherwise it is only by fortunate coincidence that in the course of a project in order to proceed students need access to specific mathematical understandings and skills that they have previously developed. This points to the interdependence of framing in relation to the criteria for legitimate contributions and the classificatory principle of the knowledge. Strong external knowledge classification can only be maintained through strong framing values in relation to the selection and sequencing of the discourse. It should not go unnoticed that the teacher's responsibility for the explicit ordering of the discourse does not exclude a weak framing of the hierarchical rules, that is the creation of rather open interaction relations between teacher and students and between the students themselves (Morais & Neves, 2009).

A radical realisation of a conservative pedagogy highlights shared competences and stresses that the acquirer is *active* in decoding and regulating a necessarily recontextualised practice. In a radical conservative pedagogy the students collectively access and participate in academically valued social practices and get introduced and used to the discourses by which academically valued practices are constituted. This would lead to acquire insights into the discourses that are the focus of critique and has the potential to reconcile a pedagogy of dissent with a pedagogy of access.

ACKNOWLEDGEMENT

We are indebted to Rico Gutstein and Kate Le Roux who, in their function as reactors to our presentation at the 6th Mathematics Education and Society conference, stimulated a critical discussion of the issues. This contribution is an elaborated version of our presentation.

REFERENCES

Bernstein, B. (1990). *The structuring of pedagogic discourse*. London, UK: Routledge.

Bernstein, B. (1996). *Pedagogy, symbolic control and identity: theory, research, critique*. London, UK: Taylor & Francis.

Bourne, J. (2004). Framing talk: Towards a "radical visible pedagogy". In J. Muller, B. Davies & A. Morais (Eds.), *Reading Bernstein, researching Bernstein* (pp. 61–74). London: RoutledgeFalmer.

Bruner, J. (1960). *The process of education*. Cambridge, MA: Harvard University Press.

Chouliaraki, L. (1996). Regulative practices in a "progressivist" classroom: "Good habits" as a disciplinary technology. *Language and Education, 10*(2&3), 103–118.

Clayton, M. (1999). What skills does mathematics education need to provide? In C. Hoyles, C. Morgan & G. Woodhouse (Eds.), *Rethinking the mathematics curriculum* (pp. 22–28). London: Falmer Press.

Cooper, B., & Dunne, M. (2000). *Assessing children's mathematical knowledge: social class, sex and problem solving*. Buckingham, UK: Open University Press.

D'Ambrosio, U. (1985). Ethnomathematics and its place in the history and pedagogy of mathematics. *For the Learning of Mathematics, 5*(1), 44–48.

D'Ambrosio, U. (2007). Peace, social justice and ethnomathematics. In B. Sriraman (Ed.), *International perspectives on social justice in mathematics education* (pp. 25–34). Monograph of The Montana Enthusiast. Montana: The University of Montana and The Montana Council of Teachers of Mathematics.

de Freitas, E. (2008). Critical mathematics education: recognizing the ethical dimension of problem solving. *International Electronic Journal of Mathematics Education, 3*(2), 79–95.

Dowling, P. (1998). *The sociology of mathematics education: mathematical myths/pedagogical texts*. London: RoutledgeFalmer.

Dowling, P. (2009). *Sociology as method: departures from the forensics of culture, text, and knowledge*. Rotterdam: Sense Publishers.

Dowling, P. (2010). Abandoning mathematics and hard labour in schools: a new sociology of knowledge and curriculum reform. In C. Bergsten, E. Jablonka & T. Wedege (Eds.), *Mathematics and mathematics education: cultural and social dimensions. Proceedings of the Seventh Mathematics Education Research Seminar* (pp. 1–30). Linköping, Sweden: SMDF.

English, L. D. (2006). Mathematical modeling in the primary school: Children's construction of a consumer guide. *Educational Studies in Mathematics, 63*(3), 303–323.

Fantinato, M. C. (2008). Teachers' practice under the ethnomathematical perspective: A study case in young and adult education. In J. F. Matos, P. Valero, & K. Yasukawa (Eds.), *Proceedings of the Fifth International Mathematics Education and Society Conference* (pp. 228–236). Lisbon, Portugal: Centro de Investigação, Universidade de Lisboa.

Fioriti, G., & Gorgorió, N. (2006). Mathematical challenge to widen excluded students' foreground. Paper presented at *ICMI Study 16: Challenging mathematics in and beyond the classroom*. Retrieved from http://www.amt.canberry.edu.au/icmis16pargfioriti.pdf.

Foshay, A. W. (1961). Knowledge and the structure of disciplines. In W. A. Jenkins (Ed.), *The nature of knowledge. Implications for the education of teachers. An invitational conference sponsored under a grant from the Uhrig Foundation, November 16–18, 1961* (pp. 28–41). Milwaukee, WI: University of Wisconsin-Milwaukee.

Frankenstein, M. (2008). Quantitative forms in arguments. In J. F. Matos, P. Valero & K. Yasukawa (Eds.), *Proceedings of the Fifth International Mathematics Education and Society Conference* (pp. 261–271). Lisbon, Portugal: Centro de Investigação, Universidade de Lisboa.

Freire, P. (1971). *Pedagogy of the oppressed*. New York: Herder and Herder.

Galbraith, P., Stillman, G., & Brown, J. (2006). Identifying key transition activities for enhanced engagement in mathematical modeling. In P. Grootenboer, R. Zevenbergen, & M. Chinnappan (Eds.), *Identities, cultures and learning spaces. Proceedings of the 29th annual conference of the*

Mathematics Education Research Group of Australasia (pp. 237–245). Adelaide, Australia: MERGA.

Gellert, U., & Jablonka, E. (2009). "I am not talking about reality": word problems and the intricacies of producing legitimate text. In L. Verschaffel, B. Greer, W. Van Dooren, & S. Mukhopadhyay (Eds.), *Words and worlds: Modelling verbal descriptions of situations* (pp. 39–53). Rotterdam: Sense Publishers.

Gerdes, P. (1990). *Ethnogeometrie. Kulturanthropologische Beiträge zur Genese und Didaktik der Geometrie.* Bad Salzdetfurth, Germany: Franzbecker.

Gerdes, P. (1998). On culture and mathematics teacher education. *Journal of Mathematics Teacher Education, 1*(1), 33–53.

Greer, B. (2008). Discounting Iraqi deaths: A societal and educational disgrace. In J. F. Matos, P. Valero, & K. Yasukawa (Eds.), *Proceedings of the Fifth International Mathematics Education and Society Conference* (pp. 293–301). Lisbon, Portugal: Centro de Investigação, Universidade de Lisboa.

Gutstein, E. (2008). Reinventing Freire: Mathematics education for social transformation. In J. F. Matos, P. Valero, & K. Yasukawa (Eds.), *Proceedings of the Fifth International Mathematics Education and Society Conference* (pp. 9–24). Lisbon, Portugal: Centro de Investigação, Universidade de Lisboa.

Hasan, R. (2001). The ontogenesis of decontextualised language: Some achievements of classification and framing. In A. Morais, I. Neves, B. Davies, & H. Daniels (Eds.), *Towards a sociology of pedagogy: the contribution of Basil Bernstein to research* (pp. 47–79). New York: Peter Lang.

Hughes, J.N. (2002). Authoritative teaching: Tipping the balance in favor of school versus peer effect. *Journal of School Psychology, 40*(6), 485–492.

Jablonka, E. (2003). Mathematical literacy. In A. J. Bishop, M. A. Clements, C. Keitel, J. Kilpatrick, & F. K. S. Leung (Eds.), *Second international handbook of mathematics education* (pp. 75–102). Dordrecht, The Netherlands: Kluwer.

Jablonka, E. (2007). The relevance of modelling and applications: Relevant to whom and for what purpose? In W. Blum, P. Galbraith, H.-W. Henn, & M. Niss (Eds.), *Modelling and applications in mathematics education: The 14th ICMI Study* (pp. 193–200). Berlin: Springer.

Jablonka, E., & Gellert, U. (2007). Mathematisation – demathematisation. In U. Gellert & E. Jablonka (Eds.), *Mathematisation and demathematisation: Social, philosophical, and educational ramifications* (pp. 1–18). Rotterdam: Sense Publishers.

Jablonka, E., & Gellert, U. (2011). Equity concerns about mathematical modelling. In B. Atweh, M. Graven, W. Secada & P. Valero (Eds.), *Mapping equity and quality in mathematics education* (pp. 223–238). New York: Springer.

Jama Musse, J. (1999). The role of ethnomathematics in mathematics education: Cases from the Horn of Africa. *Zentralblatt für Didaktik der Mathematik, 31*(3), 92–95.

Julie, C. (2002). The activity system of school-teaching mathematics and mathematical modelling. *For the Learning of Mathematics, 22*(3), 29–37.

Kajander, A., & Lovric, M. (2005). Transition from secondary to tertiary mathematics: McCaster University experience. *International Journal of Mathematical Education in Science and Technology, 36*(2–3), 149–160.

Kaplan (2003, February 10). *Taprats: computer-generated Islamic star patterns.* Retrieved from http://www.cgl.uwaterloo.ca/~csk/washington/ taprats/

Khuzwayo, H.B. (2005). A history of mathematics education research in South Africa: The apartheid years. In R. Vithal, J. Adler & C. Keitel (Eds.), *Researching mathematics education in South Africa: Perspectives, practices and possibilities* (pp. 307–327). Cape Town, South Africa: The Human Sciences Research Council.

Knijnik, G. (2000). Cultural diversity, landless people and political struggles. In A. Ahmed, J. M. Kraemer & H. Williams (Eds.), *Cultural diversity in mathematics (education): CIEAEM 51* (pp. 31–39). Chichester, UK: Horwood.

Lubienski, S. T. (2000). A clash of social class cultures? Students' experiences in a discussion-intensive seventh-grade mathematics classroom. *The Elementary School Journal, 100*(4), 377–403.

Martin, D. B. (2010). Not-so-strange bedfellows: Racial projects and the mathematics education enterprise. In U. Gellert, E. Jablonka & C. Morgan (Eds.) *Proceedings of the Sixth Mathematics Education and Society Conference*. Berlin, Germany: Freie Universität Berlin.

Martin, J. M. (1993). Genre and literacy: modelling context in educational linguistics. *Annual Review of Applied Linguistics, 13*, 141–172.

McCaslin, M., & Good, T. L. (1992). Compliant cognition: The misalliance of management and instructional goals in current school reform. *Educational Researcher, 21*(3), 4–17.

McLaren, P. (1997). Revolutionary multiculturalism: Pedagogies of dissent for the new millenium. Boulder, CO: Westview.

Moore, R., & Young, M. (2001). Knowledge and the curriculum in the sociology of education: towards a reconceptualisation. *British Journal of Sociology of Education, 22*(4), 445–461.

Morais, A., & Miranda, C. (1996). Understanding teachers' evaluative criteria: A condition for success in science classes. *Journal of Research in Science Teaching, 33*(6), 601–624.

Morais, A., & Neves, I. (2009). Textos e contextos educativos que promovem aprendizagem: Optimização de um modelo de prática pedagógica. *Revista Portuguesa de Educação, 22*(1), 5–28.

Morrell, E. (2007). Critical literacy and popular culture in urban education: Towards a pedagogy of access and dissent. In C. Clark, & M. Blackburn (Eds.), *Working with/in the local: New directions in literacy research for political action* (pp. 235–254). New York: Peter Lang.

Oakes, J. (1990). Opportunities, achievement, and choice: women and minorities in science and mathematics. *Review of Research in Education, 16*, 153–222.

Powell, A. B., & Brantlinger, A. (2008). A pluralistic view of critical mathematics. In J. F. Matos, P. Valero, & K. Yasukawa (Eds.), *Proceedings of the Fifth International Mathematics Education and Society Conference* (pp. 424–433). Lisbon, Portugal: Centro de Investigação, Universidade de Lisboa.

Scott, P. (Ed.) (1985). Ethnomathematics: what might it be? *Newsletter of the ISGEm, 1*(1). Retrieved from http://web.nmsu.edu/~pscott/isgem11.htm

Skovsmose, O. (1994). *Towards a philosophy of critical mathematics education*. Dordrecht, The Netherlands: Kluwer.

Skovsmose, O., & Nielsen, L. (1996). Critical mathematics education. In A. J. Bishop, K. Clements, C. Keitel, J. Kilpatrick, & C. Laborde (Eds.), *International handbook of mathematics education* (pp. 1257–1288). Dordrecht, The Netherlands: Kluwer.

Tanner, D., & Tanner, L. N. (1980). *Curriculum development: theory into practice* (2nd ed.). New York: Macmillan.

Teese, R. (2000). *Academic success and social power: Examinations and inequality*. Melbourne, Australia: Melbourne University Press.

Treffers, A., & Goffree, F. (1985). Rational analysis of realistic mathematics education – the Wiskobas program. In L. Streefland (Ed.), *Proceedings of the Ninth International Conference for the Psychology of Mathematics Education, Vol. II* (pp. 97–121). Utrecht, The Netherlands: OW&OC, Utrecht University.

Turner, E. E. (2003). *Critical mathematical agency: urban middle school students engage in significant mathematics to understand, critique, and act upon their world.* Unpublished Doctoral Dissertation. Faculty of the Graduate School of The University of Texas at Austin.

Valedes, J. R. (2002). The influence of social capital on mathematics course selection by Latino high school students. *Hispanic Journal of Behavioral Sciences, 24*(3), 319–339.

Van den Heuvel-Panhuizen, M. (2003). The didactical use of models in realistic mathematics education: An example from a longitudinal trajectory on percentage. *Educational Studies in Mathematics, 54*(1), 9–35.

Vithal, R., & Skovsmose, O. (1997). The end of innocence: A critique of "ethnomathematics". *Educational Studies in Mathematics, 34*(2), 131–158.

Zbiek, R. M., & Conner, A. (2006). Beyond motivation: Exploring mathematical modeling as a context for deepening students' understandings of curricular mathematics. *Educational Studies in Mathematics, 63*(1), 98–112.

Eva Jablonka
Luleå Tekniska Universitet
Sweden

Uwe Gellert
Freie Universität Berlin
Germany

CHAPTER 14

A PHILOSOPHICAL PERSPECTIVE ON CONTEXTUALISATIONS IN MATHEMATICS EDUCATION

ANNICA ANDERSSON AND OLE RAVN

INTRODUCTION

> It is because you teachers walk up front and say: This is how it is – and here you have the book. That's how it has always been since ages back. It's the book that counts, and that is how it is. It's hard to break that pattern. One has always had the book to refer too… (Katja[1], a 16-year-old upper secondary student sharing her experiences of mathematics education in an interview with Annica)

Mathematics is often taught as a subject that can be presented in a clear structure and abstract form within the confines of a textbook resembling what Cobb, Wood, Yackel, and McNeal (1992) referred to as the "school mathematics tradition". Katja quoted above is, thus, probably not unique in her description of her experiences in mathematics education. Because of the special characteristics of mathematics there might not initially seem to be any obvious reason for introducing students to the troublesome complexities of contextualisation where mathematics is, for example, used in a practical setting, a cross-disciplinary project or the like. Learning the core skills of mathematical calculations can be more than enough of a challenge for both teachers and students.

Sometimes, however, serious attempts are made to contextualise the mathematics to be learned and this is frequently done for motivational reasons (for an example, see Boaler, 1993). Students in upper secondary mathematics classrooms (at least in the Swedish context) are usually invited, through textbook problem-solving exercises, to reflect on more or less real-life situations where the mathematics learned can be applied to a predefined task. This is one way to contextualise mathematics in a school setting. But other forms with other purposes are also tried out in practice from time to time.

In this chapter, we discuss the contextualisation of mathematics[2] in a school setting, focusing on upper secondary students in Sweden that are faced with different types of contextualisations in mathematics education. In the following we explain our approach to, and perspective on, this discussion.

O. Skovsmose and B. Greer (eds.), Opening the Cage: Critique and Politics of Mathematics Education, 309–324.

"Context" in mathematics education

There is a rich body of research conducted on context/ contextualising/ contextualisation of mathematics education from a number of different perspectives. To give some examples, first the ethnomathematical movement (D'Ambrosio, 1985, 2006) should be mentioned, that strongly advocates the importance of cultural contexts in mathematics education, understanding mathematics in cultural context as mathematics "which is practiced among identifiable cultural groups such as national-tribal societies, labour groups, children of a certain age bracket, professional classes, and so on" (D'Ambrosio, 1985, p. 45)[3].

Second, a societally oriented research movement is connecting mathematics with societal contexts. Critical mathematics education (see e.g. Gutstein, 2006; Skovsmose, 2005) and the work by researchers like Atweh and Brady (2009), Frankenstein (2009), Valero (2007), Vithal (2003) and many others are examples of researchers advocating for taking such concepts as equity, social justice, empowerment, and democracy seriously in mathematics education. Research presented at the Mathematics Education in Society (MES) biennial conferences foregrounds mathematics in society from different points of view.

In this chapter, we attempt to add a philosophical argument to the many different discussions relating to context and contextualisation in mathematics education. The idea we bring to the fore is not to see the issue of using contextual situations in mathematics education from either an ethnomathematical perspective or from a critical mathematics education perspective. Instead we argue within a philosophical framework for the relevance of contextualising school mathematics education beyond the confines of textbooks. Through a philosophical theoretical perspective, and a case study highlighting the impact of different types of contextualisation techniques in teaching, we build an argument that emphasises the importance of connecting the teaching and learning of mathematics with the problems, visions, conflicts, developments, and different spheres of life beyond the mathematics textbook.

This study is, therefore, drawing on the philosophy of mathematics education, with the particular aim of producing an interpretation of how we can understand the notions of the "core" of mathematics and the "context" of mathematics in relation to mathematics education.

The aim of the chapter

The situation in Sweden, as in many countries, is that students need to take compulsory mathematics courses in connection to their major subject and their choice of study path in upper secondary school. In these required courses, mathematics is not necessarily taught as a subject that clearly involves the surrounding world of affairs or has any close relation to other sciences or subjects in the education programmes. In general, mathematics in Sweden is studied as a subject rarely connected to the students' everyday life (see, e.g., Johansson, 2006) and this rather radical choice of content in mathematics education is what we will discuss.

In this chapter, we aim for a conceptual clarification of why there are no paramount arguments to support the idea that mathematics education is best taught in isolation by being focused on abstract and isolated syntactical approaches and problem-puzzle contextualisation assignments. Instead, we argue, drawing on a human-centred conception of mathematics, that a strong argument can be made for using student-centred contextualised approaches in mathematics education. Not because it may provide a strong motivational force but because it is in line with the nature of mathematics to do so.

After the philosophical, theoretical discussion we focus on a case study relating to two different types of contextualisation approaches in an upper secondary school setting for social science students in Sweden. The first type discusses contextualisations indirectly proposed through a particular mathematics textbook used in the social science students' mathematics course. The second type relates to examples of a student-centred approach to contextualisation in the same mathematics classroom. Building on this case study, we finally discuss how a student-centred contextualisation of mathematics can, in different respects, be advantageous considering the philosophical standpoint developed in the chapter.

THE "CORE" AND THE "CONTEXT" OF MATHEMATICS

In Western philosophy of mathematics there has been a strong tendency to understand mathematics as consisting of objects that can be studied in a parallel manner to the objects of, for example, physics or chemistry. In relation to these special mathematical objects the idea has been pursued – first by Euclid and his contemporaries and, after Euclid, many other mathematicians – that there is a kernel of mathematical knowledge that is fundamental and that only exists in its abstract form. Western mathematics has been preoccupied with axiomatic mathematics in which mathematics is presented as orderly and isolated from the doings of human affairs and any practical use of mathematics. Instead, especially in the European history of ideas, mathematics has been associated with the building blocks of the universe or the principles of nature as well as the universal and logical structures of human reasoning (we refer to Skovsmose and Ravn (2011) for an in-depth development of this argument).

In mathematics education today this persisting ideology about mathematics naturally has implications for the skills that students are expected to learn in the mathematics classrooms. Mathematics is first and foremost centred on abstract reasoning and what we could refer to as the "core" of mathematics, and only secondarily related to practical life and practical problems – what could be referred to as the "context" of mathematics. In education, the "core" is associated with assignments and problems with clear-cut answers and with no fluffy sources of error or lack of information or the like in the mathematical syntax or the problem solving. As Boaler (1993) wrote, historically mathematics has been presented as a subject of "absolute truths" with one correct answer to each problem.

As described above, the presentation of mathematics to students as abstract and essentially isolated from practice hinges, in our understanding, on a particular and ancient ideology about what mathematics is, following the tradition established by Euclid and his contemporaries. In parallel, contemporary writers note that an important force in the general conception of mathematics relates to the view that mathematics consists only of computation and formulas:

> A major inertial force holding back radical reform of mathematics education is the simplistic perception of mathematics prevalent among people in general, including politicians and other policymakers. Mathematics is commonly seen as consisting essentially of computation and formulas, yielding exact and infallible answers, without relevance to everyday life, accessible only by experts, and not open to criticism. Indeed, in many respects mathematics is commonly perceived as the antithesis of human activity – mechanical, detached, emotionless, value-free, and morally neutral. (Mukhopadhyay & Greer, 2001, p. 297)

This dominant idea about mathematics is, however, contested by a more human and socio-culturally oriented conception of what mathematics is, and we intend to highlight the characteristics of a particular position of this sort. We relate this human-centred idea about mathematics to the later Wittgenstein's conception of mathematics. Here we only outline some main characteristics of his conception and refer to Ravn and Skovsmose (2007) for a detailed discussion of the human-centred orientation in Wittgenstein's philosophy of mathematics.

Wittgenstein's ideas about mathematics have their foundations in a philosophy of language as outlined in his principal work *Philosophical Investigations* (Wittgenstein, 1983). He continuously sought an explanation for how our words have meaning – how they make sense to us – and he argues that it is only through the practices of human beings that different types of signs like numbers, equations, letters, words, gestures, outbursts, etc. gain meaning. No sign has a meaning in itself, according to Wittgenstein – only through our joint use of signs in what he refers to as "language games" will a sign be given a meaning. A language game can be about cooking, needlework, politics, scientific experiments but also about finding solutions to mathematical equations, adding two numbers, proving a theorem, calculating in tens, and so on. In Wittgenstein's interpretation, mathematics is to be understood as a family of language games sharing family resemblances in a crisscross network of mathematical practices. This, however, does not mean that mathematics is a game in which you can do as you think or like. It characterises mathematics that it is exactly the family of language games where the syntax cannot be reasonably questioned. Instead, Wittgenstein refers to mathematics as the language games in which we decide on the measures that we shall use to measure the world: "What I want to say is: mathematics as such is always measure, not thing measured" (Wittgenstein, 1978, p. 201 [III-75]). Mathematics is understood as the language games whereby we negotiate which measures to use in relation to different practices. It therefore stands on a pedestal in relation to other sciences because it is in mathematics that we define what measures – which ways of calculating, which ways of deducing etc. – to use in many areas of life.

In arguing a language game oriented conception of the nature of mathematics, Wittgenstein destroys any notion of there being any mathematics that necessarily comes before other parts of mathematics – for example, in an abstract axiomatic structure. The notion of a "core" of mathematics thereby becomes problematic as referring to an entity that can firstly be reasoned about and secondarily applied in a particular practice. On the contrary, in the Wittgensteinian conception there is nothing that can be claimed to be the real or true core of mathematics. The extreme order and organisation of mathematical concepts, theorems, proofs, vocabulary, number systems etc. that we know today is a construction that makes us forget that mathematics is in fact a network of language games wherein the symbols have gained meaning from the use of the mathematical signs in different practices, or in Wittgenstein's terms, language games.

In this human-centred understanding, the idea that mathematics exists in itself is therefore rejected. Mathematics is considered a completely human enterprise in which the mathematical symbols have no other meaning than the meaning that different types of mathematical communities – in research, in the classroom, in everyday life – give them, through practice and endless repetition in the diverse settings in which we use mathematics. It is essential for the argument of this chapter that there is nothing in the nature of mathematics that forces upon us the isolation of a sort of "core" of mathematics. Importantly, this human-oriented conception of mathematics also does not advocate the primacy of the "context" of mathematics. It states that these two poles or perspectives from which mathematics can be conceived are, in fact, intertwined and intimately connected and can only secondarily be divided into different domains.

We will refrain from going deeper into the Wittgensteinian arguments for his conception of mathematics and the many philosophical discussions and questions that revolve around his work. For such discussions, we point to the sources that we have used in this section from Wittgenstein's own work and for further insight into the turning point that we believe occurred in Wittgenstein's ideas on mathematics we also refer to Shanker (1987). However, it should be clear from the above that in this human-centred understanding of mathematics it is obvious that the learning of mathematics would necessarily seem to include acquaintance with practices in which mathematics is used. And the practice of learning the syntax of mathematics from this perspective seems to have a very limited scope, teaching students only a fragment of what mathematics is all about. It resembles a learning situation in which cooking or needlework would be taught only through the principles of these crafts in textbooks and never by engaging with the real thing.

In conclusion, a human-centred philosophy of mathematics imposes the importance of another set of educational practices that are directed towards mathematical reasoning in a given practical setting – contextualisations of mathematics – as opposed to a more abstract and isolated teaching of mathematics. In the following section we present a case study from mathematics education comprising two contrasting types of contextualisation approaches in mathematics education at the upper secondary school level. These two approaches are discussed throughout against the background of the philosophical ideas presented above.

A SWEDISH CASE STUDY ON TWO TYPES OF CONTEXTUALISATION

In Sweden, students choose from different study programs when starting upper secondary school around the age of 15. Different theoretical programs provide students with an academic foundation for further studies. There is also a variety of occupational and crafts programs to choose from. In this particular study, we focus on the compulsory Mathematics A course for students who have selected a three-year theoretical social science program. The research presented in this chapter is a cut-out slice of a larger research project by Annica, a mathematics education researcher. The planning and implementation of the described projects was done in collaboration with Elin, a mathematics teacher and mathematics education co-ordinator in the Swedish upper secondary school where the research took place (Andersson, in press; Andersson & Valero, in press-a).

The Mathematics A course covers mathematical contents such as arithmetic calculation skills, geometry, algebra, statistics, and linear functions (Swedish Ministry of Education, 2000). A problem with this particular course is that the mathematical content is, as we will see, not obviously connected to other subjects in the social sciences, even if the intention as expressed in the national mathematics curriculum states otherwise:

> The subject aims at pupils being able to analyse, critically assess, and solve problems in order to be able to independently determine their views on issues important both for themselves and society, covering areas such as ethics and the environment. (Swedish Ministry of Education, 2000, original translation)

Focussing on the goals for the specific Mathematics A course, we read:

> Pupils should be able to formulate, analyse and solve mathematical problems of *importance for everyday life* and *their chosen study orientation.* (Swedish Ministry of Education, 2000, original translation, our emphasis)

Our interpretation of these ministerial guidelines is that mathematics teaching should give students mathematical knowledge and competences for taking well-grounded decisions in everyday life and to interpret the flow of information and thereby follow, understand, and participate in political discussions in society. Specifically in relation to the social science programs, we also understand the curriculum as arguing that mathematics education ought to be deeply connected to, and contextualised in relation to, the social sciences. There are several possible explanations for the Mathematics A course content not being clearly connected to the study programs. In our interpretation, this disconnect hinges on the dominant non-human-oriented idea about what mathematics is, as explained above. From the two ministerial quotations it is clear that the least problem in contextualising mathematics in the Mathematics A course is one of political backing and the laws of government relating to the contextualisation of mathematics.

Working with the ideas inspired by the later Wittgenstein, we now present and discuss two different language games for contextualising the syntax of mathematics in mathematics education in upper secondary schools. First, we address the contextualisation approach used by the students' mathematics textbook.

Second, we address a contextualisation approach in which teachers and students are active in forming a contextualisation.

The school mathematics language game of textbook contextualisation

The mathematics textbook used in the school where this case study took place is one of the five most used textbooks in Swedish upper secondary mathematics education. The textbook is, according to the particular textbook editors, designed in a way that it meets the social science students' needs particularly well (Szabo, Larson, Viklund, & Marklund, 2007). According to the mathematics education co-ordinator, Elin, this claim was one of the main reasons for the school to use this textbook. The book covers the mathematical areas to be addressed in the Mathematics A course in accordance with the Swedish national curriculum of mathematics. The tasks and exercises are on three difficulty levels, mostly related to the three different assessments and grading levels in Swedish mathematics education.

In the Swedish context, mathematics education textbooks are understood as having a unique status (Johansson, 2006). Teachers in Sweden often work close to the textbook. According to Johansson, "Textbooks influence not only *what* kind of tasks students are working with and the examples presented by the teachers but also how mathematics is portrayed in terms of concepts and the features that are related to the subject." This statement supports other findings that the authority in the classroom often resides with the textbook and the teacher (Bishop, 1999), with obvious limitations for the students to influence, and reflect on, their learning of mathematics and mathematics per se – as illustrated by Katja's comments at the very beginning of the chapter. In the first language game of contextualisation by use of the textbook assignments, the contextualised problems illustrate very well what Skovsmose (2001) referred to as "semi-reality problems" in the tradition of exercises. We exemplify this approach by discussing five textbook problems, two from the chapter on "Percentage in society" and three from the "Statistics in society" chapter. These chapters were chosen as their titles indicate a focus on mathematics in society, and thus ought to be the most relevant for social science students and connected to their chosen study program as intended in the curriculum. The problems were picked randomly, thus are neither exceptional in any way, nor different from other problem tasks in the textbook. All five are at the basic level in the book, thus all students are expected to calculate, understand, and complete them.

Percentage in society We start with three examples from the textbook's chapter on "Percentage in society" (Szabo et al., 2007, p. 121) and discuss their content in order to highlight the characteristics of this type of language game of contextualisation.

> 4202. When Anna went to Cyprus on holiday last year she paid 6530 kr. for the trip. This year the same trip would cost her 7200 kr. By how many percent has the price increased? (p. 122, our translation)

This contextualisation presupposes a background in which students are acquainted with travelling and going on holiday trips. First, this might not be the case for a number of students in Swedish society today. Second, who, if buying a holiday trip, would reflect on the percentage increase of the price in relation to the previous year? Was this particular trip with the same airline, dates, and accommodation even available the prior year? The characteristics of this example of contextualisation seems to have a strong orientation towards syntactical reasoning and a minimal context that appears to make the mathematical steps more difficult but does not close the gap between syntax and a use of it in society.

> 4207. Arthur has a pond with goldfish. One morning he discovers that a heron has been there. At the end of the week he has lost 40% of his goldfish and only 54 fish are left. How many fish did Arthur have at the beginning? (p. 123, our translation)

The questions students posed on this task were mostly not mathematical. They rather related to "real reality" or the context of the problem. They wondered: How did Arthur know it was a heron and not another bird or animal? As a teacher it is not possible to answer that question and, as Arthur is figurative, it is not possible to ask him either. The author of the textbook might know the answer, but, as illustrated by Wagner (2010), mathematics textbooks authors' voices are seldom recognized in the mathematics textbooks' content.

There are other questions a critical student might ask: Why didn't Arthur prevent the losses instead of thinking of percentage calculations? How come he does not know how many he had from the beginning when he knows he lost 40%? This example highlights that the notion of "Percentage in society" is taken to its limits and also that the context for percentage calculations can raise many questions from students when it is minimalistic and has underdetermined content.

> 1106. Lasse is helping his uncle selling fruit at the local market. As he doesn't want to calculate so much (ska slippa räkna så mycket) he asks his uncle to complete a table with what the fruit costs. His uncle starts with a table for grapes:

Weight	1 hg	2 hg	3 hg	½ kg	7 hg	1 kg
Price	1,70 kr					17,00 kr

Complete the table for Lasse. (Szabo et al. (2007), p. 10, our translation)

The contextualisation in this task is remarkable from several points of view. First of all, it is worth noting that Lasse's possibly negative feelings for mathematics are highlighted in an exercise in a mathematics education textbook (second sentence).

Either accepting or ignoring this comment in the task, the students continued wondering: Aren't the grapes very cheap? (Yes, they are in a Swedish context). Does anybody want to buy 8 hg? Or 12 hg? Here the teacher has to make a decision. Either (s)he can answer something along the lines of "you only need to complete what is asked for", or alternatively s(he) has an opportunity to challenge the student's mathematical thinking by asking "reflect on why these particular weights were chosen for the table." These are examples of decisions teachers have to take when students question the contexts of problems. A last example of a conversation in the upper secondary classroom: "Do we have to do the exercise for apples and bananas too? No, Lasse only needs a table for grapes, the rest of the fruit he enjoys calculating", and so on.

This example also highlights how many important elements of a practical scenario are left out and/or taken for granted in a short textbook contextualisation. As a result the discussions in class were not about the use of mathematics but about what were interpreted as peculiar circumstances.

Statistics in society The following two examples are from the chapter "Statistics in society" and exemplify problem-solving exercises on mean values and arithmetic averages.

> 5204. A building contractor is going to build a block of rental apartments. He analysed how big apartments the 10 people first in line for new apartments wanted and got the following answers (in number of rooms): 2,4,1,2,4,2,4,2,2,3. Which value is most appropriate to use in this context? Motivate your answer. (p. 155, our translation)

This example is far from 15–16 year old students' personal context. Actually it could be argued that it is far from anybody's context, even that of building constructors.

> 5211. When the students asked their mathematics teacher how old he was, he answered: We are five in the family and our average age is 28 years. If we count me out the average will be 21 years old. How old was the teacher? (p. 156, our translation)

The picture of a mathematics teacher answering a polite question in this, in our view, tricky way can be troublesome. It might have the effect of positioning mathematics teachers in a discourse of "mathematical nerds" which is neither desirable nor relevant in this context. These final two examples in our interpretation show that in order to have the desired syntactical content in a very short contextual assignment the context will tend to be unrealistic and sometimes even tricky. In these two cases it is, for example, not quite obvious why the problems were placed in a chapter on "Statistics in society" as their content only remotely relates to this issue.

All the examples above, in their individual ways, portray the language game of contextualised problem solving in upper secondary mathematics textbooks. It is a language game that we find has some serious flaws in its approach to mathematical practice in different contexts. It works with highly simplified and underdetermined contexts. It does not support a notion of mathematics as a network of language games that is practised in numerous ways in the private, public, or global sphere. It raises many questions in student discussions in relation to the problem context that are seemingly not meant to be raised. However, the skilled student will know how to cut all unnecessary information away from the problem (that is, the context) (as discussed by e.g. Verschaffel, Greer, & de Corte, 2000; Gellert & Jablonka, 2009; Palm, 2009) and thereby focus on, and resolve, the syntactical "core" of the problem.

The point is that the mathematical calculations requested in the textbook problems above are both challenging and relevant, but hiding them in simplistic and underdetermined contexts just gets silly with the result that they are not recognised by the students as either interesting in themselves as mathematical syntactical problems or interesting as a contextualisation of mathematics.

In conclusion, we find that when the contextualisation is developed in the portrayed manner, far removed from both the use of mathematics in society in its uncountable forms and from the context of students' lives, there is no sense in using textbook contextualisation. Following Skovsmose's (2001) vocabulary, they are semi-reality problems and can easily degenerate into non-reality problems as we have seen above.

The school mathematics language game of student-centred contextualisation

In this section we analyse an alternative language game of contextualisation in school mathematics education. We do this by presenting three examples in similar mathematical fields to those discussed above, namely percentage calculation and statistics for upper secondary school. Subsequently, we discuss this alternative approach to contextualisation. The tasks built on the idea that curriculum objectives could be achieved on all assessment and grading levels while, at the same time, the students could be part of establishing the context.

Annica developed the three examples in collaboration with Elin, the teacher of the upper secondary class in question (see Andersson, in press; Andersson & Valero, in press-a, for further details on the background, development, and assessment of the projects). This language game of contextualisations was guided by the idea that a human-centred conception of mathematics implies that the use of mathematical syntax in particular practices should be taken very seriously. Also, Boaler's emphasis on using open problems has been an important source of inspiration. She states that:

> The teaching of content mathematics within a scheme or textbook and the development of isolated process skills within an investigation are insufficient to encourage a deep and genuine mathematical understanding. There is no reason to expect students who have learned isolated strategies and atomised content throughout school to be able to combine, separate, or integrate what they have learned in "real" and demanding situations. Mathematical activities need to be open enough, not only for students to formulate strategies but to formulate their own meaning. Tasks should require that students develop an understanding of the underlying processes and the way that these link with content; in this way students will appreciate and develop an understanding of the interrelationship of the two. (Boaler, 1993, p. 16)

On the basis of the human-centred conception of mathematics and an emphasis on the need for open-ended mathematical activities in demanding situations of practice, the tasks were developed as follows.

Percentage in society The first task was formulated in the following way:

> Making your dreams come true?
>
> Reflect on something you would like to do, experience, or buy, for yourself or others, which costs so much that you need to borrow the money to cover the expenses. You have to find out how much money you require to finance the project and what repayment (including interest) the bank expects you to pay.
>
> We suggest the following: the repayments are made to the bank once a year, and you pay back the loan within five years. If this is not possible for you we will discuss that.
>
> - How much will you be paying back each year in interest? In total over the five years?
> - How much do you need to pay back in total per year?
> - What did the total cost add up to?
> - Was it worth it? Why/why not?

The students worked in pairs or in groups of three in order to facilitate discussions with peers, both about the mathematical content and the critical contextual reflections. The task was to think about what they might like to/need to borrow money for. They came up with ideas ranging from taking a study year abroad, starting a music band, a holiday trip, a trip for doing volunteer work in Africa, buying hifi equipment. They learnt how to find out about loans and interest and actual rates and calculate the cost of the repayments and the total cost. Discussions taken in the classroom are exemplified by: Was it worth it? What different lending schemes are possible and when is it appropriate to choose one kind of loan in preference to another? And what about the "quick loans" which you can get via mobile phone texts or at shops, are they smart? Why/why not?

The second task on percentage calculations was an exercise inspired by the work of Frankenstein (2008). Besides percentage calculations it also required arithmetic calculation skills relating to fractions, division, and whole and decimal numbers. A larger project on Human Rights was going on in the school. A decision was taken to relate to that topic and connect with a mathematical argumentation assignment themed on the United Nations Conventions on the Rights of the Child. The task was developed in the following way:

> **Newspaper posters with mathematical argumentation!**
>
> The task of today is, in small groups, to create a number of newspaper posters that hit people, engage people, arouse curiosity, reflections and/or emotions – with a mathematical content!
>
> The goal is for you to acquire insight into how big the penetrating power of numbers can be in advertisements and newspaper articles. There are 54 articles in "Convention on the Rights of the Child". Choose the one that interests you the most and focus on that specific one. Search and find information addressing the special children in your focus – information you consider important and want all people at the school to know about. You might want to start a debate; it might be positive information, maybe information with facts that the article has not presented – or something else. Reflect on how to present the numbers to get the message on your news poster through in the very best way.

The students found the relevant information from multiple data sources. They creatively composed different ways of showing the numbers – as percentages, fractions, whole numbers etc. Which illustration best covered the issue they wanted to address? For example, does 10 % look more or less than 10 out of 100, 1/10 – or maybe the issue can be illustrated as 90% or in a different way? If the students completed the exercise as intended they would have carried out at least as many exercises as the number of routine exercises in their textbook. The reflective questions posed after this exercise related either to the content of the chosen Children's' Rights article, or how the numbers were best exposed to get the attention of an audience. These reflections related both to creativity in using colours and pictures as well as discussing the power of the displayed numbers.

Statistics in society The third, statistical, task was a further development of a prior project (Andersson & Valero, in press-b), that, in collaboration with the school's environmental subject teacher, was now expanding, on the theme "Ecological footprints we make on earth", to a larger cross-disciplinary subject project (Andersson & Valero, in press-a). The statistical project covered statistical mathematical goals stated in both the Mathematics A and the Mathematics B courses (Swedish Ministry of Education, 2000). Opportunities were also given to reach objectives in environmental science and computer science courses and thus made it possible to work across subjects. The students were invited to conduct a survey. The survey theme was expected to relate to the "ecological footprints" we all make on earth (e.g., Wackernagel & Rees, 1996). Within this frame, students focused on such topics as transport, food, consumption, or energy and narrowed down their focus within one of these. For assessment, they were asked to create a

PowerPoint presentation or poster where they showed the data analysis and their results in appropriate diagrams and tables and with high accuracy. The students were also asked to take a stance in relation to their survey results and argue for it in their diagrams and texts. For that purpose, they were "allowed" to manipulate diagrams (but not fake them) – just as newspaper articles and advertisements do. Within this exercise, they reflected on the possibility of seeing through, or at least being skeptical about, diagrams, tables etc. presented in the media that are typically constructed from a particular political or entrepreneurial perspective that may be quite hard to identify.

DISCUSSION

In the contextualisation examples of the textbook, students were faced with a context that we judge was far from meaningful to them in addressing the issues of their reality from the perspective of mathematics. These tasks were, in our interpretation, built in a way that complicates the syntax of mathematics and which did not succeed in bringing this syntax into connection with issues either of society, as intended in the curriculum, or the students' lives in general. This language game of contextualisation takes as its starting-point the "core" mathematical problem solving, and only secondarily has the goal of constructing a framework that connects the problem solving to real life situations or "context".

In the second language game of contextualisation, the main characteristic is the structuring and formulation of the context by the teacher in a way that ensures that the mathematical topics and syntax to be learned will be in focus while at the same time a complex and relevant context can be the starting-point for doing and learning mathematics. The idea of introducing, for example, percentage calculation within the existing language game of loans and repayments is to teach mathematics in its natural habitat, so to speak, and to intertwine the "core" and the "context" of mathematics. Or, more to the point, to ensure that "core" and "context" become unnecessary terms when working with mathematics.

What are we arguing here, then? That the first language game is superfluous and that we only need the second type in school mathematics? Well, we would say yes, we actually do not need the first type of language game if we adhere to the conception of mathematics presented above. Does this also mean that training in the pure syntax of mathematics is unnecessary? Well, no, it indicates, rather, that focused training of mathematical syntax can be very helpful indeed, but then it should be presented to students as such.

The idea of confusing mathematical syntax with mathematics itself is the mistake that we have tried to use a Wittgensteinian perspective to deconstruct. Mathematics is entangled in the complexities of the other sciences, of the use of mathematics in everyday life and in newspapers, sports events, classrooms and so on. So, what we argue is that a move towards the second type of language game of contextualisations in mathematics education is not a step away from "pure" or "real" mathematics but a step towards a teaching of mathematics that is more in accord with the actual phenomenon of mathematics. Mathematics is not an isolated thing in itself. Mathematical symbols have meaning as a result of what we do with them in various types of practices, and therefore the real and pure focus of mathematics education should be mediated in contextualisations of what we (the students and the teacher) can do with them.

The human-centred conception of mathematics that we have argued here does not stand uncontested. Mathematics can of course be interpreted as an other-dimensional logical unity or even a pure formalism that has no connection to reality at all and much mathematics education is directly or indirectly based on this assumption. Here we argue that this interpretation of mathematics ought to be left behind in order to advance both our conception of mathematics and the teaching and learning of mathematics.

We have tried to present our argument for using more student-centred contextualisations as not being about the motivation of students or the question of equity in society. We have done so, not because motivation and equity are not to be considered extremely important in the education of any topic, but, first and foremost, because we want to emphasise that the nature of mathematics does not force upon us an abstract and isolated approach to the learning of mathematics.

Detached and taught in isolation, mathematics loses many of its attributes as an enormously important part of our society, culture, and science and the students lose their ability to handle complex situations where mathematics is in action. This ability is something entirely different from being able to make correct proofs or develop and solve general systems of equations (the tasks of the professional mathematician). Although governments and authorities often address the ability to handle mathematics in complex situations as a very important dimension for students to learn, we believe it will not be a strong ability in students as long as it is not trained and practised.

In this chapter, we have only started the construction of a second language game of contextualisation by contrasting it with a particular language game of school textbook contextualisations. A lot of work is still needed to develop alternative language games of mathematics education even though many attempts have been made during the years. The educational transition towards this type of alternative language games, however, is unlikely to develop without a general development in the conception of mathematics in which the core and context of mathematics are considered inseparable.

NOTES

1. All names of students in this chapter are pseudonyms.
2. Wedege (1999) distinguishes specifically between situation context and task context in mathematics education. The first meaning of context, "situation context" is by Wedege understood as social and historical matters and relations as "contexts of learning". In this paper we understand context in mathematics as "task context" in Wedege's definition: "In this sense

the word is often normatively employed, e.g. in curriculum documents as a requirement that teaching and materials shall contain 'real-life context' or 'meaningful and authentic contexts' " (pp. 206–207).

3 Andersson (2007) has challenged mathematics education with a mathematics teaching grounded in ethnomathematical theories, understanding social science students as a cultural group in line with D'Ambrosio's definition.

REFERENCES

Andersson, A. (2007). Ett kulturellt besök i matematikundervisning: En aktionsstudie med ett etnomatematiskt perspektiv [A cultural visit in mathematics education: Action research with an ethnomathematical perspective]. Unpublished Masters thesis. Malmö Högskola. Retrieved from http:// dspace.hmah.se/bitstream/handle/2043/5861/Magister2.pdf?sequence = 1

Andersson, A. (in press). A "curling teacher" in mathematics education: Teacher identities and pedagogy development. *Mathematics Education Research Journal.*

Andersson, A., & Valero, P. (in press-a). Mathematics education giving meaning to social science students. A case from Sweden. *Proceedings of II Congreso Internacional de Investigación, Educación y Formación Docente in Medelin, Colombia 26–28 Aug 2009.*

Andersson, A., & Valero, P. (in press-b). Examining a critical pedagogical discourse for agency and social empowerment. In P. Ernest & B. Sriraman (Eds), *Critical mathematics education: Theory and praxis.* Charlotte, NC: Information Age Publishing.

Atweh, B., & Brady, K. (2009). Socially response-able mathematics education: Implications of an ethical approach. *Eurasia Journal of Mathematics, Science & Technology Education, 5*(3), 267–276.

Bishop, A. (1999). Mathematics teaching and values education – an intersection in need of research. *Zentralblatt für Didaktik der Mathematik, 31*(1), 1–4.

Boaler, J. (1993). The role of contexts in the mathematics classroom: Do they make mathematics more real? *For the Learning of Mathematics.* I3(2), 12–17.

Cobb, P., Wood, T., Yackel, E., & McNeal, B. (1992). Characteristics of classroom mathematics traditions: An interactional analysis. *American Educational Research Journal, 29*(3), 573–604.

D'Ambrosio, U. (1985). Ethnomathematics and its place in the history and pedagogy of mathematics. *For the Learning of Mathematics*, *5*(1), 44–48.

D'Ambrosio, U. (2006). *Ethnomathematics. Link between tradition and modernity.* Rotterdam: Sense Publishers.

Frankenstein, M. (2008). Quantitative form in arguments. In J. F. Matos, P. Valero, & K. Yasukawa (Eds). *Proceedings of the Fifth International Mathematics Education and Society Conference.* Lisbon: Universidad de Lisboa/ Aalborg: Department of Education, Learning and Philosophy, Aalborg University.

Frankenstein, M. (2009). Developing a criticalmathematical numeracy through real real-life word problems. In L. Verschaffel, B. Greer, W. Van Dooren & S. Mukhopadhyay (Eds)., *Words and worlds: Modelling verbal descriptions of situations* (pp. 111–130). Rotterdam: Sense Publishers

Gutstein, E. (2006). *Reading and writing the world with mathematics: Toward a pedagogy for social justice.* New York: Routledge.

Gellert, U., & Jablonka, E. (2009). "I am not talking about reality". In L. Verschaffel, B. Greer, W. Van Dooren, & S. Mukhopadhyay (Eds.), *Words and worlds: Modelling verbal descriptions of situations* (pp. 39–53). Rotterdam: Sense Publishers.

Johansson, M. (2006). Teaching mathematics with textbooks: A classroom and curricular perspective. Doctoral thesis. Luleå University of Technology. Retrieved from http://epubl.ltu.se/1402–1544/2006/23/LTU-DT-0623-SE.pdf

Mukhopadhyay, S., & Greer, B. (2001). Modelling with purpose. Mathematics as a critical tool. In B. Atweh, H. Forgasz & B. Nebres (Eds.), *Sociocultural research in mathematics education. An international perspective* (pp. 295–311). Mahwah, NJ: Lawrence Erlbaum.

Palm, T. (2009). Theory of authentic task situations. In L. Verschaffel, B. Greer, W. Van Dooren & S. Mukhopadhyay (Eds.),. *Words and worlds: Modelling verbal descriptions of situations* (pp. 3–19). Rotterdam: Sense Publishers

Ravn, O., & Skovsmose, O. (2007). Mathematics as measure. In *Revista Brasileira de Historia da Matematica. Brazil: Publicação Oficial da Sociedade Brasileira de História da Matemática* (pp. 143–156.)

Shanker, S. (1987). *Wittgenstein and the turning-point in the philosophy of mathematics*. London: Croom Helm.

Skovsmose, O. (2001). Landscapes of investigation. *Zentralblatt für Didaktik der Mathematik, 33*(4), 123–132.

Skovsmose, O. (2005). *Travelling through education: Uncertainty, mathematics, responsibility*. Rotterdam: Sense Publishers.

Skovsmose, O., & Ravn, O. (2011). *Matematikfilosofi*. Aarhus: Systime.

Swedish Ministry of Education (2000). Goals to aim for in mathematics education in upper secondary school. Retrieved from http://www.skolverket.se

Szabo, A., Larson, N., Viklund, G., & Marklund, M. (2007). *Origo. Matematik kurs A för samhällsvetenskapliga och estetiska program*. Värnamo: Fälth & Hässler.

Valero, P. (2007). A socio-political look at equity in the school organization of mathematics education. *Zentralblatt für Didaktik der Mathematik, 39*(3), 225–233

Verschaffel, L., Greer, B., & De Corte. E. (2000). *Making sense of word problems*. Lisse, The Netherlands: Swets & Zeitlinger.

Vithal, R. (2003). *In search of a pedagogy of conflict and dialogue for mathematics education*. Dordrecht, The Netherlands: Kluwer.

Wackernagel, W., & Rees, E. (1996). *Our ecological footprint. Reducing human impact on the earth*. Gabriola Island: New Society Publishers.

Wagner, D. (2010). The seductive queen - mathematics textbook protagonist. In U. Gellert, E. Jablonka & C. Morgan (Eds), *Proceedings of the Sixth International Mathematics Education and Society Conference* (pp. 438–448). Berlin: Freie Universität Berlin.

Wedege, T. (1999). To know or not to know - mathematics, that is a question of context. *Educational Studies in Mathematics, 39*, 205–227.

Wittgenstein, L. (1978). *Remarks on the foundations of mathematics (3rd edition)*. Oxford: Blackwell.

Wittgenstein, L. (1983). *Philosophical investigations* (2nd edition, first published 1953). Oxford: Blackwell.

Annica Andersson
Aalborg University
Denmark

Ole Ravn
Aalborg University
Denmark

CHAPTER 15

MATHEMATICS EDUCATION AND DEMOCRATIC PARTICIPATION BETWEEN THE CRITICAL AND THE ETHICAL: A SOCIALLY RESPONSE-ABLE APPROACH

BILL ATWEH

In the mathematics education literature, the relationship of mathematics education and democratic participation has been discussed by different authors writing from various perspectives. Here, I have in mind writers who have addressed the issue directly or indirectly through their discussion of, for example, politics of mathematics education (e.g. Mellin-Olsen, 1987), critical mathematics (e.g. Frankenstein, 1983; Skovsmose, 1994), social justice (Gutstein, 2006), ethnomathematics (e.g. D'Ambrosio, 1985; Powell & Frankenstein, 1997), and equity (e.g. Burton, 2003; Secada, 1989). In this chapter I adopt an ethical perspective that, I will argue, complements these approaches by providing tools to deal with three inherent complexities encountered in linking mathematics education and democratic participation, namely: the uncertainty in the relationship, the question of power, and the elusive nature of democratic participation in globalised pluralistic times. These complexities are discussed in the first section of the chapter, followed by a discussion of an ethical approach to mathematics education based on the theorisation of ethics by the French philosopher Emmanuel Levinas. The chapter concludes by outlining an approach to mathematics education that brings the focus on democratic participation to the forefront of decisions on curriculum and pedagogy.

COMPLEXITIES IN THE RELATIONSHIP BETWEEN MATHEMATICS EDUCATION AND DEMOCRATIC PARTICIPATION

The first complexity in the relationship between mathematics education and democratic participation to be discussed here is identified by Skovsmose and Valero (2001). The authors point out that for some writers in mathematics education there appears to be an assumption of *intrinsic resonance* between mathematics and democratic participation – in the sense that more mathematical knowledge directly leads to more democratic participation. At the same time, other authors focus on an *intrinsic dissonance* between mathematics and democratic participation by pointing out that mathematics achievement can act as a "critical filter" or "badge of eligibility" that stands in the way of democratic participation by certain groups that are traditionally excluded from participation and success in mathematics education.

O. Skovsmose and B. Greer (eds.), Opening the Cage: Critique and Politics of Mathematics Education, 325–342.

In rejecting both positions, the authors call for a *critical stance* where the claims about the relationship between mathematics and democratic participation themselves should always be questioned and our practices should always be examined. In particular, Skovsmose and Valero note the increasing focus of many curriculum reform documents, in many countries, on the principle that mathematics achievement can, and should, promote citizenship and democratic ideals of society as a primary aim of mathematics education. However, they warn that:

> … it is important to realise that such aims are without much descriptive value. Even if they guide mathematics curricula, the actual mathematics education may not necessarily support the development of democratic values. Nor have such aims much prescriptive force, since what in fact prescribes the practices of mathematics education is that whole range of external factors considered as a justification for the thesis of dissonance. (p. 44)

In a later work, Skovsmose (2005), based on the writings of D'Ambrosio (1994), notes the critical role of mathematics in society that is, on one hand, intrinsically related to significant advances in knowledge and technology and on the other to most devastating instruments of war and destruction. Skovsmose calls this the "paradox of reason" and asserts that even though there is nothing intrinsically in mathematics that determines its effects, it is in the midst of – and cannot escape from – this paradox. Here I might add the dual effects of mathematics for "empowerment" and "exclusion" as further manifestations of this paradox. Skovsmose goes on to make two points that are essential for the discussion here. Firstly, the "wonders'' and "horrors'' regarding the social effects of mathematical knowledge are often *unpredictable* and *uncertain*; moreover, to add complexity, "we might be lacking any reasonable standards for judging [between them]". (p. 101)

Secondly, he rejects critical rationality as a means of providing the foundation for the necessary critique to deal with the socio-political effects of mathematics – since rationality itself has led to this paradox in the first place. Using the concepts of existential freedom and responsibilities of Sartre, Skovsmose argues that in the face of uncertainty, responsibility is expressed as concerns, and shared and discussed with others, thus forming a "critique without foundation" – in other words a critique that is not based on "logical, philosophical, political nor ethical" grounds (p. 131). In several places in the book Skovsmose presents responsibility as a way to deal with uncertainty; yet stops short of following it to the heart of the discourse on ethics – thus, using Habermas's (1998) terms, he points to the road taken here.

The second complexity in relating mathematics education and democratic participation relates to the necessary politicisation of mathematics education that this relationship implies. The concern here is not that mathematics is objective, value-free, and hence beyond politics. On the contrary, as Mellin-Olsen (1987) argues, mathematics education is political through and through. Among other reasons, it is political because it supports the ideology of objectivity (Bishop, 1998); it is associated with practices of legitimisation of social stratification

(Apple, 1992); and with patterns of colonisation (Powell & Frankenstein, 1997). However, political considerations necessarily raise the question of power. I concur with Ernest (2002) that mathematics can lead to empowerment for active citizenship. Only the cynic can disagree with the often-made claim that certain ability to understand and use mathematics is not only useful but also necessary to make many informed decisions about day-to-day affairs. Following this argument one can safely say the more ability to deal with mathematical situations somebody has the more access they have to social power. However, social power is not unproblematic. As Simmons (1999, p. 97) points out, that power "unbounded may lead to tyranny, absolute power of the strongest". Thus, increasing the capability to deal with mathematical situations might enhance civic participation of an individual, but it may also lead to increasing personal gain at the expense of the public good and, at worst, to domination of others and reduction of their opportunity for meaningful participation. Hence, relating mathematics education to democratic participation for the public good requires other considerations in order to keep democratic participation under check. Traditional views of mathematics as a system of knowledge and truths, and of mathematics education as a set of statements about desired content and means of its development, fail to provide such mechanisms. Such a role necessarily involves social values and ethical judgements.

The third complexity – rather set of complexities – in linking mathematics education with democratic participation relates to traditional understandings of democratic participation in a globalised pluralistic and new-times society (Giddens, 1990). It appears to me that an essentialist understanding of the construct of democratic participation, and of democracy itself, is becoming increasingly untenable. It remains, using a term discussed by Gallie (1956), an "essentially contested" construct that has a variety of uses in different contexts. Skovsmose (2005) relates his experience in travelling in post-Apartheid South Africa where the salient understanding of the concept of democracy was the right to vote in elections, a feature taken for granted in many Western countries. Perhaps some of the meanings and characteristics of democracy identified by Wikipedia illustrate this diversity.

1. Democracy is a political government carried out either directly by the people (direct democracy) or by means of elected representatives of the people (representative democracy).
2. Democracy includes: equality and freedom … These principles are reflected in all citizens being equal before the law and having equal access to power.
3. "Majority rule" is often described as a characteristic feature of democracy… An essential process in representative democracies is competitive elections that are fair both substantively and procedurally. Furthermore, freedom of political expression, freedom of speech, and freedom of the press are essential so that citizens are informed and able to vote in their personal interests.

The article goes on to list 12 different forms of the term "both in theory and practice ... [that are] not exclusive of one another: many specify details of aspects that are independent of one another and can co-exist in a single system". (Democracy, Wikipedia, undated).

Here, I note some observations about these articulations that are representative of wide understandings of the construct. To start with, they all seem to be based on the ideology of individualism and nationalism. In other words, they refer to democratic participation by independent but equal citizens in a particular nation-state. This understanding does not take into account our increasing awareness that nation-states are composed of a variety of minority groups that resist identification with the identity and values of the majority. Many have standards of social organisation that are not based on individualistic identities but rather as members of family and social groups (Brubaker & Cooper, 2000).

I am thinking here of the Australian Aboriginal people and their struggle to have group ownership of land and to have their traditional law exist in parallel with the European law of the majority. I also have in mind many migrant communities who value family affiliations and traditions as well as their individual choices. To add to the complexity, members of these minorities do not necessarily have a single identity that allows them to speak with a single voice. In an increasingly globalised world, identities are not unitary and fixed; rather they are fluid and multiple (Butler, 1990). Hence, regarding individual identities *without* a consideration of their social identities is untenable for democratic participation. Similarly, it is futile to treat them as *only* members of a group since social groups consist of individuals with varying histories, needs, and interests.

Further observations about the traditional understandings of democracy and democratic participation relate to their implied ontological and epistemological foundation on natural rights and freedoms. In particular, the rights and freedoms enshrined in the United Nations Declaration of Human Rights are widely used as bases for both social organisations policy and their contestations by dissident groups in many liberal regimes around the world. However, Heller (1992) raises questions about the ontological character of these rights and their theoretical nature. She notes that they are not descriptions of reality – hence they are "fictions" (p. 351). Similar understanding is presented by Lakoff and Johnson (1980) where they interpret abstract constructs as metaphors. In his recent and controversial book *Whose freedom? The battle over America's most important idea* (Lakoff, 2006), he outlines how the debate on freedom between liberals and conservatives in the USA can be constructed as based on distinct family metaphors where the conservatives' view of freedom is based on the "strict father" metaphor while the liberal discourse seems to be based on the "nurturing parent" model.

Similarly, K. Roth (2007) investigated the foundation of democratic participation on epistemological grounds and found it problematic. Knowledge of the other may lead to acceptance inclusion, but, by the same token, it may lead to indoctrination and oppression. Roth claims that such knowledge may be necessary but not sufficient. As the writings of Popkewitz (2004) show, disciplinary knowledge acts as inscription of the child and controls her/his way of thinking and behaviour. In that article, Popkewitz goes on to critique the discourse of democratic participation itself as means of disciplining social participation as a compliance to social functions and structures. Undoubtedly he has in mind here the conservative understanding of the construct referred to by Lakoff. Hence, with the lack of solid ontological and epistemological bases, the rights and freedoms that are assumed to be behind democratic participation are a set of political and ethical principles to guide social organisation and actions. As I will elaborate below, rather than seeking ontological and epistemological foundations of democracy, Levinas posits ethics as the foundation of being and knowledge and as a basis for politics. Here, I don't take democratic participation as based on the humanist construction that "we are all born free". Rather, I understand freedom as being based on ethics, which in turn is based on responsibility towards the other. In other words, we are free because of our responsibility to the other, not the other way around.

The last observation about the above articulations relates to the deconstruction of the term "democracy" by Derrida (1997). Democratic participation is intrinsically based on agency of each citizen being and acting within a collective. But it is also based on balancing this participation by single citizens with the participation of others in the collective. This balancing inevitably leads to consideration of the agent's voice as a single voice among others – in other words, not as an individual but as a number. To quote Derrida: "there is no democracy without respect for irreducible singularity, or alterity, but there is no democracy without the 'community of friends', without the calculations of majorities, without identifiable, stabilizable, representable subjects, all equal" (p. 22). Hence, there is no democracy that empowers citizens to participate without limiting such participation. However, this does not imply that democracy is an empty construct. Derrida goes on to talk about "democracy-to-come", not as a new form of democracy, but as an affirmation of it as an essential ideal – albeit it cannot be reached.

All these observations that problematise the understanding of the term "democratic participation" imply complexities that need to be considered in establishing the relationship of mathematics and democratic participation. Similarly to the "paradox of reason" discussed by Skovsmose, they call for critique that, if it will not determine action, will, at least, allow for reflection on action.

ETHICS AND CRITIQUE

In the quotation above, Skovsmose calls for a "critique without a foundation". He also acknowledges that putting the concerns stemming from a sense of responsibility in the public arena avoids the accusation of "relativism" (p. 132) often raised against some postmodern perspectives. Here, I interpret this stance as an avoidance of privileging a unique foundation for the critique rather than a call for no foundation at all – since every concern, and the reaction to it from others, has some basis, whether rational, legal, political, ethical or otherwise. Accepting the limitation of each of these perspectives to provide an exhaustive foundation for critique of mathematics education, they can be used as a basis for a "reaction" to a critical situation.

In particular, social justice has often been used as an argument to provide critique of mathematics education. However, social justice itself, as a foundation of critique, raises its own problems. As Young (1990) reminds us, the principles of social justice are not theorems, or laws, rather they are claims that one group makes of others, and hence, the notion of social justice itself is contested (Gallie, 1956; Rizvi, 1998). Further, Simmons (1999), quoting Kant, claims that social justice to one group may imply social injustice to another group outside our immediate concern. A contested social justice depends on discourse and language, and hence it is inherently "violent" in the sense discussed by Derrida (in Critchley, 1992). Political considerations in general, and social justice in particular, are under threat of reducing the individual to merely being a member of a species. By saying social justice is violent, I do not understand it here as being evil to be overcome. Rather it is inherently open to the possibility of violence and, hence, needs to be kept under questioning and in need of another foundation to deal with its conflicting claims. As I will argue below, the political, while not reducible to ethics, requires ethics as a foundation of its decisions (Simmons, 1999).

In another context (Atweh & Brady, 2009), I argued that the discussion of *ethics* is raised in mathematics education literature very infrequently, and that this silence is paralleled by the avoidance of discussion of ethical questions in most traditions of Western philosophy. With the rise of scientific rationality, ethics is often associated with questions of morality, dogma, codes of behaviour, and legal imperatives and is often seen as belonging to the domain of metaphysics rather than philosophy proper. Cohen (2005) explains this avoidance of ethical discussion in philosophy as a fear of moralising, preaching, and questions of values in philosophical discourses that are mainly focused on ontology rather than meaning. Similarly, in Western thinking, there is a movement away from essentialist thinking represented in the universality of ethical principles (Christie, 2005) and their foundation on rationality as established by philosophers such as Kant. As Levinas (1969, 1997) maintains, philosophy is mainly concerned with questions of being (ontology) and knowledge (epistemology). The discussions of being and knowledge are achieved by reducing the Other to the same (Critchley, 2002).

K. Roth (2007) notes that the relationship between ethics and knowledge is not new. Going back to the philosophical and ethical discourses of Socrates who established the primacy of the knowledge of the good over the knowledge of the truth, Cohen raises the question "has the philosopher abdicated responsibilities?" (p. 39). However, this avoidance of dealing with ethical discourse is slowly dissolving. As Critchley (2002) indicates, it was only in the 1980s that the word "ethics" came back into intellectual discourse after the "antihumanism of the 1970s" (p. 2). Further, the post-ontological philosophical writings of Levinas (1969, 1997) have been accredited by the re-introduction of ethics within philosophy by establishing ethics as the First Philosophy.

However, discourse of ethics is not unitary (Giroux, 1987). K. Roth (2007) identifies some alternative approaches to constructing ethical decisions. A utilitarian approach bases ethical decisions on consequences of action – in other words, an action or knowledge is neither good nor bad by itself; its ethical value depends on what it leads to. A deontological approach identifies principles for ethical duties regardless of consequences. More recent feminist writers developed an ethics of care approach, which focuses on principles for emotions and virtues that are morally relevant. Discourse ethics establishes ethical claims on the same basis as claims of truth and fact – that is, on argumentation and the logic of communicative action; hence ethical norms can only be justified intersubjectively through the processes of argumentation between individuals in a dialectic manner. For Levinas, ethics is before any philosophy and is the basis of all philosophical exchanges. It precedes ontology, "which is a relation to otherness that is reducible to comprehension or understanding" (Critchley, 2002, p. 11). This relation to the Other that precedes understanding he calls "original relation". Critchley points out that Levinas's original contribution to ethics is that he does not see ethics as a pre-determined set of principles that can be used to made decisions about particular instances of behaviour. Rather it is an adjective that describes a relationship with the Other that precedes any understanding and explanation. Using a phenomenological approach, Levinas argues that to be human is to be in a relationship to the other, or more accurately, in a relationship *for the other*. This relation is even prior to mutual obligation or reciprocity. W. R. Roth (2007) argues that this original ethical relationship discussed by Levinas consists of an "unlimited, measureless responsibility toward each other that is in continuous excess over any formalization of responsibility in the law and stated ethical principles".

In his later work, Levinas (1997), in response to Derrida's claim that the encounter with the other is "violent" if it is based on language and discourse, introduced the distinction between *saying* and the *said* in the face-to-face encounters with the Other. Further he located the initial encounter with the Other as based on saying which precedes the ontological said. Simmons (1999, p. 88) explains "Prior to the speech act, the speaker must address the Other, and before the address is the approach of the other or proximity". Importantly for our purposes here, Levinas places ethics in the *saying* and politics and social justice in the realm of the *said*. He argues that *peace* is in the *saying* and the *said* is necessarily open to the possibility of violence. Using this distinction, Levinas demonstrates how ethics and politics are necessarily independent; however, each needs the other. Ethics, which is the encounter with the Other, needs politics since the Other is not singular – there are many others. On the other hand, politics needs ethics since politics is always open to the possibility of excesses and needs to be kept in check.

Here, I propose that a conception of ethics is necessarily implicated with any critique of the relationship between mathematics and democratic participation. Further, I propose that the inclusion of an ethical/responsibility perspective – in particular an understanding of ethics not as a set of specific codes of behaviours, but as basic inescapable responsibility to the other – in that critique assists in dealing with the uncertainties and complexities discussed above.

Giroux (1987) points to a paradox facing many radical educational theories that often posit "moral" indignation about social and political justices and yet have "failed to develop a moral and ethical discourse upon which to ground its version of society and schooling" (p. 9). He further adds that, without such discourse, it is not possible for critical education to "move from criticism to substantive vision" (p. 9). He calls for an ethical discourse that transcends both the essentialist constructions of ethics from the right – that may lead to standardisation of being and conduct – on the one hand, and constructions of certain "free-floating" forms of postmodernism – that may lead to pragmatism and relativism – on the other.

Arguably, every complexity identified above gives rise to situations where the choices need to be made and outcomes critiqued and where the outcomes are neither pre-determined nor simple. Nevertheless, choices still have to be made – one hopes responsibly. However, every critique consists of a judgement about "what is good". Hence, it enters an ethical discourse. Skovsmose was right in noting that rationality is limited to providing the foundation of the needed critique or reflection on action. However, placing ethics prior to (in both temporal and precedence meanings) rationality (and philosophy, politics and law for that matter) allows rational thinking itself to be used as one among many bases for critique while it is itself kept under check by the sense of responsibility towards the other.

TOWARDS SOCIALLY RESPONSE-ABLE MATHEMATICS EDUCATION

In my introduction, I referred to different perspectives that support and enhance democratic participation by various segments of the population. Here, I do not make the assumption that these different perspectives are necessarily in accord with each other in their theoretical foundations or in their implications for practice in mathematics education. I do make the observation, though, that behind many of their concerns is the lack of distribution of power and access to mathematics and, in particular, making mathematics empowering to the less advantaged in society. Without a doubt, this is a worthwhile endeavour that should remain, with urgency I might add, at the forefront of our collective consideration in the field. What I propose in this section is that an ethical dimension to the above dissenting discourses to mainstream mathematics education would increase the possibility of achieving the role of mathematics education as a tool to increase democratic participation by the marginalised social groups – and also by the whole society. It also provides that any approach to mathematics education would be self-reflective and critical as to its assumptions and practices.

Here, I put forward a vision of mathematics education based on ethics and, in particular, on the concept of responsibility. Elsewhere (Atweh & Brady, 2009), I suggested that in current political discourse the demand for responsibility, or more often for its synonym "accountability", is an increasing concern in educational policy and practice. However the term is used with a variety of meanings. Responsibility is often presented as a requirement or duty that restricts (as in, it is the teachers' responsibility to cover the curriculum), as well as enables (as in, evaluating students' learning is the teachers' responsibility), or sometimes in the placement of blame (as in, who is responsible for the students' lack of achievement?). As Christie (2005) suggests, when it comes to responsibility or ethics, it is possible to "work with and work against" (p. 240) the construct at the same time. In other words, I adopt a critical stance on the concept, its usefulness, and its limitations. To distance the approach to responsibility proposed here from these legal and rationalistic understandings, I will suggest a slight change in the term "responsibility". Puka (2005) suggests that a great contribution to ethics is the feminist distinction between responsibility and "response-ability" (for diverse feminist stances with respect to Levinas, see Chanter, 2001). Response-ability highlights the ability to respond to the demands of our own well-being – hence it focuses on agency – and the ability to respond to the demands of the other – hence doing that responsibly. This is similar to the observation that W. R. Roth (2007, p. 5) makes that:

> … etymologically [responsibility] derives from a conjunction of the particles *re-*, doing again, *spondere*, to pledge, and *–ble*, a suffix meaning "to be able to." Responsibility therefore denotes the ability to pledge again, a form of re-engagement with the Other who, in his or her utterances, pledges the production of sense. Each one, on his or her own and together, is responsible for the *praxis* of sense, which we expose and are exposed to in transacting with others.

In the following two sections, I briefly outline some implications of a Socially Response-able Mathematics Education (SRaME) as they relate the two of the three message systems of schooling: curriculum and pedagogy (Bernstein, 1971).

Implications of Social Response-ability for the curriculum

The dominance in school mathematics of content needed for careers that are seen as mathematically based – mainly science and engineering – does not promote democratic participation and, perhaps, is a residue of times when few students finished high school and went to university. Notwithstanding the importance of jobs in science and engineering for social technological development, only a few students end up in such careers. The approach to mathematics taken here is that *all* students need considerable amount of mathematics for effective citizenship in the increasingly mathematised world of today – albeit different type of mathematics depending on their interests, capacities and career choices. Hence, a *utilitarian* approach to mathematics falls short of developing a response-able student. As Ernest (2002) argues, a critical approach to mathematics and citizenship is needed. This ethical response-ability discussion applied to mathematics education posits the primary aim of mathematics education as enabling the response-ability of students in their current and future lives as citizens. Here I will discuss two implications for the curriculum of mathematics that promotes democratic participation.

Firstly, an SRaME implies a shift of focus on what is central in mathematics education. Curriculum documents around the world often contain lists of outcomes or topics in mathematics that students are expected to cover in their progression from year to year of school. It is customary to present this content in strands along the lines of number, algebra, geometry, probability, and statistics. At times this content is articulated as concepts, skills, and procedures. Lastly, most new curriculum documents focus on applications and problem solving as important aspects to be developed with students. Undoubtedly, such topics dominate the majority of classroom time and assessment instruments that teachers utilise. However, many curriculum documents also articulate aims or outcomes that cut across the different topics. For example, the Western Australian Curriculum Framework (Curriculum Council of Western Australia, 1998) identifies "working mathematically" as an important rationale for mathematics education. The document states that:

> Every student needs to develop an awareness of the nature of mathematics, how it is created, used and communicated, for what purposes, and how it both influences and is influenced by the things we believe and the values we hold. (p. 179)

Further, it lists some specific outcomes that students need to demonstrate. In particular:

> Appreciate that mathematics has its origins in many cultures, and its forms reflect specific social and historical contexts, and understand its significance in explaining and influencing aspects of our lives.
>
> Show a disposition to use mathematics to assist with understanding new situations
>
> Choose mathematical ideas and tools to fit the constraints in a practical situation, interpret and make sense of the results within the context and evaluate the appropriateness of the methods used.

For many teachers, however, these outcomes are problematic in the sense they don't define particular content nor do they easily lend themselves to particular ways of assessment. The Western Australian Curriculum Framework itself asserts that working mathematically is not an area that needs to be targeted in assessment directly. Hence, the tendency of many mathematics classroom practices is to pay lip service to this aspect of mathematics. Further, the focus on the set of what might be seen as "core mathematics" skills and understandings is further encouraged through the increasingly high stakes multiple-choice national numeracy tests that are conducted every two years in the country. Lastly, for most mathematics teachers, the curriculum is very crowded. Often working mathematically is seen to be desirable, but not an essential addition to the curriculum, and remains of secondary importance.

An SRaME approach that aims to increase democratic participation requires that a shift be made away from mere content and procedures into problem solving, modelling, and applications. Further, while it is usual to find applications in mathematics from science and the natural world of the student, applications from the student's social life often remain neglected. Social applications in mathematics are often seen as contrary to rigorous mathematics that is needed for higher studies and often dealt with in special less academic courses targeting students designated as less able. However, this binary might be counter-productive by denying the opportunity and the ability to develop their generalised abstractions of mathematical concepts and procedures to the majority of students taking the so-called social or practical mathematics. Further, in spite of the rhetoric of curriculum documents, and the assurance by many teachers that the two streams deal with equally valuable mathematics – albeit for different needs – for many students a hierarchy of values exists between them, resulting in higher status for the formal academic mathematics.

However, not every focus on applications and modelling guarantees the development of a socially response-able curriculum. As Warnick and Stemhagen (2007) point out:

> If acquiring a mathematical worldview means that students begin to see how the subject applies to the problems of everyday life, it does make sense to say that the mathematization of experience is an important goal of mathematics education. At the same time, though, we argue that students should also recognize the limits of the mathematical language game, and that mathematics education should play a part in fostering this recognition. We move toward this goal by exposing the relationship between a mathematical worldview and a technological worldview (p. 305).

Secondly, an SRaME approach implies a shift of sequencing in the development of mathematical knowledge and its application. The common practice in many mathematics classrooms is that students develop mathematical understandings and skills before they are able to apply them in problem solving. Hence mathematical knowledge is often presented as decontextualised and abstract. This approach often leads students into asking, "Why are we studying this?" and to students switching off mathematics before real and interesting applications are encountered. Mathematics education that promotes democratic participation must aim at not only developing mathematical knowledge and skills, but also knowledge and skills about the real world of the students. The approach promoted here is for the use of real-world activities that promote students' learning about their social world *while* they are learning mathematics and, at the same time, their learning about mathematics *while* they are engaging with real-world activities. Moreover, there has to be a balance between these two areas of learning. An SRaME teacher always needs to ask what mathematics, higher order mathematics in particular, is learnt by such activities and what significant learning about the social world is anticipated. In particular, they need to raise the question about the mathematics itself, its assumptions, power, and limitations as a result of these activities. These stances are consistent with the approaches promoted by critical mathematics, ethnomathematics, and social justice approaches. What an SRaME approach adds to this is the raising of issues of social responsibility with students as they engage in learning to read and write the world through mathematics (Gutstein, 2006).

Implications of Social Response-ability for pedagogy

In this context I understand pedagogy in the sense discussed by Lingard (2005) who, using Bernstein (1971) elaboration, states that pedagogy goes beyond mere teaching methods or instructional techniques to include teachers' interpersonal competencies for interacting with students as well as contextual considerations and questions of power relationships enacted in the classroom. Here, I consider three implications for a Socially Response-able Mathematics Education to pedagogy.

Firstly, an SRaME approach stipulates relationships between teachers and students in the classroom that are not common in traditional practices. Neyland (2004) demonstrates how in mathematics education the demand for accountability or responsibility, as portrayed in the worldwide push towards standards and testing, reflects a "scientific management" rationality that posits institutions and norms as the cause of ethical behaviour. Using Levinas's writings, he goes on to argue that such institutions externalise and mechanise ethical behaviour and thus "sometimes erode a primordial ethical relation between people" (p. 517). In this context, a focus on ethical responsibility shifts the focus of interactions between students and teachers from technical and system demand considerations to an encounter between two human beings, and while it is not totally free from *system* demands (Habermas, 1987), it allows for teachers' decision making based on the interest of the student. It implies a collaborative and mutually respectful classroom environment where the participants are constructed as co-learners, an environment to which Vygotsky and Freire aspired. In working towards SRaME, the teachers and students develop a new relationship of co-inquirers or co-learners in contrast to the traditional construction of expert and novice. In such real life activities, while the teacher is not the source of knowledge about what needs to be changed, the students need support in identifying these needs and in negotiating change. As Atweh and Bland (2005) point out, in their reflection on one such project, there needs to be a balance between the teachers abiding by their duty of care and minimizing the risk of student failure, and thus limiting students' agency, on one hand, and their willingness to take risks by maximising students input, on the other.

Secondly, an SRaME approach implies new understandings of what constitutes knowledge in mathematics classes. This understanding of the Socially Response-able pedagogy is in harmony with some of the tenets of constructivism (Von Glasersfeld, 1991), a position that constructs the learner as an active participant in the development of their own knowledge. Further, it posits the student and the teacher in a "reflexive" relationship developing contextualised knowledge wherein neither party can claim a monopoly of expertise. However, since such mathematical knowledge is to be used in social and political contexts, questions of values and ethical decisions about possible action must necessarily arise. This approach is perhaps more aligned with critical constructivism as discussed by Kincheloe (1995), who claims that:

> Critical constructivists ... ask what are the forces which construct the consciousness, the ways of seeing of the actors who live in it. ... Critical constructivism concerns the attempt to move beyond the formal style of thinking which emerges from empiricism and rationalism, a form of cognition that solves problems framed by the dominant paradigm, the conventional way of seeing. (p. 88)

Hence, SRaME activities should do more than attempt to achieve students' *engagement with the mathematics* learning by giving them real-world examples of the content. Students should also engage with the world situation being investigated. Through the SRaME activities, students engage in critical reflection about the assumptions behind the mathematics developed as well as the assumptions behind social practices being investigated. Lastly, through these activities a sense of an empowered agency is developed to reflect ethically on various possible lines of action and to actively listen to alternative points of views. Hence, the call here is for an interdisciplinary approach to mathematics education and the willingness to deal with controversial topics in which debate and difference of opinion and human interests are part of the equation rather than nuisance variables. This approach is in direct conflict with the view of mathematics as an abstract, decontextualised, and value-free system of knowledge.

Thirdly, an SRaME approach implies a socially just pedagogy that necessarily raises the question of inclusion of marginalised groups of students in the study of mathematics. Education is often discussed as the most effective solution to addressing disadvantage in society and between societies. After at least fifty years of development and reform in education, it is important to raise the question as to whether education has been able to address this challenge. Perhaps Basil Bernstein (1971) was correct in his conclusion that schools do not compensate for society. However, there is some good news. A wide-ranging review of the effects of educational interventions aiming to alleviate disadvantage show that increasing quality teaching does contribute to improving opportunities for marginalised groups of students (Hayes, Mills, Christie & Lingard, 2005). This research shows that quality education assists *all* students; however, as Christie (2005) comments, "it is for the most disadvantaged children that improvements in school quality will make the most difference in achievement" (p. 245). Further, out of all the school factors that affected students' achievement the most important was the teacher. Hence good teaching "can make a difference, but not *all* the difference" (Hayes et al., p. 178). The danger of exclusion is not in challenging disadvantaged and underachieving students to higher intellectual quality, but in "dumbing down" the curriculum for them – thus locking them into marginalisation and disempowerment.

These conclusions, however, should not be taken to imply that a focus on quality automatically results in equity. The authors go on to discuss Productive Pedagogy as a framework for reflection on pedagogy to ensure it focuses on both quality and equity. The Productive Pedagogy framework consists of four main categories with each divided into several subsections:

Intellectual quality
Connectedness
Supportive classroom environment
Recognition of difference

An ethical response-ability places the primacy of ethical considerations in the teacher-student pedagogical encounter. There are two dangers in this encounter that might erode ethical response-ability for the student. First, to deal with the students as individuals with no regard for their gender, ethnicity, or socioeconomic background – factors that are demonstrably related to student achievement in mathematics – is to relate to an "abstract" student. Not only is this a recipe for failure – it is also dehumanising and is unethical, as argued by Neyland (2004). Similarly, the other extreme of seeing a student *only* as being of a particular gender, ethnicity, or social status is equally counterproductive. Such stereotyping also limits the possibility of an authentic encounter with the real *Other*. An ethical encounter attempts to be open to any possibility that exposes itself and responds to the students' needs and aspirations rather than in a stereotypical fashion. In supporting the students' response-ability a teacher can provide the opportunity to develop the high intellectual quality to the maximum of the students' needs and capacities. This is consistent with Vithal and Skovsmose's (1997) argument that a focus on the background of the student can obscure and hinder a focus on the *foreground* that sees possibilities as to what the student can be rather than a focus on where they have come from.

CONCLUDING REMARKS

Increasingly, educational policy and curriculum discourse around the world are being constructed in terms of citizenship and democratic participation. For example, In Australia, the Ministerial Council on Education, Employment, Training, and Youth Affairs, consisting of all the state ministers of education along with the Federal government, issued the Melbourne Declaration (Australian Government, 2008) which forms the basis of the current attempts in the country to develop its first National Curriculum. The declaration identifies two goals of education:

Goal 1: Australian schooling promotes equity and excellence

Goal 2: All young Australians become

- successful learners
- confident and creative individuals
- active and informed citizens. (p. 7)

The identification of "active and informed citizens" as a main goal of education may be taken as a commitment for social transformation as discussed above. In this chapter I explore a type of mathematics education that is likely to contribute to the aim of "active and informed citizens".

I began this chapter by problematising the relationship between mathematics education and democratic participation. Building on the observations by authors such as D'Ambrosio and Skovsmose, this relationship cannot be assumed and needs to be scrutinised with careful critique. The experience of several mathematics educators within movements such as ethnomathematics, critical mathematics, and socially just mathematics education have provided valuable critiques of traditional mathematics education and demonstrated that mathematics teaching in formal education can contribute to both empowerment of individuals and groups of students as well as enhance social justice in many societies. The approach taken here complements these perspectives by presenting ethics as the basis for this critique.

I conclude this discussion by making the assertion that ethics is not an add-on to the concerns in mathematics education. It lies at the very foundation of every decision in the field. It is reflected in identifying the aims of mathematics education, in making decisions about practices of teaching, learning, and assessment. It raises questions of inclusion and exclusion. However, it is not deterministic in a sense that following simple rules or principles ensures ethical conduct. As Foucault (1983) famously warned us, “everything is dangerous”. But Foucault added, “If everything is dangerous, then we always have something to do. So my position leads not to apathy but to hyper- and pessimistic- activism” (pp. 231–232). By the same token, an ethical approach to mathematics education calls for taking risks, albeit with a great sense of responsibility, and to be constantly vigilant about the outcomes of our actions. Ethics invites us, rather compels us, to a continuous and exhaustive sense of engagement with the welfare of the other.

REFERENCES

Apple, M. (1992). Do the standards go far enough? Power, politics, and practice in mathematics education. *Journal for Research in Mathematics Education, 23*(5), 412–431.

Atweh, B., & Bland, D. (2005). Mathematics through/for understanding social life: Productive Pedagogies meets critical mathematics. In M. Goos, C. Kanes & R. Brown (Eds.), *Proceedings of the 4th International Mathematics Education and Society Conference*. Brisbane: Griffith University.

Atweh, B. & Brady, K. (2009). Socially response-able mathematics education: Implications of an ethical approach. *Eurasia Journal of Mathematics, Science & Technology, 5*(3), 267–276.

Australian Government (2008). *The Melbourne declaration on the educational goals for young Australians*. Victoria: Ministerial Council on Education, Employment, Training and Youth Affairs.

Bernstein, B. (1971). *Class, codes, and control.* London: Routledge and Paul.

Bishop, A. (1998). *Mathematical enculturation: A cultural perspective on mathematics education.* Dordrecht, The Netherlands: Kluwer.

Brubaker, R., & Cooper, F. (2000). Beyond “identity”. *Theory and Society, 29*, 1–47.

Burton, L. (Ed.). (2003). *Which way social justice in mathematics education?* London: Praeger.

Butler, J. (1990). *Gender trouble: Feminism and the subversion of identity.* New York: Routledge.

Chanter, T. (Ed.). (2001). *Feminist interpretations of Emmanuel Levinas.* University Park, PA: Pennsylvania State University Press

Christie, P. (2005). Towards an ethics of engagement in education in global times. *Australian Journal of Education,* 49(3), 238–250.

Cohen, R. (2005). Levinas, Plato, and ethical exegesis. In J. Bloechl & J. Kosky (Eds), *Levinas studies: An annual review (Vol. 1)* (pp. 37–50). Pittsburgh, PA: Duquesne University Press.

Critchley, S. (1992). *The ethics of deconstruction: Derrida and Levinas,* Oxford: Blackwell.

Critchley, S. (2002). Introduction. In S. Critchley & R. Bernasconi (Eds.), *The Cambridge companion to Levinas* (pp. 1–32). Cambridge: Cambridge University Press.

Curriculum Council of Western Australia. (1998). *Curriculum framework for kindergarten to year 12 in Western Australia.* Perth: Author.

D'Ambrosio, U. (1994). Ethnomathematics, the nature of mathematics and mathematics education. In P. Ernest (Ed.), *Mathematics, education and philosophy: An international perspective* (pp. 230–241). London: Falmer.

D'Ambrosio, U. (1985). Sociocultural basis for mathematics education. In M. Carss (Ed.), *Proceedings of the Fifth International Congress on Mathematics Education* (pp. 1–6). Boston: Birkhäuser.

Derrida, J. (1997). *Politics of friendship*. London: Verso,

Ernest, P. (2002). What is empowerment in mathematics education? In P. Valero & O. Skovsmose (Eds.), *Proceedings of the 3rd International MES conference* (pp. 1–12). Copenhagen: Centre for Research in Learning Mathematics.

Foucault, M. (1983). On the genealogy of ethics. In H. Dreyfus & P. Rabinow (Eds.), *Michel Foucault: Beyond structuralism and hermeneutics* (pp. 231–232). Chicago: University of Chicago Press.

Frankenstein, M. (1983). Critical mathematics education: An application of Paulo Freire's epistemology. *Journal of Education, 165*(4), 315–339.

Gallie, W. B. (1956). Art as an essentially contested concept. *The Philosophical Quarterly, 6*(23), 97–114.

Giddens, A. (1990). *The consequences of modernity*. Cambridge: Polity.

Giroux, H. (1987). Schooling and the politics of ethics: Beyond liberal and conservative discourses. *Journal of Education, 169*(2), 9–33.

Gutstein, E. (2006). *Reading and writing the world with mathematics: Toward a pedagogy for social justice*. New York: Routledge.

Habermas, J. (1987). *Theory of communicative action: Volume two: Lifeworld and system*. Boston, MA: Beacon Press.

Habermas, J. (1998). *The philosophical discourses of modernity* (F. Lawrence, Trans.). Cambridge, MA: MIT Press.

Hayes, D., Mills, M., Christie, H., & Lingard, B. (2005). *Teachers and schooling making a difference: Productive Pedagogies, assessment and performance*. Sydney: Allen & Unwin.

Heller, A. (1992). Rights, modernity and democracy. In D. Cornell, M. Rosenfeld & D. G. Carlson (Eds.), *Deconstruction and the possibility of justice* (pp. 346–460). New York: Routledge.

Kincheloe, J. (1995) *Teachers as researchers: Qualitative inquiry as a path to empowerment*. London: Falmer.

Lakoff, G., & Johnson, M. (1980). *Metaphors we live by*. Chicago: University of Chicago Press.

Lakoff, G. (2006). *Whose freedom? The battle over America's most important idea*. New York: Farrar, Straus and Giroux.

Levinas, E. (1969). *Totality and infinity: An essay on exteriority* (A. Lingis, Trans.). Pittsburgh, PA: Duquesne University Press.

Levinas, E. (1997). *Otherwise than being or beyond essence* (A. Lingis, Trans.). Pittsburgh, PA: Duquesne University Press.

Lingard, B. (2005). Socially just pedagogies in changing times. *International studies in sociology of education, 15*(2), 165–186.

Mellin-Olsen, S. (1987). *The politics of mathematics education*. Dordrecht, The Netherlands: Reidel.

Neyland, J. (2004). Rethinking curriculum: An ethical perspective. In B. Barton, K. Irwin, M. Pfannkuch & M. Thomas, (Eds.*), Mathematics education in the South Pacific: Proceedings of the 25th Annual Conference of the Mathematics Education Research Group of Australasia*. Auckland: MERGA.

Popkewitz, T. (2004). The alchemy of the mathematics curriculum: Inscriptions and the fabrication of the child. *American Educational Research Journal, 41*(1), 3–34.

Powell, P. & Frankenstein, M. (Eds). (1997). *Ethnomathematics : Challenging Eurocentrism in mathematics education*. Albany, NY: State University of New York Press.

Puka, B. (2005). Teaching ethical excellence: Artful response-ability, creative integrity, character opus. *Liberal Education, 91*(3), 22–25.

Rizvi, F. (1998). Some thoughts on contemporary theories on social justice. In B. Atweh, S. Kemmis, & P. Weeks (Eds.), *Action research in practice: Partnerships for social justice in education* (pp. 47–56). London: Routledge.

Roth, K. (2007). Education for responsibility: Knowledge, ethics, and liberation. In K. Roth & L. Gur-Zeev (Eds.), *Education in the era of globalisation* (pp. 105–122). Dordrecht, The Netherlands: Springer.

Roth, W. R. (2007). Solidarity and the ethics of collaborative research. In S. Ritchie (Ed.), *Research collaboration: Relationships and praxis* (pp. 27–42). Rotterdam: Sense Publishers.

Secada, W. (1989). *Equity in education.* Philadelphia: Falmer.

Simmons, W. (1999). The third: Levinas' theoretical move from anarchical ethics to the realm of justice and politics. *Philosophy and social criticism*, *26*(6), 83–104.

Skovsmose, O., & Valero, P. (2001). Breaking political neutrality: The critical engagement of mathematics education with democracy. In B. Atweh, H. Forgasz & B. Nebres (Eds.), *Sociocultural research on mathematics education* (pp. 37–55). Mahwah, NJ: Lawrence Erlbaum.

Skovsmose, O. (1994). *Towards a philosophy of critical mathematics education.* Dordrecht, The Netherlands: Kluwer.

Skovsmose, O. (2005). *Travelling through education: Uncertainty, mathematics, responsibilities.* Rotterdam: Sense Publishers.

Vithal, R., & Skovsmose, O. (1997). The end of innocence: A critique of "ethnomathematics". *Educational Studies in Mathematics, 34*, 131–157.

Von Glasersfeld, E. (Ed.). (1991), *Radical constructivism in mathematics education.* Dordrecht, The Netherlands: Kluwer.

Warnick, B., & Stemhagen, K. (2007). Mathematics teachers as moral educators: The implications of conceiving of mathematics as a technology. *Journal of Curriculum Studies. 39*(3), 303–316.

Young, I. M. (1990). *Justice and the politics of difference*. Princeton, NJ: Princeton University Press.

Bill Atweh
Science and Mathematics Education Centre
Curtin University of Technology, Western Australia

CHAPTER 16

TOWARDS A CRITICAL MATHEMATICS EDUCATION RESEARCH PROGRAMME?

OLE SKOVSMOSE

Mathematics education is a momentous social system with socio-political, economic, and cultural impacts. It can be of relevance for a globalising learning economy; it can be an essential ingredient of productivity within a neo-liberal economic order. Mathematics education may provoke both exclusion and suppression; it might exercise a gender bias; it may operate as a secret weapon of Western imperialism, as indicated by Alan Bishop (1990); it may operate as part of cultural colonisation, as observed by Ubiratan D'Ambrosio (2006).[1] It could also be that mathematics education, when organised in an adequate form, ensures empowerment and provides a basis for critical citizenship and social justice. To me all such observations indicate that mathematics education could make a difference: for the better or for the worse. As mathematics education can be acted out in many different ways, it has no a priori "essence". It could serve very different social functions. This implies that, basically speaking, mathematics education becomes what it is doing. Its "essence" is produced as it becomes being acted out.

This is a basic observation in taking steps towards formulating a *critical mathematics education research programme*. In the title of this chapter, I have written "towards" and added a question mark. This is certainly necessary. Some of the authors that I am referring to in the following have talked about critical mathematics education, but very many refer to frameworks as, for instance, "mathematics education for social justice", "pedagogy of conflict and dialogue for mathematics education", or "ethnomathematics". I try to formulate some general concerns that might run across these different frameworks, and in this way I hope to take steps towards a critical mathematics education research programme.[2]

One could think of critical mathematics education in terms of both theory and practice. It can refer to certain ways of theorising about mathematics education, also to certain ways of practising this education. Certainly there can be close interaction between theory and practice, and in the following I do not make a big point of distinguishing. Sometimes I might refer to theoretical perspectives, sometimes to practical issues. This has implications for the notion of "research" when talking about a "research programme". Apparently, research refers more directly to theory than to practice. However, I use the notion of research in a broad way. It can refer to exploration, as when we talk about teachers trying to find a different way of introducing a certain topic; it can refer to teachers' research of their own practice; it can refer to different forms of collaborative inquiries; it can

O. Skovsmose and B. Greer (eds.), Opening the Cage: Critique and Politics of Mathematics Education, 343–368.

refer to a pedagogical imagination applied when identifying alternative educational possibilities; yet again, it can refer to research in an academic sense of the word. I use "research" in all these different ways when I talk about a critical mathematics education research programme.

I will restrict myself to considering the following three issues in moving towards such a programme. Here the notion of "programme" refers first of all to concerns (and not to statements about what to do).[3] That many more issues could be considered is all too obvious.

(1) *Exploring the variety of sites for teaching and learning mathematics.* The teaching and learning of mathematics takes place in very many different settings around the world. A certain "prototypic mathematics classroom" seems, however, to have dominated mathematics education research. I find it important that critical mathematics education challenges the dominance of the discourse created around this prototype, or stereotype. Mathematics education can operate in very many different ways, depending on the site. One can consider how a particular mathematics education may function for immigrant students in Denmark, for Indian students in Brazil, for students from a favela environment. There is no context-independent interpretation of how mathematics education might function. It is important to critical mathematics education to address the variety of sites for teaching and learning mathematics and abandon the paradigmatic assumptions that emerge from research that has embraced stereotypes. Thus, it is a concern of critical mathematics education to address critically any form of research concerning mathematics education.

(2) *Exploring the variety of forms of mathematics in action.* It becomes important to broaden the school-centred discourse on mathematics education and to address the very different out-of-school practices that include mathematics. Mathematics can be brought in operation in very many different practices. One can think of professional practices as accounting, engineering, brick building, etc. Mathematics makes part of technical settings as statistical packages, medical cost-benefit calculations, decision procedures in marketing, strategies for automatisation in production, etc. However, often mathematics is integrated within work practices in a form that is not transparent to people involved in the professional practice. It is important to address such variety of practices to provide a critical investigation of how mathematics might operate and to investigate possible relationships between out-of-school mathematical practices and how mathematics might be contextualised in a learning environment. Thus, it is a concern of critical mathematics education to address critically all forms of mathematical rationality, whatever practices they might make part of.

(3) *Exploring the variety of educational possibilities.* When the variety of sites for teaching and learning mathematics and the variety of forms of mathematics in action become related, we reach deep into the socio-political dimension of mathematics education. How may a mathematics education ensure an empowerment? Is it possible to work for social justice through mathematics education in an unjust society? What could be the meaning of empowerment and social justice? Could discrimination, cultural exclusion, and economic oppression also be acted out through mathematics education? Certainly such questions are in

urgent need of clarification. They call for the formulation of notions and visions that might be helpful for exploring educational possibilities. This exploration is an important part of a critical mathematics education research programme. Thus, it is a concern of critical mathematics education to address critically any form of educational practice by locating it among possible alternatives.

In the following three sections I will address these issues regarding research, mathematical rationality, and educational practice. That many other concerns are crucial as well for critical mathematics education is all too obvious; nevertheless, here I concentrate on only three. In the conclusion, I turn to the point that any critical research programme will include a profound uncertainty. In other words, self-critique is a decisive part of critical mathematics education. No critical approach can establish itself in any self-proclaimed stability. This means that one should not think of a critical mathematics education research programme as something to be implemented. It is, instead, an expression of concerns that continuously have to be reformulated, developed, and reconsidered. It is an expression of profound uncertainties.

THE VARIETY OF SITES FOR LEARNING MATHEMATICS

The research discourse in mathematics education has been dominated by a stereotype that I refer to as the *prototype mathematics classroom*. This classroom is well equipped, and the students are willing to learn. Due to this stereotype, research may develop conceptions of learning mathematics that are blind to the many conflicts that often form part of the context of learning. Research grounded in the prototype mathematics classroom becomes a defining element of a broadly accepted research tradition within mathematics education. However, I find it important for critical mathematics education to look beyond the prototype mathematics classroom.

Taking a look at the world around, we find many schools without electricity. Schools might be missing all kind of equipment, while the students might be missing schoolbooks. Many schools are located in violent neighbourhoods, where students might fear gangs operating in the vicinity. Poverty dominates many sites for learning. Poor neighbourhoods are found everywhere around the world, as globalisation includes ghettoising as an integral part. Immigration is a world-wide phenomenon that influences the learning situation of students. Contrary to all this, the prototype classroom stays homogeneous. However, what has been characterised as a prototype mathematics classroom belongs to a small minority of the sites for learning mathematics.[4]

Much research discourse in mathematics education appears to include a prototype-bias.[5] What we could call the prototype-bias of research in mathematics education concerns a whole set of underlying paradigmatic priorities and presumptions. This bias makes part, not only of research priorities, but also of the developed conceptual framework; for instance, in the way students' motives are formulated, experiences of meaning are analysed, and forms of contextualisation are addressed. Not only does the prototype classroom represent a minority of the world's classrooms, it also signifies a discursively constructed classroom, where the students could be expected to behave in a particular way. Paola Valero (2002) has characterised such students as "schizo-mathematics-learners", who do not seem to express any real-life interest, but appear like the "epistemic subject" as characterised by Jean Piaget in his elaboration of a genetic epistemology (see Beth and Piaget, 1966). When such an idealised learner has been conceptualised, it becomes possible to sort out what data may count as relevant, and what not, for a further understanding of learning mathematics. If we consider the total number of transcripts that over time have been presented to the research community (in all kinds of reviewed research publications), it appears that a high degree of (self-) censorship must have operated. Only a few cases give voice to clearly obstructive or violent students. Strong paradigmatic criteria of relevance must have existed, establishing the mathematics classroom in a stereotypical format.

It is easy to point out economic causes for this bias. If we consider that research in mathematics education is a costly affair, we should expect that it addresses issues that can obtain funding. We can expect progress in research that produces results that can be transformed into learning materials, textbooks, computer programs, etc. precisely for an affluent context. Research in mathematics education is sensitive to the system of the demand and supply scheme.[6] Reconsidering the economic perspective, there is nothing surprising (nor much praiseworthy) in the bias of mathematics education. One can also point out causes emerging from paradigmatic priorities, for instance about what to count as valid research.[7] Let me finally emphasise that I find it important that *also* the prototype mathematics classroom is researched as part of critical mathematics education. This is an important site for learning mathematics, although only one site amongst many others.

A non-prototype classroom could have an overwhelming number of students, it could be located in a poor neighbourhood, it could be ruined by violence. The non-prototype classroom could be located in cultural settings, which according to paradigmatic research norms could be counted as "foreign". Insecurity could dominate a non-prototype classroom. We find classrooms located in almost war-like situations. Jeanne Albert and Intisar Natsheh (2002) refer to the Middle East Children's Association (MECA) that was established in 1996 jointly by Israeli and Palestinian educators to promote the peace process. In 2001, MECA started a group for mathematics teachers in elementary schools. Albert and Natsheh refer to practices so different from what we normally address in research: "Israeli teachers who work in schools whose neighbourhood is being shelled reported that the children quickly became accustomed to the situation, and the lessons progressed as before." (2002, p. 127). This observation surprises me. Furthermore, they observed that the Palestinian teachers found that the students' motivation was low: "The students complained of the lack of connection between mathematics and their 'real life'." (2002, p. 75). This observation, however, certainly does not surprise me.

Let me address a couple of examples more carefully, that are less extreme in being non-prototypical. Eric Gutstein (2003) describes and analyses mathematics education for social justice in an urban, Latino school in the USA.[8] This example shows what a close interaction, if not a unity, could mean between research and practice. He taught a seventh grade mathematics class from November 1997 to the end of the school year, and then he moved with the class to eighth grade in 1998–1999. The school, referred to as the Diego Rivera School (pseudonym), is located in a working class, Mexican and Mexican-American community in Chicago.

The Diego Rivera School is, in many ways, a typical Chicago public school, and the students must meet all the required tests. Students wear uniforms, which reduces the economic pressure on families, and it also helps to keep the school in "neutral territory" with respect to gangs. That is, certain clothing styles and colours are associated with particular street gangs that are present in the Rivera neighbourhood. Most students live in the neighbourhood and attend the neighbourhood high school but there is a dropout rate of over 50%. The class with 26 students that Gutstein was teaching was demographically representative for the school. All students were from Latino, immigrant, working-class families. About half were born in the USA, and the rest in Mexico except for one student from the Dominican Republic and another whose family was from Puerto Rico. Spanish was the first language of all students, and all but one was fluent in English as well.

I consider the class to exemplify a non-prototype classroom for several reasons, and let me mention three. First, the school is located in a poor area (although not in an extremely poor one), which has implications for the resources that are available in the school, for instance in terms of computers and access to the Internet. The poverty of the neighbourhood could have implications for the students' possibilities of doing homework and for getting support at home. An important issue with Rivera families is that many parents either do not speak English well enough to help their children with homework or were themselves denied educational opportunities in Mexico, and thus have little formal schooling. Second, I consider the classroom to be non-prototypical because it contains immigrant students. In the Diego Rivera School, the immigrant students are a majority within the school and within the neighbourhood, while being a minority within society as a whole. Naturally, there are many cases, in Denmark for instance, where immigrant students make up a minority in both school and neighbourhood. In both cases I would, however, talk about non-prototype mathematics classrooms. One particular issue for immigrant students has to do with the opportunities they experience in the socio-political context as being their realistic opportunities in future life. They might experience a restricted set of opportunities, even when they belong to a local majority, as is the case of students from the Diego Rivera School. Third, I consider the classroom from the Diego Rivera School to be non-prototypical, since the experience of violence is part of the daily reference for many students. I do not simply think of violence as necessarily being experienced directly by the students, but as being so present in their environment that it becomes part of the way students makes sense of what they are doing and learning in school. Thus students at the Diego Rivera School are quite familiar with stories of undocumented relatives (those without legal immigration papers) "crossing over" (i.e., sneaking in) to the USA, or of "la migra", the USA Immigration and Naturalization Service, conducting raids on factories or farms to look for, and round up, undocumented workers who are known to toil in the sweatshop, non-unionised conditions of semi-slavery. These forms of violence permeate the consciousness of Rivera students. Naturally, research on a non-prototype mathematics classroom can focus on very many issues: the violence, poverty, immigration, and discrimination in general. However, in his presentation of Diego Rivera School, Gutstein describes, first of all, in what sense it is possible through mathematics education to develop a *conscientização*, to which I will return to in the third section.

Let us move from Chicago to Barcelona. When a student with a foreign cultural background enters a classroom, it appears from general observation that some difficulties do occur. One broadly accepted explanation for this is that immigrant students bring along norms that do not harmonise with the prevailing norms of the classroom. Thus, it could be assumed that cultural conflicts will arise. This interpretation, however, has been seriously questioned in the studies of Núria Gorgorió and Núria Planas (2005), who examine schools with many immigrants in neighbourhoods of Barcelona. They identify different norms that regulate classroom practices, such as "in this class we work collaboratively and people must help each other". Another norm is that "the contextualisation of the mathematics task should be considered seriously". Such norms stimulate practices that are open to explore situated mathematics; they facilitate group work and inquiry cooperation. In their study, they investigated classrooms that, according to the teachers, were ruled by such norms.

In their study, Gorgorió and Planas concentrate on two immigrant students, Ramia and Esmilde. As an outcome of the investigated episodes both students decided to become non-participants. An exclusion has taken place. Ramia did not accept the classroom norms. She restricted her considerations, with respect to a certain task, to purely mathematical aspects of proportionality, which implies that she calculated that 6.66 eggs should be applied in a recipe, and that this was the proper result. Here the teacher and other students disagreed and referred to the norm of contextualisation – contextualisation has to be taken seriously, and it does not make sense to use 6.66 eggs when cooking food or baking. However, Esmilde, also an immigrant student, was well aware of the norm of contextualisation, and he explicitly referred to it when a mathematical task concerned the density of the population in a neighbourhood familiar to him. He found that his knowledge was relevant to the solution of the problem. However, in this case both teacher and the other students claimed that the task concerned proportionality, and that there was no need to make any more fuss about this. The norm of contextualisation was abandoned for a while. The teachers were different in the two cases, but, as emphasised by Gorgorió and Planas, they were both open and supported the idea of situated learning in mathematics.

The interesting observation by Gorgorió and Planas is that the pattern of explanation (that immigrant students bring along norms that do not harmonise with the prevailing norms of the classroom) does not seem to apply generally. This brings us to a more bitter interpretation of the observation. The point is not simply that immigrant students bring different norms to the classroom, and in this way cause conflicts. It might very well be that immigrant students struggle to become members of the classroom community by observing prevailing norms. Other forms of exclusion might be operating. These forms need not have much to do with the fact that immigrant students do not observe such norms. They may do it, or they may not – either case could mean exclusion. It might happen that the teacher and students who are familiar with the norms choose to cancel the norms for a while in such a way that exclusion can be maintained. This might indicate that causes for exclusion are produced from within the established classroom community, and that they need not be sought in some divergent norms of the immigrant student. Reasons might be found in a stereotyping of "immigrant", and this stereotyping might be exercised through particular ways of administering recognised norms.[9]

Through these short visits to Chicago and Barcelona, I have only addressed particular examples of non-prototype classrooms.[10] The obtained insight in processes of possible empowerment and of how processes of inclusion and exclusion might operate cannot easily be identified if we concentrate on the prototype classroom. Therefore, I find it important that a critical mathematics education research programme should emphasise the importance of developing insight into the variety of socio-political, cultural, and economic contexts of mathematics education.

THE VARIETY OF FORMS OF MATHEMATICS IN ACTION

Mathematics forms part of many different work contexts, often in an implicit way, and as mathematical thinking, theories, and techniques are developing in all possible directions, the very notion of mathematics is developing as well. We cannot expect mathematics to represent any unity. "Mathematics" is an open concept that can form part of different language games and discourses. Mathematics can be implicated in different practices, without any shared "essence" in the forms that it takes. We can count numerous and very different activities as being mathematical. Examples are: The calculation of change at the baker's shop; solving cubic equations for homework; searching for a more efficient algorithm of prime factorisation; investigating the functioning of a robot arm by using matrix calculus; reading statistical figures; estimating the risks connected to the construction of an atomic power plan; planning the cheapest route for going on holiday at the beach; estimating how much to leave in tips at a restaurant; constructing the roof of a hut; weaving baskets; knitting a pullover; carrying out the planning of bridge construction; making a time schedule for a conference.

We can find mathematics everywhere. Or can we? We could extend this list much further by activities like: Watching somebody doing knitting; eating bread from the baker's shop; copying the solutions of cubic equation from a friend; forgetting to pay tips, etc. To call such activities mathematical seem to include a Herculean stretching of the concept of "mathematics". Nevertheless, the mathematics-is-everywhere discourse opens new perspectives for addressing mathematics in action.[11]

It is possible to find many different practices of mathematics in action, which I will try to illustrate.[12] And, as emphasised, we cannot assume that we are dealing with the "same" underlying mathematics. Keeping this in mind, I continue in the following to talk about mathematics.

A range of studies demonstrates that advanced forms of mathematics constitute part of many technological enterprises, also forms of mathematics that have been designated as pure. Mathematics makes an integral part of computer technology and of all forms of processing information and knowledge. As a particular example, one can just think of modern cryptography, which is a highly developed mathematical discipline. Applications of mathematics are pervasive in modern warfare. And in economic management as well. This applies to all businesses, as well as to governmental programmes. Any tax system is a tremendous form of mathematics brought into action. One can also consider the mathematics of surveillance and control, where automatic procedures are constituting new panopticons. Uwe Gellert and Eva Jablonka (2009) have explored the notions of mathematisation and demathematisation. Mathematisation refers to the phenomenon that many social practices have becomes structured through mathematics-based technological systems. We would not even be able to imagine what daily life without such systems operating might be like. At the same time, a demathematisation is taking place, as it is possible (if not necessary) to operate with the technology built up through mathematical packages without being aware of the particular content of the packages.

With reference to many such examples one can address the hegemony of certain forms of mathematics, and show how mathematics makes part of forms of oppression (see, for instance, Greer and Mukhopadhyay, Chapter 10, this volume). Mathematical rationality is a questionable rationality. It can operate in dubious or malevolent technological, economic, and military contexts. On the other hand, mathematical rationality can also operate in ways that are considered most benevolent. As with mathematics, so also the rationality of mathematics is without essence. We have to do with a rationality that is always in need of critique. This realization applies to all forms of mathematics: academic mathematics, engineering mathematics, ethnomathematics, etc.

We see mathematics in operation, often in an implicit form, in many work-processes. Tine Wedege (2002) has made many observations about mathematics at work.[13] As an example, she observed how the person responsible for loading an airplane has to take into account certain numbers which indicate how well-balanced the plane is before take off. The person has to make estimations and judgments based on figures and numbers. However, most often in such cases people do not think of themselves as doing any mathematics; we are dealing with a demathematisation. Many ethnomathematical studies have investigated practices that include mathematical structures and principles, and D'Ambrosio has not only provided the initial formulations of ethnomathematics but also ensured an ongoing elaboration, refinement, and generalisation of the ethnomathematical approach (see, for instance, D'Ambrosio 2010a, 2010b).

Mathematics contributes to our daily life, for example when information and decisions are presented with reference to numbers and figures. Statements from experts are expressed each and every day on television and in newspapers. Numbers could refer to proposed investments. The "necessity" of introducing some economic restrictions can be justified by certain calculations. Systems of salary and pensions reflect degrees of inflation and schemes of productivity. Taxes have to be paid, and the tax level is a permanently discussed issue. It has an impact on how each and every one of us may manage their daily-life situation. It seems that the whole forum of democratic debate is deluged by numbers, figures, and statistics. Experts can be interviewed, and they can express their opinion in public. A classic study was made by Jean Lave (1988), investigating mathematics at the supermarket, which appears to be an explicit example of mathematics in a practice of consumption.

We find many groups of people living in the fringe area of the informational economy. This economy, as organised according to dominant economic interests and priorities, excludes huge groups of people. Some groups of people seem to be swallowed up by the "black holes of informational capitalism".[14] People can be marginalised, if not excluded. Anyway, they are involved in practices including mathematics, for example in selling and buying. Madalena Santos and João Filipe Matos (2002) explored the mathematical practices of the *ardinas*, who are described as the young boys between 12 and 17 that sell newspapers in the streets of Praia, the capital of the Republic of Cabo Verde. Diana Bocasante (2009) investigated the mathematics that made part of the life of children whose existence was linked to the recycling of waste from the city.[15]

Being aware of how mathematics operates in many different practices is important in order to understand the complexity of mathematical rationality and the variety of mathematics in action. Such actions could mean domination and oppression. They could exercise the hegemony of academic mathematics. However, there are no general characteristics to be associated to mathematics in action. Like any action, so mathematics in action can have such qualities as being promising, relevant, surprising, devastating, spectacular, useful, expensive, valuable, risky, misleading, dangerous, etc. And this applies to whatever kind of mathematics that might have been brought into action, whether the most advanced provinces of academic mathematics or some traditional forms of cultural practices. Practices including mathematics have to be investigated in all their diversities. Critique is needed.

Being aware of the variety of practices brings the notion of articulation into focus. I interpret articulation as referring to any issue relating insight, competence, understanding, notions, techniques, tools, etc. operating in one context to "similar" insight, competence, understanding, notions, techniques, tools, etc. operating in a different context. Briefly, articulation concerns the bridging between different practices. When the variety of in-school and out-of-school practices that might include mathematics (in one or another interpretation of the notion) expands enormously, the number of articulation issues also grows. How to express the relationship between these different practices? What use might we make of this bridging in an educational setting? Articulation has to do with the production of meaning. When relationships are made visible between what is happening in the classroom and some practices outside school in which the students might become involved, a resource for students' construction of meaning has been established.

Questions about articulation in mathematics education have been formulated in terms of "transition" with reference to school organisation. For instance: How to ensure a smooth transition between the mathematics of secondary school and the mathematics of the upper secondary school? Such a transition problem has to do with the relationship between different parts of an overall curriculum. However, this issue is only a minor one within a network of articulation issues, and is one reason that I prefer to use the broader notion of articulation instead of transition, which includes a connotation of one-way direction. The students are, or could in later life become, involved in many different activities and practices that include mathematics, and it becomes an educational challenge to consider what articulation might be possible. Articulation problems could concern students from a small city in Denmark, an African township, a metropolis. Such problems emerge everywhere, but their significance depends on the context. Even when we address the "same" mathematical issues, say solving quadratic equations, the articulation problem might be very different for different groups of students. In particular, when we search outside the prototype mathematic classroom, we easily come to see how bridging points towards a huge variety of socio-political issues.

One example of the articulation issue was addressed by Marta Civil and Rosi Andrade (2002). The project they describe primarily took place in schools located in working class, Mexican-American communities in the USA. The children could be recent immigrants, or their parents could be immigrants. Some children were bilingual, some Spanish monolingual, some English monolingual. In the project, they tried to relate out-of-school mathematics with in-school mathematics. Establishing such relationships may ensure meaning to activities in the classroom, it may serve an "empowerment" of students. However, to establish such a relationship is problematic, which was also emphasised by Civil and Andrade. One element in the project was household visits. Here teachers could become aware of the children's' situation and learn in what tasks the children might be involved when at home. Such tasks could contain elements of mathematics. A mathematical archaeology, consisting of analysing a situation or a practice for its mathematical elements, could be carried out.[16] Such an archaeology could be helpful for reflecting on how to introduce a mathematical topic in school and what examples to use as illustrations. Certainly, difficulties were experienced be the teachers: "The transformation of household knowledge into pedagogical knowledge for the classroom is not easy. As much as we enjoyed the wealth of information that comes out of these household visits, we find ourselves constantly wondering about connection to the teaching of mathematics in school" (Civil and Andrade, 2002, p. 156).[17]

The aim of the project was not to *bring* home-mathematics to school but to *bridge* home-mathematics and school-mathematics. Home mathematics is one out-of-school site for mathematics. Mathematics at the workplace is another site, and in the project, the household visits were supplemented by interviews with a mechanic, a carpenter, a welder, a construction worker, and a seamstress. Could there, one way or another, be mathematics integrated in the practices in question? Could mathematics become excavated? Could some of the activities be read mathematically? Again, the difficulty of carrying out a mathematical archaeology was experienced.[18]

Reconsidering the very many different sites for learning mathematics (of which the prototype mathematics classroom is only one) as well as the very many sites of mathematics practices, we see that the articulation problem can take very many forms. Clarifying possibilities for articulation could be a sense-making activity for many students. In particular, one can think of bridging between activities in school and practices that could make part of the students' foregrounds.[19] The variety of sites for the teaching and learning of mathematics and the many forms of mathematics in action have to be addressed through a critical mathematics education research programme. In particular it becomes important to investigate the different forms of bridging between out-of-school practices with school practices. This is crucial for establishing meaning in mathematics education.

THE VARIETY OF EDUCATIONAL POSSIBILITIES

We could think of learning mathematics in general as preparation for any social practice rich in mathematics, explicit or implicit, in what we could call a functional or accommodating way. Mathematics education could regiment and discipline students. Alternatively, mathematics education could bring about competencies which can be described in terms of, for instance, Mündigkeit, empowerment, conscientização, mathemacy, ubuntu and botha, social justice, autonomy, equity, etc.[20] To explore what this could mean constitutes part of a critical mathematics education research programme. In this way, critical mathematics education addresses the question: What could it mean to try to establish an education for social justice in an unjust society? It fact it could turn out to be an impossible task, as any such attempt may be swallowed up by dominant socio-economic power structures. However, I do not look for any *a priori* answer to this question. I see it as an ongoing challenge, keeping in mind that, in a just society, an education for social justice would be unnecessary; conversely, teaching for social justice in, and through, mathematics education must be seen in the context of the greater struggle within an unjust society.

Much research in mathematics education is not addressing the variety of functions that could be exercised through mathematics education. In this sense it demonstrates "blindness", maybe by assuming that mathematics education contains an intrinsic value. This essentialism assumes that there is a positive value in mathematics education guaranteed by the very fact that this education addresses mathematics. Such an assumption will ensure that mathematics educators can operate as ambassadors of mathematics, with the conviction that they are acting on behalf of a good cause.

Let me point out some of the ways in which mathematics education can operate in a discriminating way. I will make remarks about (1) discrimination in terms of (lack of) resources, (2) racism, (3) sexism, (4) discrimination in terms of language, and (5) discrimination in terms of what is referred to as "ability".[21]

First, mathematics education presupposes investment. Computers enter the (prototype) classroom, and often they are celebrated as ensuring a new powerful learning environment. The computer, with the proper software, can engage students in mathematical activities. They can develop their creativity, they can conduct experiments and explorations, and they can construct mathematical knowledge. Computers can ensure motivation as well as "learning efficiency". So goes the celebration. What is seldom under discussion, however, is the implication of this observation for the majority of the world's children, who learn mathematics in classrooms without any computer in sight. Are they left behind? Do we have to deal here with a new form of social exclusion? A computer-can-be-taken-for-granted perspective pervades the discussion of technology in mathematics education. Naturally, there is no problem that particular studies of the use of technology in mathematics education assume that this technology is available, but the taken-for-granted perspective turns into a devastating research bias, if the majority of studies assume the perspective as a natural given.[22] To make the story brief: Mathematics education and poverty is not a topic explored widely in research in mathematics education. But it is an essential topic, as discrimination of learning opportunities are caused by an unequal distribution of resources.[23]

Second, it is not difficult to find examples of sheer racism exercised through mathematics education, especially when we consider the role of education throughout the apartheid era in South Africa. Other, possibly more indirect, expressions of racism are revealed when, as suggested by Wenda Bauchspies (2005), we consider to what degree learning, and in particular the learning of mathematics, can mean colonisation. There are many studies that have emphasised that learning mathematics in a particular form could serve as a suppression of an existing form of thinking. Munir Fasheh (1993, 1997) has talked about the "occupation of the mind" and related issues, and Herbert Khuzwayo (2000) has developed in further detail what this could mean for interpreting the operation of mathematics education during the apartheid era in South Africa (see also Fasheh, Chapter 3, this volume). With particular reference to the USA context, Danny Martin (2009) has discussed in what sense mathematics education could become a liberating force in the lives of black children (see also Martin & Gholson, Chapter 9, this volume).

Third, sexism, or the issue of gender, has been addressed in mathematics education for a longer period. Mathematics can be interpreted as a language giving access to power, to technology, to job opportunities. Statistics have documented the unequal distribution of men and women with respect to mathematics-dense studies and later jobs. It is apparent that mathematics education includes or materialises discrimination. Gender issues are analysed through a range of studies, and they have been broadened in an important way by Gelsa Knijnik (1996), with reference to the Movimento Sem Terra (Landless People Movement) in Brazil.

In that study, she focussed on a group of people suffering from poverty and social exclusion. Within such a group, the gender issue can assume definite shapes that could hardly be grasped by research set within a traditional framework (see also Knijnik and Wanderer, Chapter 8, this volume). In this way, gender studies become multidimensional by considering interactions with class, culture, and ethnicity.

Fourth, the language issue includes many socio-political controversies. As an example, according to formal regulations the language spoken in schools in Barcelona must be Catalan, while Spanish is forbidden. However, Catalan represents a middleclass culture, and students in poorer parts of Barcelona often come from other parts of Spain or from other Spanish-speaking countries. Although these students' mother tongue is Spanish, the teacher has to address them in Catalan. To be forced to learn (mathematics also) in a language different from one's mother tongue could amount to oppression. At the same time it could make available new life opportunities. Thus, mastering Catalan could turn out to be very useful for the students later in life. The conflict between the cultural suppression provoked by having to learn in a language different to one's mother tongue, and the possible advantages of accepting to be taught in the dominant language, has to be addressed with reference to the particular context. The conflict is, for instance, present in South Africa with 11 official languages. What is the price and what is the gain of learning mathematics in English?[24] The language issue is also an important aspect of many ethnomathematical studies.[25] Finally, I should mention that according to the dominant educational policy in Denmark, the approach to immigrant students is to teach them Danish as fast as possible. The issue of cultural oppression is simply ignored.

Fifth, discrimination in terms of ability could take the form of elitism in mathematics education. In many cases, it has been argued that it is important to differentiate between students according to their so-called "abilities" (certainly the notion of ability is a problematic one). Differentiation could turn into elitism when groups of students are treated differently according to their seemingly different capacities for learning mathematics and when those who have become labelled "best" are best resourced. Elitism might be seen as "justified" in economic terms by claiming it is most profitable to invest in the seemingly better students. But if we consider education as a human right, then this appeal to economic productivity as an underlying principle for an unequal distribution of learning possibilities appears absurd.

So, many studies emphasise that mathematics education is domesticated by many discriminating factors. One might even claim that mathematics education provides a flexible medium for dominant socio-economic power to act through.[26] Let us, anyway, try to formulate features of an educational imagination, which I also find forms part of critical mathematics education.

The many identified problematic functions of mathematics educations do not, nevertheless, exclude that one could try to look for alternatives. And in order to search for alternative it is useful to be aware of some notions that signify aspirations and visions, and whish might help in searching for possibilities. Notions like conscientização, mathemacy, critical citizenship and dialogue, as referred to previously, may inspire an exploration of educational possibilities. Exploration of educational possibilities may draw from the *pedagogical imagination* included in such notions. And certainly, one can imagine many other notions that support a pedagogical imagination making part of a critical mathematics education. I see it as a concern of critical mathematics education to explore such imagination.[27]

In his work at the Diego Rivera School, as referred to previously, Gutstein is inspired by the notion of *conscientização* that was explored by Paulo Freire (1972). This notion refers to the power of reading the world as being open to change. Reading the world, drawing on mathematical resources, means, according to Gutstein, to use mathematics to:

> … understand relations of power, resource inequities, and disparate opportunities between different social groups and to understand explicit discrimination based on race, class, gender, language, and other differences. Further, it means to dissect and deconstruct media and other forms of representation and to use mathematics to examine these various phenomena in one's immediate life and in the broader social world and to identify relationships and make connections between them. (Gutstein, 2003, p. 45)

This effort can be expressed as a concern for supporting the students' development of mathematical literacy or of mathemacy.[28]

The notion of *mathemacy* refers to some potentials of mathematics education. Sticking with this concept, it is complex and contains tensions, if not contradictions. It is a contested concept. The notion cannot be pinned down within a well-elaborated definition.[29] As a consequence, there is no recipe waiting for how to organise a practice that might support the development of mathemacy. This is an important point to consider when attempts are made to develop a practice of mathematics education with a critical dimension. In any case, very many examples of what it might mean to develop a mathemacy have been suggested. In Alrø and Skovsmose (2002), we describe a project in the classroom that addresses the uncertainties that accompany information put in numbers and statistics.[30] The project was about problems connected with salmonella-infected eggs. In the project, eggs took the form of empty film cases. Such black cases could easily be opened and checked. When the project started a whole population of eggs was brought into the classroom in a trolley containing 500 "eggs", of which 50 were infected by salmonella. These numbers were known to every student in the classroom. Thus, 450 eggs contained a healthy yolk, in the form of a yellow piece of plastic, while the remaining 50 contained a blue piece of plastic, indicating salmonella infection. The students worked in groups, and their first task was to select a sample of 10 eggs each from the trolley. Through this work, they became aware of the fact that a sample far from always reveals the "truth" about the population from which it is drawn. The students tried to find explanations for this observation: Could it be that the eggs in the trolley were not mixed up well enough? Or are samples rather unreliable messengers regarding properties of the whole population from which they are drawn? Such considerations point towards a more profound problem, namely that, in almost all real-life situations, we know nothing about the whole population except what is revealed through samples. This applies to any form of quality control of a product. In this way the project opened up to a broader discussion of the reliability of samples and of statistics. The overall point was to address the point that information provided by numbers might be far from reliable. To me considerations about reliability are an essential feature of developing mathemacy.[31]

The notion of *critical citizenship* may also direct a search for educational possibilities. The project City Planning, as describes by Denival Biotto Filho (2008), addressed the different features that makes part of city planning. A principal reference for the project was Rio Claro, a city in the interior of São Paulo State. As part of the project it became revealed that only 53% of the water that was delivered into the water supply system in fact became registered and paid for by the customers. Information about the disappearance of water could be expressed verbally, but putting it in numbers makes it possible to reflect in a more systematic way on the efficiency of the water supply system. One could consider, first, if the information is correct. How are the amounts of delivered and received water in fact calculated? Do some customers not figure in the system of measuring? One can also start addressing the possibility of locating the problem. Are there some ways of tapping water from the system without being measured? Are the water-supply systems leaking? Are there ways of estimating the water supply for specific neighbourhoods? Could the disappearance of the water be related to the amount of time the water supply system has been in service, which certainly might vary from neighbourhood to neighbourhood? One could also start considering the possible improvement of the water supply system. What could be the maximum percentage of delivered water that could possibly be registered by the customers (assuming that 100% would be a practical impossibility)? What is the situation in other cities? With an estimation of the optimal percentage, one could start gauging the yearly gain of having the 53% system repaired. Such a gain could in turn be compared to the cost of repairing the system. All such reflections can be carried out, and Biotto Filho's main conclusion based on the project City Planning is that mathematics is an important resource for formulating, strengthening, and specifying a broad variety of socio-political and economic reflections. This example illustrates that educational possibilities might be approached by considering what could be an interpretation of critical citizenship with reference to mathematics. As the same time the project illustrates important features of mathemacy.

The notion of *dialogue* may also be helpful in a searching for educational possibilities. It plays an important role in Renuka Vithal's (2003a) formulation of a pedagogy of conflict and dialogue within a post-apartheid South Africa. Through these two notions, she emphasises that the context in whish the education takes place is filled with conflicts, and that dialogue has to play an important role. Vithal is studying how different projects become acted out in the classroom in a historical Indian school that now included several black students from the neighbourhood. The students perceived the contextualised projects in very different ways, and, working with the contextualised projects, clashes of culture occurred in the classroom. The point is not to prevent or "eliminate" such clashes, but to see them as starting points of a dialogue. The context Vithal was addressing is of no prototypical format, and it is important for critical mathematics educations to explore possibilities in a variety of situations. Still, one should not forget that prototypical situations also need to be explored for alternatives. Thus, an indoctrination of dominant socio-economic priorities and an adaptation to the demands of the labour market of today's globalised capitalist order may be deeply rooted in the school mathematics tradition as this appears in its prototypical format.[32]

Searching for possibilities makes part of a critical mathematics education research programme. It is a task that concerns educational practice. It also concerns research methodology. It certainly concerns the interaction between theory and practice. It might even be seen as a way of eliminating the distinctions between theory and practice. Searching for educational possibilities and trying to formulate an educational imagination is a collective affair.[33]

CONCLUSIONS: UNCERTAINTIES

The lack of essence in mathematics as well as in mathematics education, to which I referred previously, produces uncertainty. This uncertainty concerns the whole research programme in critical mathematics education. It is a political programme, but the uncertainty eliminates possibilities for the programme to identify any theoretical strongholds. In this way, there is an important difference between the notion of critique as used in the expression "critical mathematics education research programme" and the modern conception of critique.

This conception was clearly expressed through *Critique of Pure Reason* where Immanuel Kant wanted to clarify the *a priori* conditions for obtaining knowledge, and also to outline what could not be known. He wanted to address our basic epistemic conditions through critical investigations. In this way, he wanted to establish a solid foundation for knowledge. The existence of such a foundation was also assumed to exist by the approaches that tried to associate certainty with scientific knowledge. Karl Marx elaborated the notion of critique to address not only theories of economy, but also the very economic structures themselves. In this way, Marx established critique as much more than an epistemic enterprise of a Kantian format. He established it as a political endeavour. Different as they might be, these two forms of critique share the assumption that it is possible to establish the critical activity on a solid foundation. By associating critique with certainty, both Kant and Marx contributed to a modern conception of critique.

The notion of critique got further developed through Critical Theory. It incorporated a strong inspiration from Marx, but it also included a new orientation. Critical Theory started out from a Marxist approach, as represented by Carl Grünberg, the first director of the Institute für Sozialforschung in Frankfurt. Max Horkheimer, who was to follow Grünberg as director of the Institute, introduced a broader perspective on what could be called critical social studies. A critical approach to the whole Enlightenment was formulated in the *Dialectic of Enlightenment* from 1944 by Max Horkheimer and Theodor W. Adorno. Another astonishing contribution of Critical Theory can be found in the studies made by Walter Benjamin. In *The Arcade Project*, he opened for a radically different approach to critical studies. Benjamin's studies of modern society took an almost surrealistic format by pasting together ideas, observations, and formulations. Apparently no coherent critical perspective was guiding this artistic-creative process. I see *Dialectics of Enlightenment* and *The Arcade Project* as transgressing the modern conception of critique. The ties between critique and certainty got loosened.

Through Michel Foucault's work the critique of modernity became underpinned by a profound archaeology of knowledge.[34] Foucault revealed how what might appear as a solid scientific conceptual framework includes notions, priorities, and interests of suspicious character. The Modern outlook, embracing thrusts of scientific knowledge and the modern conception of critique, became revealed, not as a universal, but as a particular (and peculiar) outlook. It could be questioned in many ways. These observations have an impact on the very notion of critique.

It appears that we have to give up the possibility of formulating critique as based on any solid foundation. This observation is crucial for any form of critical activity that tries to step out of the epistemic project that formed part of the Modern outlook, whatever the protection established by Kant or by Marx. One cannot associate critique with any form of certainty, whether critique is epistemic or political. This also applies to critique with respect to education. When formulating any research programme of critical mathematics education it becomes important to observe that the notion of critique is open to radical changes. Certainty has been established through layers of assumptions linked to Modernity. But when such assumptions become deconstructed, critique turns into an uncertain activity. A critique becomes tentative and preliminary. It becomes fragile.[35]

This fragility becomes exposed when the many different sites for teaching and learning are investigated; it becomes exposed when the variety of sites for mathematics in action becomes explored; and it becomes exposed through the variety of concepts through which one may direct a pedagogical imagination, such as conscientização, mathemacy, social justice, and dialogue. Such notions are explosive, as any attempt to clarify them brings us to use other concepts just as complex as the ones we set out to clarify. Thus, trying to clarify what social justice might mean could bring about a discussion of democracy, citizenship, empowerment, equity, etc. As soon as we have to do with explosive concepts, we realize that any attempt to provide a conceptual clarification ends up in a proliferation of issues each of which might be as complex to clarify as the initial issue. This does not mean, however, that one needs to avoid using such notions. They play important roles in a critical mathematics education research programme. At the same time they represent a profound uncertainty, an integral part of the programme.

Finally, it needs to be observed that these concluding remarks have implications for the very notion of "programme". Talking about a research programme may include assumptions about the possibility of pointing out directions and formulating general perspectives with respect to what to research, how to research, and why to research. And that it is possible to do so with a degree of well-argued certainty. However, the observations about uncertainty might have brought this chapter outside the scope of its title. Instead of research programme it might be more appropriate just to talk about preoccupations for critical mathematics educations. Maybe only the final question mark of the title should remain.

ACKNOWLEDGEMENTS

This paper is partly based on Skovsmose (2006), which represented my contribution to one of the survey teams for the 10th International Congress on Mathematics Education in Copenhagen, 2004. I want to thank Paulus Gerdes, Gelsa Knijnik, and Miriam Godoy Penteado for suggestions for improving the paper.

NOTES

1 See Apple (2000) for exploring connections between standards, markets, and inequality in education. See also Brown (2010) and Greer (Chapter 4, this volume).

2 There are many sources of inspiration for formulating such a programme. Let me just mention some general references: Alrø, Ravn and Valero (Eds.) (2010); Ernest, Greer and Sriraman (Eds.) (2009); Greer, Mukhopadhyay, Powell, and Nelson-Barber (Eds.) (2009); and Sriraman (Ed.) (2008); and some particular studies: Anderson (2010); Frankenstein (2010); Jacobsen (2010); Lesser (2010); Pais (2010); Planas and Civil (2010). Povey (2010); Rasmussen (2010); Sánchez and Blomhøj (2010); and Valero and Stentoft (2010). See also D'Ambrosio (1992) for a formulation of an ethnomathematical research programme.

3 For a presentation of some concerns of critical mathematics educations, see, for instance, Skovsmose (2010a).

4 This observation can be underpinned with many statistics. See, for instance, UNESCO (2000).

5 A preliminary empirical underpinning of this point can be developed along the same lines as the survey presented in Skovsmose and Valero (2002b).

6 For comments on the political economy of mathematics education, see Skovsmose and Valero (2002a).

7 If we try to move beyond prototypical settings, we face issues like the "disruption for data" and the notion of generalisability and validity have to be rethought. See, for instance, Valero (2004, 2007); Valero and Zevenbergen (Eds.) (2004); Vithal and Valero (2003), Vithal (2007), and Atweh et al., (Eds.) (2007) for a broader look at mathematics education research.

8 Gutstein has provided me with supplementary information for the present description. See also Gutstein (2006, 2008, 2009, Chapter 1, this volume).

9 However, Gorgorió and Planas (2005) do not leave us here. They suggest that negotiation of norms might be a useful strategy for coping with these situations. By a negotiation, norms and their justification can be made explicit.

10 See also Skovsmose and Penteado (2011).

11 By calling an activity mathematical, it may be "colonised" as a "domain" for mathematics education. Dowling (1998) has emphasised this in his critique of ethnomathematics. Knijnik (1996) also addresses this issue. For a broader discussion for mathematics in context, see also Restivo and Collins (2010).

12 For examples of mathematics in action see, for instance, Christensen, Skovsmose and Yasukawa (2009); Skovsmose (2005, 2009a); Skovsmose and Yasukawa (2009); and Yasukawa, Skovsmose, and Ravn (Chapter 12, this volume). For broader critical perspectives on mathematics, see, for instance, Ravn (2010); Baber (2010); and Skovsmose (2010a, 2010b).

13 See also Yasukawa and Brown (Chapter 11, this volume).

14 A formulation coined by Castells (1998, p. 162).

15 See also Knijnik and Bocasante (2010); Vithal (2003b, 2009); and Mesquita (2004).

16 See Skovsmose (1994) for a discussion of mathematical archeology in an educational context.

17 See also Civil (Chapter 5, this volume) and Baber (Chapter 6, this volume). For a critical investigation of contextualisation in mathematics education see Andersson and Ravn (Chapter 14, this volume).

18 Any such archaeology runs the risk of presuming what is questioned: mathematics is found in the practice, often to the great surprise of those who are involved in the practice, and the mathematics found might well be the mathematics included in the perspective of those who interpret the practice.

19 See, for instance, Baber (2007); Skovsmose, Alrø and Valero in collaboration with Silvério and Scandiuzzi (2008); and Skovsmose, Scandiuzzi, Valero and Alrø, H. (2008). See also Lindenskov (2010).

20 The German notion *Mündigkeit* has been used broadly in the educational discourse. The Portuguese notion conscientização has been used by Freire to indicate a crucial aspect of literacy. The Zulu term *ubuntu* and the Sotho term *botha*, have both achieved educational significance through the work of Bopape (2002). The notion of social justice has been explored in many contexts, see, for instance Burton (Ed.) (2003); Gates (2006); and Frankenstein (1995). See also Pais (Chapter 2, this volume) for a critical approach to research on equity.

21 For more general observations about power and mathematics education, see Christensen, Stentoft and Valero (2007) and Valero (2007, 2009).

22 See, Penteado and Skovsmose (2009).

23 For an exploration of this issue, see Berliner (2005).

24 The complexities of such multilingual issues are addressed by many; see for instance, Setati (2005). The hegemony of English in the USA context has been analysed by Macedo, Dendrinos and Gounari (2003). See also Setati and Planas (Chapter 7, this volume).

25 See, for instance, Barton (2008).

26 See also Skovsmose (2008).

27 The notion of pedagogical imagination can be related to the notion of sociological imagination as proposed by Wright Mills (1959). Explorations of educational possibilities are found in: Appelbaum with Allen (2008), who discuss a variety of examples; Araújo (Ed.) (2007), who has brought together a set of examples which shows a close connection between ethnomathematical and the critical mathematics outlook; Gutstein (2006), who has investigated possibilities with particular reference to a Freirean framework; Mora (Ed.) (2005), who presents critical mathematics education in a variety of contexts, both theoretical and practical; Almeida (2010) and Corlu (2010, who explore the notion of democracy with respect to mathematics education; Appelbaum (2010), who explores sense and representation in elementary mathematics; and Greer (2008), who discusses how numbers and figures enter public discourses. See also Frankenstein (1989, 1998) for a range of examples, as well as Jablonka and Gellert (Chapter 13, this volume) and Atweh (Chapter 15, this volume).

28 A careful clarification of the notion of "mathematical literacy" is presented by Eva Jablonka (2003). See also Jablonka (2010); Johnston and Yasukawa (2001); and Mukhopadhyay and Greer (2009), on education for statistical empathy.

29 See Skovsmose (2005) for a presentation of mathemacy and Chronaki (2010) for a fascinating exploration of the notion.

30 The example was developed in collaboration with Helle Alrø, Morten Blomhøj, Henning Bødtkjer, and Mikael Skånstrøm .

31 The project continued and addressed the notion of responsibility, which is also an important feature of mathemacy.

32 See also Alrø and Johnsen-Høines (2010) for an exploration of possibilities where dialogue plays a crucial role. See Alrø and Skovsmose (2002) for presentation of dialogue as a crucial notion in mathematics education.

33 A presentation of what is could mean to research possibilities is found in Skovsmose (2009b). See also Vithal (2010); and Yasukawa (2010).

34 See, for instance, Foucault (1989, 1994).

35 See Skovsmose (2009a) for an exploration of "doubt" and Ernest (2010) for an explicit critique of critical mathematics education.

REFERENCES

Albert, J. & Natshed, I. (2002). Using mathematics to break down stereotypes. In P. Valero & O. Skovsmose (Eds.) *Proceedings of the Third International Mathematics Education and Society Conference* (pp. 126–129). Copenhagen, Roskilde, Aalborg: Centre for Research in Learning Mathematics, Danish University of Education, Roskilde University, and Aalborg University.

Almeida, D. F. (2010). Are there viable connections between mathematics, mathematical proofs, and democracy? *Philosophy of Mathematics Education Journal.* Special issue on Critical Mathematics Education, Vol. 25.

Alrø, H., & Johnsen-Høines, M. (2010). Critical dialogue in mathematics education. In H. Alrø, O. Ravn & P. Valero (Eds.), *Critical mathematics education: Past, present and future* (pp. 11–21). Rotterdam: Sense Publishers.

Alrø, H., Ravn, O., & Valero, P. (Eds.). (2010). *Critical mathematics education: Past, present and future.* Rotterdam: Sense Publishers.

Alrø, H., & Skovsmose, O. (2002). *Dialogue and learning in mathematics education: Intention, reflection, critique.* Dordrecht, The Netherlands: Kluwer.

Alrø, H., Skovsmose, O., & Valero, P. (2009). Inter-viewing foregrounds: Students' motives for learning in a multicultural setting. In M. César & K. Kumpulainen (Eds.), *Social interactions in multicultural settings* (pp. 13–37). Rotterdam: Sense Publishers.

Anderson, A. (2010). Can a critical pedagogy in mathematics lead to achievements, engagement, and social empowerment? *Philosophy of Mathematics Education Journal.* Special issue on Critical Mathematics Education, Vol. 25.

Appelbaum, P. (2010). Sense and representation in elementary mathematics. *Philosophy of Mathematics Education Journal.* Special issue on Critical Mathematics Education, Vol. 25.

Appelbaum, P., with Allen, D. S. (2008). *Embracing mathematics: On becoming a teacher and changing with mathematics.* New York: Routledge.

Apple, M. W. (2000). Reform through conservative modernization: Standards, markets, and inequality in education. In J. Boaler (Ed.), *Multiple perspectives on mathematics teaching and learning* (pp. 243–259). Westport, CT: Ablex.

Araújo, J. de L. (Ed.). (2007). *Educação matemática crítica: Reflexões e diálogos.* Belo Horizonte, Brazil: Argumentum.

Atweh, B., Keitel, C., Calabrese Barton, A., Vistro-Yu, C., Borba, M., Vithal, R., & Gough, N. (Eds.) (2007). *Internationalisation and globalisation in mathematics and science education.* New York: Springer.

Baber, S. A. (2007). *Interplay of citizenship, education and mathematics: Formation of foregrounds of Pakistani immigrants in Denmark.* Unpublished PhD thesis. Aalborg: Aalborg University.

Baber, S. A. (2010). Mathematics from the perspective of critical sociology. In H. Alrø, O. Ravn, & P. Valero (Eds.), *Critical mathematics education: Past, present and future* (pp. 23–29). Rotterdam: Sense Publishers.

Barton, B. (2008). *The language of mathematics: Telling mathematical tales*. New York: Springer.

Bauchspies, W. K. (2005). Sharing shoes and counting years: Mathematics, colonalisation and communication. In A. Chronaki & I. M. Christiansen (Eds.), *Challenging perspectives on mathematics classroom communication* (pp. 237–259). Greenwich: Information Age Publishing.

Benjamin, W. (1999). *The Arcades Project*. Cambridge, MA: Harvard University Press.

Berliner D. C. (2005). *Our impoverished view of educational reforms*. [Based on a presidential invited speech given at the meetings of the Americal Educational Research Association]. Retrieved from http://www.educationanddemocracy.org/Resources/Berliner.pdf

Beth, E. W., & Piaget, J. (1966). *Mathematical epistemology and psychology*. Dordrecht: Reidel.

Biotto Filho, D. (2008). *O desenvolvimento da matemacia no trabalho com projetos*. Masters dissertation. Universidade Estadual Paulista (UNESP), Instituto de Geociências e Ciências Exatas, Campus Rio Claro.

Bishop, A. J. (1990). Western mathematics: The secret weapon of cultural imperialism. *Race and Class, 32*(2), 51–65.

Bocasante, D. M. (2009). A gente não quer só comida: Processos educativos, crianças catadoras e sociedade de consumidores. Masters dissertation. Universidade do Vale do Rio dos Sinos, Programa do Pós-Graduação em Educação.

Bopape, M. (2002). *Mathematics school based in-service training (SBINSET): A study of factors contributing towards success or failure of SBINSET in the South African school context*. Doctoral dissertation. Aalborg, Denmark: Aalborg University.

Brown, T. (2010). Cultural continuity and consensus in mathematics education. *Philosophy of Mathematics Education Journal*. Special issue on Critical Mathematics Education, Vol. 25.

Burton, L. (Ed.). (2003). *Which way social justice in mathematics education?* Westport, CT: Praeger.

Castells, M. (1998). *The information age: Economy, society and culture. Volume III, End of millennium*. Oxford: Blackwell.

Christensen, O. R., Skovsmose, O., & Yasukawa, K. (2009). The mathematical state of the world: Explorations into the characteristics of mathematical descriptions. In B. Sriraman & S. Goodchild, (Eds.), *Relatively and philosophically earnest: Festschrift in honor of Paul Ernest's 65th Birthday* (pp. 83–96). Charlotte, NC: Information Age Publishing.

Christensen, O. R., Stentoft, D., & Valero, P. (2007). Power distribution in the network of mathematics education practices. In E. de Freitas & K. Nolan (Eds.), *Opening the research text: Critical insights and in(ter)ventions into mathematics education* (pp. 131–146). New York: Springer.

Chronaki, A. (2010). Revisiting mathemacy: A process-reading of critical mathematics education. In H. Alrø, O. Ravn & P. Valero (Eds.), *Critical mathematics education: Past, present and future* (pp. 31–49). Rotterdam: Sense Publishers.

Civil, M., & Andrade R. (2002). Transitions between home and school mathematics: Rays of hope amidst the passing clouds. In G. de Abreu, A. Bishop & N. C. Presmeg (Eds.), *Transitions between contexts of mathematical practices* (pp. 149–169). Dordrecht: Kluwer.

Corlu, M. S. (2010). A historical analysis of democracy in mathematics and mathematics education in European cultures. *Philosophy of Mathematics Education Journal*. Special issue on Critical Mathematics Education, Vol. 25.

D'Ambrosio, U. (1992). Ethnomathematics: A research programme on the history and philosophy of mathematics with pedagogical implications. *Notices of the American Mathematics Society, 39*, 1183–1185.

D'Ambrosio, U. (2006). *Ethnomathematics: Link between traditions and modernity*. Rotterdam: Sense Publishers.

D'Ambrosio, U. (2010a). Mathematics education and survival with dignity. In H. Alrø, O. Ravn & P. Valero (Eds.), *Critical mathematics education: Past, present and future* (pp. 51–63). Rotterdam: Sense Publishers.

D'Ambrosio, U. (2010b). Ethnomathematics: A response to the changing role of mathematics in society. *Philosophy of Mathematics Education Journal*. Special issue on Critical Mathematics Education, Vol. 25.

Dowling, P. (1998). *The sociology of mathematics education: Mathematical myths/pedagogic texts*. London: Falmer.

Ernest, P. (2010). The scope and limits of critical mathematics education. In H. Alrø, O. Ravn & P. Valero (Eds.), *Critical mathematics education: Past, present and future* (pp. 65–87). Rotterdam: Sense Publishers. (Also published in *Philosophy of Mathematics Education Journal*. Special issue on Critical Mathematics Education, Vol. 25.)

Ernest, P., Greer, B. & Sriraman, B. (Eds.) (2009). *Critical issues in mathematics education*. Charlotte, NC: Information Age Publishing.

Fasheh, M. (1993). From a dogmatic, ready-answer approach of teaching mathematics towards a community-building, process-orientated approach. In C. Julie, D. Angelis & Z. Davis (Eds.), *Political dimensions of mathematics education 2: Curriculum reconstruction for society in transition* (pp. 15–19). Cape Town: Maskew Miller Longman.

Fasheh, M. (1997). Mathematics, culture, and authority. In A. B. Powell & M. Frankenstein (Eds.), *Ethnomathematics: Challenging Eurocentrism in mathematics education* (pp. 273–290). Albany, NY: State University of New York Press.

Foucault, M. (1989). *The archeology of knowledge*. London: Routledge. (First French edition 1969.)

Foucault, M. (1994). *The order of things: An archaeology of the human sciences*. New York: Vintage Books. (First French edition 1966.)

Frankenstein, M. (1989). *Relearning mathematics: A different third R – Radical maths*. London: Free Association Books.

Frankenstein, M. (1995). Equity in mathematics education: Class in the world outside the class. In W. Secada, E. Fennema & L. Adajian (Eds.), *New directions for equity in mathematics education* (pp. 165–190). Cambridge: Cambridge University Press.

Frankenstein, M. (1998). Reading the world with maths: Goals for a critical mathematical literacy curriculum. In P. Gates (Ed.), *Proceedings of the First International Mathematics Education and Society Conference* (pp. 180–189). Nottingham: Centre for the Study of Mathematics Education, Nottingham University.

Frankenstein, M. (2010). Critical mathematics education: An application of Paulo Freire's epistemology. *Philosophy of Mathematics Education Journal*. Special issue on Critical Mathematics Education, Vol. 25.

Freire, P. (1972). *Pedagogy of the oppressed*. Harmondsworth: Penguin Books.

Gates, P. (2006). The place of equity and social justice in the history of PME. In A. Gutérrez & P. Boero (Eds.), *Handbook of research on the psychology of mathematics education: Past, present and future* (pp. 367–402). Rotterdam: Sense Publishers.

Gellert, U., & Jablonka, E. (2009).The demathematising effect of technology: Calling for critical competence. In P. Ernest, B. Greer & B. Sriraman (Eds.), *Critical issues in mathematics education* (pp. 19–24). Charlotte, NC: Information Age Publishing.

Gorgorió, N., & Planas, N. (2005). Social representations as mediators of mathematics learning in multiethnic classrooms. *European Journal of Psychology of Education, 20*(1), 91–104.

Greer, B. (2008). Discounting Iraqi deaths: A societal and educational disgrace. In J. F. Matos, P. Valero & K. Yasukawa (Eds.). *Proceedings of the Fifth International Mathematics Education and Society Conference I-II* (pp. 293–301). Centro de Investigação em Educação, Universidade de Lisboa, and Department of Education, Learning and Philosophy, Aalborg University.

Greer, B., Mukhopadhyay, S., Powell, A. B., & Nelson-Barber, S. (Eds.) (2009). *Culturally responsive mathematics education*. New York: Routledge.

Gutstein, E. (2003). Teaching and learning mathematics for social justice in an urban, Latino school. *Journal for Research in Mathematics Education, 34*, 37–73.

Gutstein, E. (2006). *Reading the world with mathematics*. New York and London: Routledge.

Gutstein, E. (2008). Building political relationships with students. In E. de Freitas & K. Nolan (Eds.), *Opening the research text: Critical insights and in(ter)ventions into mathematics education* (pp. 189–204). New York: Springer.

Gutstein, E. (2009). Possibilities and challenges in teaching mathematics for social justice. In P. Ernest, B. Greer & B. Sriraman (Eds.), *Critical issues in mathematics education* (pp. 351–373). Charlotte, NC: Information Age Publishing.

Horkheimer, M., & Adorno. T. W. (2002). *Dialectic of enlightenment.* [Translated by John Cumming]. New York: Continuum. (Originally published 1944).

Jablonka, E. (2003). Mathematical literacy. In A. J. Bishop, M. A. Clements, C. Keitel, J. Kilpatrick & F. K. S. Leung (Eds.), *Second international handbook of mathematics education* (pp. 75–102). Dordrecht: Kluwer.

Jablonka, E. (2010). Reflections on mathematical modelling. In H. Alrø, O. Ravn & P. Valero (Eds.), *Critical mathematics education: Past, present and future* (pp. 89–100). Rotterdam: Sense Publishers.

Jacobsen, L. J. (2010). Embedding mathematics in the elementary teacher education curriculum network. *Philosophy of Mathematics Education Journal.* Special issue on Critical Mathematics Education, Vol. 25.

Johnston, B., & Yasukawa, K. (2001). Numeracy: Negotiating the world through mathematics. In B. Atweh, H. Forgasz & B. Nebres (Eds.), *Sociocultural research on mathematics education* (pp. 279–294). Mahwah, NJ: Lawrence Erlbaum.

Kant, I. (1933). *Critique of pure Reason.* [Translated by Norman Kemp Smith]. London: MacMillan. (Original German edition 1791).

Khuzwayo, H. (2000). *Selected views and critical perspectives: An account of mathematics education in South Africa from 1948 to 1994.* Unpublished doctoral dissertation. Aalborg, Denmark: Aalborg University.

Knijnik, G. (1996). *Exclução e resistência: Educação matemática e legitimidade cultural.* Porto Alegre, Brazil: Artes Médicas.

Knijnik, G., & Bocasante, D. M. (2010). Do ofício da pesquisa: Pílulas neoaforistica [About the craft of research: Neoaphoristic pills]. In H. Alrø, O. Ravn & P. Valero (Eds.), *Critical mathematics education: Past, present and future* (pp. 101–120). Rotterdam: Sense Publishers.

Lave, J. (1988). *Cognition in practice.* Cambridge: Cambridge University Press.

Lesser, L. M. (2010). The necessity of equity in teaching statistics. *Philosophy of Mathematics Education Journal.* Special issue on Critical Mathematics Education, Vol. 25.

Lindenskov, L. (2010). Students' curriculum in critical mathematics education. In H. Alrø, O. Ravn & P. Valero, (Eds.), *Critical mathematics education: Past, present and future* (pp. 121–131). Rotterdam: Sense Publishers.

Macedo, D., Dendrinos, B., & Gounari, P. (2003). *The hegemony of English.* Boulder, CO: Paradigm Publishers.

Martin, D. B. (2009). *Mathematics teaching, learning, and liberation in the lives of black children.* New York: Routledge.

Mesquita, M. (2004). O conceito de espaco na cultural de criança em situatição de rua: Um estudo etnomatemático. In J. P. M. Ribeiro, M. do C. S. Domite & R. Ferreira (Eds.). *Etnomatemática: Papel, valor e significado* (pp. 125–136). São Paolo: Zouk.

Mora, D. (Ed.). (2005). *Didáctica crítica, educación crítica de las matemáticas y etnomatemática: Perspectivas para la transformación de la educación matemática en América Latina.* La Paz: Gidem.

Mukhopadhyay, S., & Greer, B. (2009). How many deaths? Education for statistical empathy. In P. Ernest, B. Greer & B. Sriraman (Eds.), *Critical issues in mathematics education* (pp. 169–189). Charlotte, NC: Information Age Publishing.

Pais, A. (2010). Portrait of an influence. In H. Alrø, O. Ravn & P. Valero (Eds.), *Critical mathematics education: Past, present and future* (pp. 133–144). Rotterdam: Sense Publishers.

Penteado, M. G., & Skovsmose, O. (2009). How to draw with a worn-out mouse? Searching for social justice through collaboration. *Journal for Mathematics Teacher Education, 12*(3), 217–230.

Planas, N., & Civil, M. (2010). Discourse processes in critical mathematics education. In H. Alrø, O. Ravn & P. Valero (Eds.), *Critical mathematics education: Past, present and future* (pp. 145–159). Rotterdam: Sense Publishers.

Povey, H. (2010). Teaching for equity, teaching for mathematical engagement. *Philosophy of Mathematics Education Journal*. Special issue on Critical Mathematics Education, Vol. 25.

Ravn, O. (2010). Step beyond mathematics: Investigations in the philosophy of mathematics. In H. Alrø, O. Ravn & P. Valero (Eds.), *Critical mathematics education: Past, present and future* (pp. 171–181). Rotterdam: Sense Publishers.

Rasmussen, P. (2010). The critical perspective on education and on mathematics education. In H. Alrø, O. Ravn & P. Valero (Eds.), *Critical mathematics education: Past, present and future* (pp. 161–169). Rotterdam: Sense Publishers.

Restivo, S., & Collins, R. (2010). Mathematics and civilisation. *Philosophy of Mathematics Education Journal*. Special issue on Critical Mathematics Education, Vol. 25.

Sánchez, M., & Blomhøj, M. (2010). The role of mathematics in politics as an issue for mathematics teaching. *Philosophy of Mathematics Education Journal*. Special issue on Critical Mathematics Education, Vol. 25.

Santos, M., & Matos, J. F. (2002). Thinking about mathematical learning with Cabo Verde *Ardinas*. In G. de Abreu, A. Bishop & N. C. Presmeg (Eds.), *Transitions between contexts of mathematical practices* (pp. 81–122). Dordrecht: Kluwer.

Setati, M. (2005). Teaching mathematics in a primary multilingual classroom. *Journal of Research in Mathematics Edcuation, 36*, 447–466.

Skovsmose, O. (1994). *Towards a philosophy of critical mathematics education*. Dordrecht: Kluwer.

Skovsmose, O. (2005). *Travelling through education: Uncertainty, mathematics, responsibility*. Rotterdam: Sense Publishers.

Skovsmose, O. (2006). Research, practice, uncertainty, and responsibility. *Journal of Mathematical Behavior, 25*(4), 267–284.

Skovsmose, O. (2008). Mathematics education in a knowledge market In: E. de Freitas & K. Nolan (Eds.), *Opening the research text: Critical insights and in(ter)ventions into mathematics education* (pp. 159–174). New York: Springer.

Skovsmose, O. (2009a). *In doubt: About language, mathematics, lnowledge and life-worlds*. Rotterdam: Sense Publishers.

Skovsmose, O. (2009b). Researching possibilities. In M. Setati, R. Vithal, C. Malcolm & R. Dhunpath (Eds.), *Researching possibilities in mathematics, science and technology education* (pp. 105–119). New York: Nova Science Publishers.

Skovsmose, O. (2010a). Critical mathematics education – in terms of concerns. In B. Sriraman, C. Bergsten & S. Goodchild (Eds.), *The first sourcebook on Nordic research in mathematics education* (pp. 671–682). Charlotte, NC: Information Age Publishing.

Skovsmose, O. (2010b). Mathematics: A critical rationality. *Philosophy of Mathematics Education Journal*. Special issue on Critical Mathematics Education, Vol. 25.

Skovsmose, O. (2011). *An invitation to critical mathematics education*. Rotterdam, The Netherlands: Sense Publishers.

Skovsmose, O., & Penteado, M. G. (2011). Ghettoes in the classroom and the construction of possibilities. In B. Atweh, M. Graven, W. Secada & P. Valero (Eds.), *Mapping equity and quality in mathematics education* (pp. 77–90). New York: Springer.

Skovsmose, O., Alrø, H., & Valero, P. in collaboration with Silvério, A. P., & Scandiuzzi, P. P. (2008). "Before you divide you have to add": Inter-viewing Indian students' foregrounds. In B. Sriraman (Ed.), *International perspectives on social justice in mathematics education*. The Montana Mathematics Enthusiast Monograph 1 (pp. 209–230). Charlotte, NC: Information Age Publishing.

Skovsmose, O., Scandiuzzi, P. P., Valero, P., & Alrø, H. (2008). Learning mathematics in a borderland position: Students' foregrounds and intentionality in a Brazilian favela. *Journal of Urban Mathematics Education, 1*(1), 35–59.

Skovsmose, O., & Valero, P. (2002a). Mathematics education in a world apart – where we are all together. In P. Valero & O. Skovsmose (Eds.), *Proceedings of the Third International Mathematics Education and Society Conference* (pp. 1–9). Copenhagen, Roskilde, Aalborg: Centre for Research in Learning Mathematics, Danish University of Education, Roskilde University Centre, Aalborg University.

Skovsmose, O., & Valero, P. (2002b). Democratic access to powerful mathematical ideas. In L. English (Ed.), *Handbook of international research in mathematics education* (pp. 383–407). Mahwah, NJ: Lawrence Erlbaum.

Skovsmose, O., & Yasukawa, K. (2009). Formatting power of "mathematics in a package" A challenge for social theorising? In P. Ernest, B. Greer & B. Sriraman (Eds.), *Critical issues in mathematics education* (pp. 255–281). Charlotte, NC: Information Age Publishing.

Sriraman, B. (Ed.). (2008). *International perspectives on social justice in mathematics education*. The Montana Mathematics Enthusiast, Monograph 1. Charlotte, NC: Information Age Publishing.

United Nations Educational, Scientific, and Cultural Organization (2000). *Education for all: Statistical assessment, 2000*. Paris: Author. Retrieved from http://unesdoc.unesco.org/images/0012/001204/120472e.pdf

Valero, P. (2002). *Reform, democracy, and mathematics education: Towards a socio-political frame for understanding change in the organization of secondary school mathematics*. Doctoral Dissertation. Copenhagen: Department of Curriculum Research, The Danish University of Education.

Valero, P. (2004). Postmodernism as an attitude of critique to dominant mathematics education research. In M. Walshaw (Ed.), *Mathematics education within the postmodern* (pp. 35–54). Greenwich, CT: Information Age Publishing.

Valero, P. (2007). In between the global and the local: The politics of mathematics education reform in a globalized society. In B. Atweh, C. Keitel, A. Calabrese Barton, C. Vistro-Yu, M. Borba, R. Vithal & N. Gough, (Eds.), *Internationalisation and globalisation in mathematics and science education* (pp. 421–439). New York: Springer.

Valero, P. (2009). What has power to do with mathematics education? In P. Ernest, B. Greer & B. Sriraman (Eds.), *Critical issues in mathematics education* (pp. 237–254). Charlotte, NC: Information Age Publishing.

Valero, P., & and Stentoft, D. (2010). The "post" move of critical mathematics education. In H. Alrø, O. Ravn & P. Valero (Eds.), *Critical mathematics education: Past, present and future* (pp. 183–195). Rotterdam: Sense Publishers.

Valero, P., & Zevenbergen, R. (Eds.). (2004). *Researching the socio-political dimensions of mathematics education: Issues of power in theory and methodology*. Dordrecht: Kluwer.

Vithal, R. (2003a). *In search of a pedagogy of conflict and dialogue for mathematics education*. Dordrecht: Kluwer.

Vithal, R. (2003b). Teachers and "street children": On becoming a teacher of mathematics. *Journal of Mathematics Teacher Education, 6*(2), 165–183.

Vithal, R. (2007). The "uncivilised" scientist. In U. Gellert & E. Jablonka (Eds.), *Mathematisation – demathematisation: Social, philosophical, sociological and educational ramifications* (pp. 95–105). Rotterdam: Sense Publishers.

Vithal, R. (2009). Researching, and learning mathematics at the margin: From "shelter" to school. In P. Ernest, B. Greer, & B. Sriraman (Eds.), *Critical issues in mathematics education* (pp. 475–484). Charlotte, NC: Information Age Publishing.

Vithal, R. (2010). Democratising mathematics education doctoral research teaching and learning: Undoing the North-South divide. In H. Alrø, O. Ravn & P. Valero (Eds.), *Critical mathematics education: Past, present and future* (pp. 197–207). Rotterdam: Sense Publishers.

Vithal, R., & Valero, P. (2003). Researching mathematics education in situations of social and political conflict. In A. J. Bishop, M. A. Clements, C. Keitel, J. Kilpatrick & F. K. S. Leung (Eds.), *Second international handbook of mathematics education* (pp. 545–591). Dordrecht: Kluwer.

Wedege, T. (2002). Numeracy as a basic qualification in semi-skilled jobs. *For the Learning of Mathematics, 22*(3), 23–28.

Wright Mills, C. (1959). *The sociological imagination.* Oxford: Oxford University Press.
Yasukawa, K. (2010). Educating critical mathematics educators: Challenges for teacher educators. In H. Alrø, O. Ravn & P. Valero (Eds.), *Critical mathematics education: Past, present and future* (pp. 209–224). Rotterdam: Sense Publishers.

Ole Skovsmose
Aalborg University, Denmark
State University of São Paulo, Brazil

CHAPTER 17

OPENING THE CAGE? CRITICAL AGENCY IN THE FACE OF UNCERTAINTY

OLE SKOVSMOSE AND BRIAN GREER

One can talk about art criticism, critical review of scientific papers, negative and destructive critique, positive and helpful critique. One can talk about critical thinking and about a critical situation. The notion of critique can form part of all kinds of discourse. It is a ubiquitous notion.

The notion of critique also permeates lofty philosophical discourses. Descartes built critique from a starting-point of universal doubt, while Kant gave the notion a formidable staging in *Critique of Pure Reason* (1993, originally 1791) as being the main means for clarifying the conditions of obtaining knowledge. Marx added a new dimension to the concept when he elaborated a critique of the political economy. The notion went through further transformations when the Frankfurt School elaborated its explicitly named Critical Theory. Since then critical education, including critical mathematics education, has made further proliferations of the notion of critique.

It is not easy to get hold of "critique", but before we try to do so, let us say a few words about how we proceed. The notion of critique has been integrated with the idea that it is possible to provide a solid basis for a critical activity. It has been claimed that a critique can be justified; that it can be rational; that it can be conclusive. Such claims characterise what we refer to as the modern notion of critique. This is a notion that has been associated with "grand narratives" in philosophy, epistemology, and politics. Thus, the modern notion of critique has been in search of an Archimedean fixed point that could provide critique with certainty. However, that aspiration might be an illusion. It might be that it is impossible to locate a solid foundation for any critical activity. This realization means that "critique" starts drifting in a new direction. From being located in a well-organised conceptual landscape, "critique" might be transposed into an alien wasteland without any fixed point in sight. To apply a simplified rhetoric, we might say that the notion of critique is moving into a post-modern condition, abandoning the "grand narratives" that were celebrated as part of the modern outlook. But what does that mean for critical education, and for critical mathematics education in particular? However, now this introduction is running far ahead of the story, so we had better just begin.

CRITIQUE AS A LOGICAL ENDEAVOUR

The power of rationality has become associated with procedures of deduction. Statements can be related through deduction, as some statements can be logical

O. Skovsmose and B. Greer (eds.), Opening the Cage: Critique and Politics of Mathematics Education, 369–386.

consequences of others. The profound trust in rationality emerges from the general conviction that it is possible to identify a few axioms from which one can deduce all true statements about a given domain. The logic of deduction ensures that if the axioms are true then all deduced theorems will be true as well. The process of deduction digs out a network of channels through which a stream of truths will reach even the most remote corners of a deductively organised theory. The only thing necessary is to ensure the truth of the axioms. Here the deductive device cannot be of any use, as deduction can only begin after the axioms have been established. Intuition is needed, but if the axioms are sufficiently simple, then intuition is assumed to be reliable, "self-evident".

Within the domain of geometry, Euclid had demonstrated the power of an axiomatic approach. His presentation achieved a paradigmatic influence in epistemology. It was crucial to Descartes' trust in rationality, and for his overall project of establishing knowledge as necessary truths (see Descartes 1993, 1998, 2001). This trust in rationality assigned a particular role to critique, namely to prepare the scene for deduction. All forms of beliefs, assumptions, superstitions, everyday knowledge, assumed theoretical knowledge, had to be swept away. The whole critical approach, as suggested by Descartes, took the form of a universal doubt. What was possible to doubt could not be true by necessity; consequently, it should be considered false. Later on, when deductive patterns had been organised, it might turn out that some of the statements that have been put aside, due to the universal doubt, would come to take up positions in the deductive system. In this way, they could be rehabilitated and nominated as being true with certainty.

Universal doubt is a powerful critical activity, and one has to remember that this doubt was formulated within a historical context dominated by the church's celebration of faith, and faith with certainty. Faith pervaded by doubt had no value. In fact, doubt represented a heresy that needed to be severely published. Descartes was a declared Christian (atheism as a possible philosophical position was only to be formulated later), but many of his formulations confronted any form of dogmatism, including religious, and Cartesianism was later to develop into radical forms of atheism. Universal doubt challenged everything that possibly could be doubted, and no dogma was unassailable by this doubt. A universal doubt was a profound, secular, anti-dogmatic device. It was a human doubt as, through doubting, human beings established themselves as sovereign. There was no external authority that by any decree of dogma could eliminate a doubt. Only a strict rational deduction could confront a doubt. Certainty had to be based on human rationality, and, according to Descartes, such certainty was crucial for establishing knowledge. In this way Descartes formulated critique as a universal human activity of a logical nature.

This interpretation of critique initiated a long tradition that understands critical thinking as examinations of arguments. To be critical means to address all logical aspects of an argument: Does the conclusion, in fact, follow from the premises? Could there be some hidden assumptions implicit in the argument? What extra assumptions would make the argument valid? The operation of critical thinking as an examination of arguments has also been developed in educational contexts, and one finds many proposals for how to develop critical thinking through educational initiatives.

CRITIQUE AS AN EPISTEMIC ENDEAVOUR

For the Enlightenment, the aim of creating an *Encyclopaedia* was to provide an overview of all forms of knowledge and to make it available for anybody. The *Encyclopaedia* was assembled as part of the fundamental conviction that knowledge is crucial for human welfare in the broadest sense of the term. In order to ensure progress, one has to struggle against ignorance and dogmatism, the former referring to lack of knowledge, and the latter to wrongly assumed knowledge. Both phenomena were to be challenged by the *Encyclopaedia*, which should serve as the ultimate resource book for knowledge.

This meant that it became of tremendous importance to clarify: What is knowledge? And what are the criteria for establishing knowledge? Descartes tried to demonstrate how it is possible to establish knowledge as a body of certain truth, but empiricism reveals huge problems in this approach. Many forms of knowledge do not seem to be certain, such as those established through empirical observations. It became urgent to provide a deeper characteristic of knowledge and the conditions for obtaining it. Such demands opened a new terrain for critical investigations, and due to his *Critique of Pure Reason*, Kant became a principal philosopher of the Enlightenment.

As a university teacher Kant covered an impressive range of topics, including geography, natural sciences, history, modern physics as formulated by Newton. However, as a philosopher, Kant addressed knowledge in a different way. He was concerned about knowledge as a general phenomenon. He wanted to look behind all instantiations of knowledge and identify the universal conditions for obtaining knowledge. His critical philosophy aimed at providing a foundation for knowledge as well as an identification of the boundaries of all possible knowledge. While Descartes, through a universal doubt, tried to clear the ground for the building up of knowledge, Kant investigated the whole topology of knowledge. Through his critique Kant tried to study knowledge *a priori* to any particular instantiation of knowledge.

Following the inspiration from Kant, critique is much more than a logical endeavour, clarifying the validity of proposed structures of argumentation. Following Kant, critique turns into an examination of the whole epistemological scenario, of universal epistemic human conditions. One can see the whole Enlightenment as an educational approach. People need to be enlightened, which will ensure wellbeing and progress, and the contents of the enlightenment have to be defined in terms of knowledge. One can see the overall examination of knowledge and of conditions for obtaining knowledge as supporting attempts to identify a universal content for education.

CRITIQUE AS A POLITICAL ENDEAVOUR

Through the work of Marx, a new dimension was added to "critique". Marx critiqued political economy, addressing it in its theoretical formulations, but also its real-life manifestations. In this way, Marx elaborated critique to be not only a logical or epistemological endeavour, but also political. From being an object of reflection and theory, critique became a real-life activity as well.

Marx was a firm believer in progress, and even, indeed, the possibility of identifying the logic of progress. This logic, however, has a different ontological status than any logic of deduction. To Marx, the logic of progress is the logic of reality. It refers to forms of material changes by being the logic of economic progress. This crushing dialectical logic has brought about the destruction of feudalism and made space for capitalism which, compared to feudalism, expresses a tremendous progress according to Marx's analysis. But the logic of real progress also includes the destruction of capitalism, making space for true socialism.

According to Marx, a critical activity gets its essence from this deep logic of progress, which ultimately takes the form of a class struggle. In this way, a critical activity takes a completely new meaning compared to Kant's profound critique. While Kant established critique as an investigation of the general epistemic human condition, Marx elaborated critique as an investigation of the human condition in its material, economic, and political formats. And not only that: he also integrated critique as part of the struggle to change these conditions.

Within orthodox Marxist interpretations, any progressive political work must serve the working class in its class struggle. This struggle was identified as the nucleus of the dialectical progress, which had to be completed in order for the universal dialectics to overcome capitalism and open the way for a true socialism. Supporting the working class in their necessary struggle was the principal critical act. Thus, orthodox Marxism is guided by the idea that Marx's identification of the logic of reality provides basic guidelines for a critical activity. To be critical means to associate oneself with the dialectical logic of progress.

One interpretation of critical education emerged directed from such Marxist orthodoxy. A critical education was an education for class struggle, which in turn received many different interpretations. One was provided by Makarenko, who developed his educational ideas with close reference to his educational practice, working with excluded and criminalised young people in the early period of the Soviet Union. Makarenko's interpretation of critical education is rich in detailed observations; at the same time he always kept a Marxist orthodoxy within convenient reach (Makarenko, 1955).

Later, Critical Theory provided a new direction in critical education by transcending dogmatic Marxist assumptions. The idea of education for class struggle was replaced by less well-defined ideas, such as education for empowerment and *Mündigkeit* (roughly translatable as "maturity"). Critical education was still concerned with making real differences, but the question of what differences to make was no longer answered through a Marxist orthodoxy.

The proliferation of critical education continued, and different types of activism came to form part of critical education, for instance by addressing many forms of oppression. The *Routledge International Handbook of Critical Education* (Apple, Au, & Gandin, 2009) provides an overview by addressing issues such as racism, sexism, feminism, critical media education, education for social justice, and research as action.[1] However, this broad spectrum not only represents a development of critique as a political enterprise, it also brings us to a different understanding of critique.

CRITIQUE AS A CERTAIN ENDEAVOUR

There are many differences between Descartes' formulation of critique as a logical activity, Kant's presentation of critique as an epistemological undertaking, and Marx's claim that critique is a political enterprise. There is also an important similarity. These critical activities all represent preoccupations with obtaining certainty, in one way or another.

Descartes presented critique in the form of a universal doubt that could serve to clear the ground for establishing all forms of knowledge as being true with certainty. Kant wanted to outline a general topology of knowledge, and in particular he wanted to demonstrate how knowledge with certainty was possible, at least within some domains.[2] Marx wanted to establish critique as a political force that allied with the laws of real progress. Marx was not in favour of any form of revolt, which he interpreted as a random protest, not bringing about any genuine progress.[3] Instead he wanted to fine-tune critique to the laws of economic development as he had identified them through studies of capitalism.

In all three cases, we see that a critical activity tries to form an alliance with certainty by searching for a foundation upon which one can carry out a critique. In general terms, during Modernity, which we recognise as a specifically European historical category, critique and certainty became united in a congenial relationship. This unity may, however, have come to an end. As a consequence it becomes important to investigate what will become of critical enterprises, if the critique-certainty alliance is abolished and the modern perspective is left behind.

DESTRUCTIONS OF FOUNDATIONS

Some important observations throw doubt on the possibility of critique achieving certainty. Many of these can be related to "the problem of language". Throughout Modernity, most philosophers somehow took language as a transparent tool for expressing ideas, thoughts, theories, and knowledge. Language was what it appeared to be. It represented epistemic solidity. One can just think of the work of Kant, who wrote hundreds of pages about the conditions for obtaining knowledge. He explored the vast terrain of pre-knowledge. But in this courageous exploration of no-knowledge land he used the German language fluently and affluently. In this way, he demonstrated a blind confidence in language. He never addressed the possibility that language, including the way in which he himself used language, might include metaphysical assumptions that could undermine his whole critical project. Language was simply at hand, independent of how deep into no-knowledge land one had travelled. Language was taken for granted as the solid foundation for a critique and a reliable companion for rationality.

That language could be much more than a transparent tool was formulated by the Romantic Movement, which also opposed rationalism in many other ways. Language became interpreted as a carrier of cultural and metaphysical profundities. Language was not just a means for expression; it was itself an expression of a range of phenomena. It condensed historical experiences and the culture of people. It was an extract of the soul of the people. The Romantic Movement combined this interpretation of language with a celebration of what the language was expressing. In poetic form language revealed a world of collective experiences, representing the identity of the people.

Like the Romantic Movement, Nietzsche saw natural language as an expression of a historic evolution, but he could not take part in any celebration of language. Instead he found that language incorporates and perpetuates all kinds of misunderstandings, superstitions, metaphysical assumptions, etc. During history, all sorts of presumptions are engraved in the grammar of language, and automatically repeated whenever language is used. Nietzsche was concerned about all metaphysical assumptions that had been pumped into linguistic expressions. The notions of God, soul, eternity, etc. constitute fabulous fictions, compared to which any Disney World appears as a minor dream world. Also, philosophical discourses were swelling with conceptual balloons, and Nietzsche was careful in pointing out that Kant's grand notions, such as "thing in itself", contained nothing but hot air. Nietzsche's interpretation of language opened the way for a profound critique of language as well as a critique of rationality itself.[4]

With his pneumatic drill, Nietzsche smashed to pieces anything that had been presented as epistemic foundations. Nietzsche found that whatever a philosopher presents as universal and well-justified is just an expression of his or her personal hypocrisies mixed up with language-engraved superstitions. Thus, Nietzsche inserts a tremendous dose of subjectivity into any philosophical system. There is no generality to be found here, just hot air from the philosopher.

Was Nietzsche critical? For sure, he was giving up the dream of providing any absolute foundation for anything. Language was mischievous, and so was any philosophical system. It seems impossible to identify a foundation for any critical activity. Nietzsche gave critique an anarchistic format, leaving no connection between critique and certainty. But could critique adopt an anarchistic guise? Does Nietzsche's acid bath dissolve all features of critique, turning it into a pure power game? Are we losing the possibility of talking about, say, a well-justified critique?

What, then, could be the meaning of critical education? If there are no "directions" to be determined, there can be no use of a notion like "improvement". There are no guidelines for educational practices to be extracted from critical investigations. However, before we dig too deep into this possible conclusion, we will take a new look at the notion of critique.

CRITIQUE WITHOUT LIMITS?

As mentioned in the Introduction, "critique" is derived from the Greek word *krinein*, which, among other connotations, relates to "separating". This indicates that one could try to think of critique as drawing a limit and thereby separating. Thus, Descartes was concerned about the distinction between what could be proved and not proved; Kant's showpiece in limiting was between what could be known and not known; while Marx separated what was progressive from what was reactionary.

Nietzsche trampled on any such separations. But when limits are erased, will any form of certainly – whether it is of logical, epistemic, or political nature – be erased as well? And with the loss of certainty, do the grounds for action also disappear? Nietzsche walked in front of other big-foot tramplers: Foucault trampled on the separation between knowledge and power; Derrida on any form of dualistic thinking; while Lyotard trampled on all kind of grand narratives.[5] So what will happen if we join the vigorous post-modern trampling? Would the implication be that any critical activity will take an anarchistic form? Could critique take any direction, serve any interest, form part of any power game? Or would a heavy trampling bring about a profound new critical approach? Let us try to condense the issues into a single question: *Should we try to eliminate any kind of limits?* We want to answer this question with both a "yes" and a "no".

Let us start with the "yes". All kinds of limits that have been formulated as basic to any critical activity have turned out to be temporary, including the separations provided by Descartes, Kant, and Marx. Nietzsche, Foucault, Derrida, and Lyotard inserted a temporality into any form of separations. Any distinction between good and bad, truth and false, just and unjust, progress and regress, beauty and ugliness is pointed out as contingent. As a consequence it becomes an illusion to associate critique with some *fixed and immutable limits* (or with any Archimedean fixed point, to use a different terminology). This conclusion is what the tramplers have helped us to understand.

The contingency of limitations became obvious when language, from being assumed a solid foundation for critique, became revealed as a most suspicious fabrication. Limitations are expressed in language, which is a carrier of complex time-dependent metaphysical layers. No limits step out from history. This insight is important for formulating any critical approach. So one answer is "yes": a heavy trampling is important for a critical approach. It is important to point out the contingency of any assumed limits.

Now over to our answer "no". If there are no limits, we enter an empty space, which is the home ground for absolute relativism and anarchism. If there are no recognized limits, everything appears the same. Without limitations there are no reasons to protest, no reasons for any resistance. In the beginning of *The Rebel*, Camus describes how, at a certain moment, a slave, who had taken orders all his life, decided that he cannot continue like this any more. He turned around and faced his oppressor. In this way, the slave affirmed that "there are limits". He confronted "an order of things which oppresses him with the insistence on a kind of right not to be oppressed beyond the limit that he can tolerate" (Camus, 1991, p. 13). The slave personified the idea that creating a limit is a basic critical act.

If we trample along with the vigorous tramplers, no limits might be left. However, when the slave faced his oppressor he *made* a limit. The limit might not exist in advance, but it is possible to establish a limit, to negotiate a limit, to construct a limit. Setting a limit can be seen as crucial for any critical activity, yet need not be an expression of any certainty. This constructive act might have nothing to do with logical certainty, epistemic transparency, or political correctness. Setting a limit can be an expression of desperation as well as of hope. It can be an expression of lived-through weaknesses. It can be tentative. It can be deeply uncertain. It is contingent; nevertheless, it is crucial for a critical act.

Most importantly, we are arguing that giving up the quest for an absolutely coherent and non-contradictory set of rules to govern our actions by no means implies paralysis. Primo Levi (1989, p. 175), in an essay in which he speaks of the ethical responsibilities of scientists (on which subject he was eminently qualified to speak) wrote that "we know the world is not black and white and your decision may be problematic and difficult: but you will agree to study a new medicament, you will refuse to formulate a nerve gas".

Sen (2009) presents a theory of justice that maintains the centrality of critique and reasoning while abandoning the illusion of a theory of justice that maintains, for example, that there is one right course of action in a given situation. He also makes clear that we do not need a comprehensive theory of justice in order to make ethical decisions, and to act, in particular circumstances. (An example is the group of reformers who judged slavery to be wrong and took action against it – they did not need a grand theory of justice in order to act.) As such, Sen's "idea of justice" is more nuanced, more open to uncertainty, than that of Rawls (1971).

So, "yes", we should trample along with the tramplers, and even our small-feet trampling might help to reveal as temporary some limits that are claimed to be made in certainty. And "no", we should not try to eliminate any forms of limitations: setting a limit, contingent as it will be, is important for a critique. Moreover, setting a limit is always accompanied by the possibility of resetting the limit.

UNCERTAINTY IN MATHEMATICS AND MATHEMATICS EDUCATION

What may come from this yes-no aporia? Do such lofty reflections bring us anywhere? Instead of trying to answer such questions directly, let us jump to the domain of mathematics education, and make some observations.

As we discussed in the introduction, mathematics has been a domain in which people have sought certainty. However, the foundational crises and epistemological shifts of the 19th and 20th centuries have undermined faith that absolute certainty can ever be found, even in mathematics. In the face of the resultant crisis, various attempts to protect the feeling of certainty have been made. Russell and Whitehead made a valiant attempt to establish mathematics upon a solid rock of logic. The extent, as well as the futility, of their diligence is revealed when, in the first edition of *Principia Mathematica*, we read on page 379 the following: "From this proposition it will follow, when arithmetical addition has been defined, that $1 + 1 = 2$." As mentioned in the introduction to this book, Russell later abandoned the hope of finding certainty, even in mathematics.

According to logical positivism, formal languages would save us from the metaphysical swamp accompanying any form of natural language. Thus, Carnap, a hardliner of logical positivism, acknowledged that Nietzsche had done an important job through his critique of natural language. In contrast to a dubious natural language, Carnap presented formal languages as a metaphysics-free symbolism. As a consequence, mathematics can serve as the language of science. By means of mathematics, observations can be presented in a neutral and objective way, and the analyses of observations and the building-up of theory can be completed without relaying on questionable assumptions. By means of mathematics, scientific processes come to represent nothing but a pure form of rationality.

However, doubt about natural languages can be broadened to apply to any form of language, including formal languages. It can also be applied to mathematics, which, too, can be considered a discourse (or more precise a range of discourses that show different forms of family resemblances).

We may also suggest, as Eco (1995, p. 311) has commented on computer languages, that the language used in doing mathematics is not a full language "because [its] syntax, though rigorous, is simplified and limited, and [it] remains parasitic on the natural languages which attach meanings to [its] empty symbols". On computer languages, he comments that they are:

> … *a priori*, in that they are based not on the rules that govern the surface structures of natural languages, but rather, ideally, on a presumed deep grammar common to all natural languages. They are … philosophical because they presume that this deep grammar, based on the laws of logic, is the grammar of thought of human beings and machines alike. They also exhibit the two limitations inherent in philosophical *a priori* languages: (1) their rules of inference are drawn from the western logical tradition, and this may mean, as many have argued, that they reflect little more than the basic grammatical structures common to the Indo-European family of languages; (2) their effability is limited; that is, they are capable of expressing only a small proportion of what any natural language can express.

Eco goes on to dismiss attempts in Artificial Intelligence to create a perfect language that covers all the complexity of a natural language as succeeding "in solving certain problems only through imposing ad hoc solutions, which work only for local portions of the range of action of natural languages" (p. 312).

From being commemorated as an instantiation of critical thinking, mathematics becomes recognised as a suspicious fabric in need of being addressed through a critique. In particular, the whole discussion of mathematics in action challenges any celebration of mathematical rationality. (For a discussion of mathematics in action see, e.g., Skovsmose, 2009; Yasukawa, Skovsmose, & Ravn, Chapter 12, this volume). As with any other language, mathematics includes metaphysical assumptions. Such assumptions can take the form of a mechanical world-view that becomes imposed on what is assumed to be "just" described. What is presented in a mathematical format – a natural phenomenon, an economic connection, a sociological correlation, a technological possibility – takes up a particular format due to the *mathematical* format of the presentation. Mathematics imposes a structure on what becomes captured mathematically.

A second attempt to preserve the sanctity of mathematics is to posit a universal set of core structures. The project of the Enlightenment celebrated all kinds of knowledge. A general educational implication of this celebration is to establish a school curriculum that presents, as carefully as possible, the domain of knowledge. Of course, asking who should determine the canon brings us directly into the nexus of knowledge and power (e.g. Apple, 2003).

Jerome Bruner's (1960) suggestion to pay attention to the structures of science can be seen as an updated version of the Enlightenment project. As knowledge is rapidly proliferating, no curriculum can keep up with this development as long as one tries to solve the problem of knowledge expansion by means of some add-on devices. One cannot create a digestible curriculum through processes of compression. But, according to Bruner, one can extract basic structures, and though processes of extraction one could try to reach an all-embracing curriculum.

This belief was the idea of the New Math Movement according to which the mathematical curriculum should represent structural distillations of mathematics. The structural approach took as a given that mathematics itself is of value, and that the educational task is to effect a transmission based on certain principles for extracting what to transmit. As a consequence, teachers of mathematics could serve as ambassadors or missionaries of mathematics.

The most elaborated attempt to provide the definitive structural essentialisation of mathematics was famously attempted by the Bourbaki group of mathematicians. In terms of mathematics education, this attempt was influential in various manifestations under the umbrella term "Modern Mathematics". Piaget famously believed he could discern an alignment between the structures posited in his theory of cognitive development and those of Bourbaki. Freudenthal (1973, p. 46) commented, with typical forcefulness, as follows:

> The most spectacular example of organizing mathematics is, of course, Bourbaki. How convincing this organization of mathematics is! So convincing that Piaget could rediscover Bourbaki's system in developmental psychology. Poor Piaget! He did not fare much better than Kant, who had barely consecrated Euclidean space as a "pure intuition" when non-Euclidean geometry was discovered! Piaget is not a mathematician, so he could not know how unreliable mathematical system builders are. ... Mathematics is never finished – anyone who worships a certain system of mathematics should take heed of this advice.

The idea of a curriculum as being a distillation of a pre-established body of knowledge has been repeated many times, for instance also in the back-to-basics approach. The overarching idea is that when the extraction of the curriculum in done in a proper way, then the mathematics education so developed would serve its ultimate aims. However, this essentialist idea has become questioned. Mathematics education is a social endeavour, and as such it may serve very different functions in society.

A prominent, politically active, mathematician has declared that "mathematics isn't a continuum. There's mathematics and there's non-mathematics and there's absolutely nothing in between" (interview with James Milgram at: http://www.baltimorecp.org/newsletter/BCPnews_fall06.htm). Even if that were true in any useful sense, someone has to decide how the separation is to be made, and there is scant agreement on that, even among mathematicians. The question *What is mathematics, really?* (Hersh, 1997) has no agreed answer, but we can establish limits and, with Hersh, we take it as self-evident that mathematics "must be understood as a human activity, a social phenomenon, part of human culture, historically evolved, and intelligible only in a social context" (and see the other criteria Hersh (1997, Chapter 2) proposed should be satisfied by any philosophy of mathematics).

From our perspective, no reasonable topography of "mathematics" – most certainly no reasonable program for mathematics education – defines mathematics in action as lying outside the border. Further, the characterization of mathematics as a human activity entails consideration of mathematical practices other than academic mathematics. We question the dominance of school mathematics by academic mathematics – in the extreme, seeing the central purpose of school mathematics as primarily the reproduction of mathematicians. A less extreme position, with which we likewise disagree, is that all children should learn considerable formal mathematics (up to symbolic algebra, for example), which is rather like having every child in a country taught to play a sport with the aim of producing a top-class national team.

More generally, we react against the tendency towards formalization of mathematics in schools that makes it irrelevant in students' (and teachers') lives. To an extent this tendency has been alleviated in recent decades by inclusion in curricula of elements of "applied mathematics", notably probability and data handling, and mathematical modelling (Barbosa, 2006; Blum, Galbraith, Henn, & Niss, 2007). Barbosa (2006) points out that in Brazil the modelling movement is closely associated with ethnomathematics "the mathematics of cultural groups" (D'Ambrosio, 2006). So, as well as applied mathematics, there is engineering mathematics, craft mathematics, everyday mathematics, and so on. All these forms have to be addressed critically.

The ethnomathematical approach has formulated a critique of the monopoly of academic mathematics. It has been emphasised that, in order to preserve the culture of indigenous people, it is important to acknowledge the complexities of indigenous knowledge, including their mathematics. These practices make up an important aspect of their culture and should make up an important aspect of their curriculum as well. In this way, indigenous students can experience meaningfulness and relevance of learning. In a profound way, ethnomathematics provides a critique of the rationality that is included in academic mathematics.

However, ethnomathematics itself represents a rationality. Ethnomathematics provides a discourse, and like any other form of mathematical discourse, it provides a way of seeing and doing. It might include a range of metaphysical assumptions. It might open up possibilities, it might include obstructions. In an interview with indigenous students in Brazil, one emphasised that he wanted to study medicine, and in order to do so he saw the importance of studying regular school mathematics (see Skovsmose, Alrø, and Valero, in collaboration with Silvéri and Scandiuzzi, 2008). He saw the health problem as being a principal problem in Indian communities, and he wanted, as a qualified doctor, to work in the community. The point is that a celebration of ethnomathematics should not overlook the important point that ethnomathematics also might include severe limitations. A student who has been subjected to a carefully elaborated ethnomathematical curriculum will not easily get opportunities for entering a range of further education, such as medicine.

This remark brings about further observations. It might be that formal mathematical rationality is not simply a problematic rationality. It could be applied in all possible ways and come to serve very different, also contradictory, ends. Thus many medical achievements are in fact obtained by seeing the human body as a machine. Many remarkable technological achievements are deeply engraved in mathematical approaches. One only needs to think of the tremendous technological machinery that is brought into use for identifying and treating cancer.

One should not think of academic mathematics and ethnomathematics as being able to substitute for each other. They represent different forms of rationality. They include different possibilities and different forms of assumptions, and have different limitations on their range of applicability. They are engraved in different forms of discourses, and they might bring about different forms of actions. Both forms of rationality are in need of critique.

We should also briefly mention mathematics education as a research field (Skovsmose, Chapter 16, this volume). As Kilpatrick (1981) pointed out, research in this field is characterized by "reasonable ineffectiveness" for the simple reason that improving mathematics education is a human problem, not a technical problem. Yet logical positivism and its relatives continue to reappear and currently – most extremely, but by no means exclusively – in the United States, there is an attempt to confine research in mathematics education within a methodological straightjacket. The National Mathematics Advisory Panel represents the *reductio ad absurdum* of this strain (Greer, Chapter 4, this volume): "Whereof one cannot do randomized experiments, thereof one must be silent."

DIGGING WHERE WE STAND: WHEN THE CRITIQUE HITS THE *REALPOLITIK*

One aim of this book is to perturb mathematics education, a term intended to cover both what happens in classrooms and other sites of learning mathematics, and the field of research and theory related thereto. Activity in both arenas is in dire need of perturbation, of the questioning of unquestioned assumptions, the dead hands of inertia and tradition resting on practices and institutions, falling into "the trap of the same", and being trapped in epistemological cages, as D'Ambrosio puts it. Furthermore, we seek to deconstruct denial of its political nature, the stance of spurious neutrality.

The "grand narratives" that have been written together with Modernity need to be perturbed. Some of these stories have taken the form of critical theories of rationality, epistemology, politics developed with aspirations of establishing solid foundations. Thus, critical approaches have been searching for an Archimedean fixed point that will provide critical activities with certainty. Critical education, at least in some of its expression in Europe and the USA (Apple et al., 2009) has been drawing on "grand narratives", but at the same time it is an expression of perturbations. Perturbation is important and critical mathematics education, as outlined in Bookend 1, needs to bring this perturbation further. This activity is what the authors of this book have pursued.

Other grand narratives from the Modern library are expressions of Western hegemony. They need to be challenged, and they have been challenged, through the recognition of cultural diversities developing in many arenas, including subaltern studies (Apple & Buras, 2006), linguistic diversity (Macedo et al., 2003; Phillipson, 2009; Skutnabb-Kangas, 2000), literature and other arts (Said, 1994), the politics of scholarship and research (Canagarajah, 2002; Smith, 1999). One theme of this book is that such grand narratives need to be challenged by counter-narratives in, and through, critical mathematics education also (we have noted our surprise and disappointment that a recent handbook of critical education (Apple et al., 2009) does not refer to critical mathematics education and "mathematics" does not appear in the index).

We maintain that equipping people to deal with mathematics in action, by developing mathematics as a critical tool (Gutstein, Chapter 1, this volume; Mukhopadhyay & Greer, 2001) should be a central purpose for mathematics education. Closely allied to this position is that people working in mathematics education – and those putting mathematics into action – have societal and ethical responsibilities (Atweh, Chapter 15, this volume). This responsibility concerns all forms of mathematics and mathematics education. As emphasised through the discussion of mathematics in action, we find mathematics brought into action in very many different contexts: ethnomathematical settings, daily life environments, business, commerce, marketing, construction, engineering, technology, research. In all such contexts, one faces social and ethical responsibilities. In this sense, a critical approach is important with respect to all forms of mathematics education, not least with respect to advanced university education in mathematics.

In the interplay between reflection and action, a point at issue is illustrated in the chapters by Pais and Gutstein in this volume. It is possible with reference to at least some formulations of Marxism to claim that a critical educational practice, serving functions of justice and empowerment, is in fact an illusion. If educational problems are caused by dominant economic structures, how then to solve educational problems without first changing those dominant economic structures? How to ensure genuine educational improvements without first making them possible? How could critical activities on a micro scale make sense, if critical activities on a macro scale have not been completed? (See Pais, Chapter 2, this volume)

Our point is that this way of thinking somehow assumes the big story of Marxism: that critical investigations need to observe the grandiose limitation between what is progressive and what is not. It maintains the Marxist idea that social changes have to take place in a proper order. But if we abandon this assumption, then there evaporates as well the idea that micro-critique does not make (much) sense if no macro-critique has been conducted in advance. If any Marxist-inspired analysis is without any "objective" foundation, then there are no justified rankings within the field of critical activities, and critical activities at any level *might* make sense. This conclusion also applies to mathematics education. In this field, as well as in any other educational context, it *might* make sense to search for small-scale, local improvements.

Let us just listen to Lupes, one of Eric Gutstein's students, who participated in a programme of mathematics education for social justice. The education took place in Chicago, and one could assume that, of all places in the world, this would be one where one could apply the argument that it is impossible to establish an education for social justice due to the formidable social, political, and economic constraints. Anyway, Lupes states: "With every single thing about math that I learned came something else. Sometimes I learned more of other things instead of math. I learned to think of fairness, injustices and so forth everywhere I see numbers distorted in the world. Now my mind is opened to so many new things. I'm more independent and aware. I have learned to be strong in every way you can think of it." (Lupes, quoted from Gutstein, 2003, p. 37).

There is a huge uncertainty related to mathematics education. What could be the role of mathematics education in a post–apartheid South Africa for a white student coming from a wealthy neighbourhood? For a black student from a township? What could mathematics education mean for marginalised young people living in huge European cities? What does mathematics education mean to children of immigrants, as for instance immigrants from Pakistan living in Copenhagen? What role could mathematics education have for young people from Indian communities in Brazil? What role could mathematic education have for Lupes's friends? Answers to such questions remain open. There are no philosophical systems, no rules, no algorithms to absolve us from the responsibility of making moral judgments and acting on them.

In reading this book, the reader will have become aware of many open questions, unresolved tensions, and outright contradictions. It is a book that aims to celebrate global diversity of forms of life, mathematical practices, culture, and educational systems, yet, in a response to *realpolitik*, it is written entirely in English. In many places, explicitly or implicitly, there arise conflicts: between indigenous mathematics and universal formal mathematics; between cultural integrity and access to power and economic opportunity; between attempting to relate mathematics in classrooms to children's lived experience and recognizing the essentially artificial nature of schooling; between "playing the game" and "changing the game"; between macro-critique of contemporary capitalism and micro-critique of carrying on the struggle in a local context (as encapsulated in the slogan "think globally, act locally").

Central to the book is the dialectic between reflection and action that we embody in the concept of critical agency. The genesis of critical agency lies in the ability and disposition to imagine that things can be different. What we present is "just" words on paper, our collective reflections, yet they are agentive in attempting to perturb, open up, and radicalize mathematics education, both as a set of cultural practices that occupy significant parts of the formative years of children and as a scholarly field. We attempt to expose the ideological substrates, the archaeological layers on which these hegemonic activity systems are built. In so doing, we insist that mathematics education be accorded the pivotal place it merits within the ambit of critical education and political analysis. This endeavour demands the picking apart of rhetoric and slogan systems, couched in the discourse of propaganda and advertising. It also demands an unmasking of the limitations and claims of self-defined "rationality" within mathematics as an intellectual activity but, more critically, when people act using mathematics and when formalistic mathematical casts of mind promote positivistic and dehumanized world-views.

In the spirit of self-critique, we acknowledge that we have inevitably been shaped by our cultural backgrounds, two different varieties of European. As we have lived and worked in mathematics education, and more generally, for decades, we have developed a degree of greater sensitivity to other world-views. We also acknowledge that we are vulnerable to a charge of Utopianism, to which we plead "guilty". One rational response to the current situation in mathematics education, and to the state of the world in general, would be to withdraw into hedonism. Well, we prefer to make an effort to express our indignation at injustice, and at intellectual dishonesty and complacency. The parrot in Tagore's parable represents a learner being force-fed with information and being forced to conform. For too many people, their experience of school mathematics is personally, emotionally, and intellectually dehumanizing. It does not have to be like that.

NOTES

1 It should also be noted that this international handbook includes two heavy biases. The whole wave of critical education that emerged in Europe during the late 1960s but was not published in English is ignored. Ironically, thus, the hegemony of English is reflected in this international handbook. Furthermore, critical education as related to content matter issues is ignored, in particular in relation to critical mathematics education.

2 Thus a principal element in his critique was to point out that it was possible to obtain certainty, not only with respect to analytical statements, as for instance "all bachelors are unmarried", but also with respect to also synthetic statements, as for instance "all events have a cause".

3 Thus it is no surprise that Marx strongly criticised any form of anarchism, as, for instance, formulated by Proudhon, who also provided a strong social critique at the time.

4 See, for instance, Nietzsche (1974, 1998, 2003). It should be noticed also that Nietzsche used the whole metaphorical power of natural language to make the point that natural language is mischievous. So Nietzsche's critical approach also demonstrated a vicious circularity. Similar comments apply to constructivists who speak and write at length about the limitations of verbal communication.

5 See Foucault (1977, 1989, 1994); Derrida (1974, 1993); and Lyotard (1984). As an aside it could be remarked that Lyotard's story, in making an end to all grand narratives, is itself is a story. Not a small one, we think.

REFERENCES

Apple, M. W. (Ed.). (2003). *The state and the politics of knowledge*. New York/London: RoutledgeFalmer.

Apple, M. W., Au, W., & Gandin, L. A. (Eds.). (2009). *The Routledge international handbook of critical education*. New York/London: Routledge.

Apple, M. W., & Buras, K. L. (Eds.). (2006). *The subaltern speak*. New York: Routledge.

Barbosa, J. C. (2006). Mathematical modelling in classroom: a socio-critical and discursive perspective. *Zentralblatt für Didaktik der Mathematik, 38*, 293–301.

Blum, W., Galbraith, P., Henn, H.-W., & Niss, M. (Eds.). (2007). *Modelling and applications in mathematics education: The 14th ICMI study*. New York: Springer.

Bruner, J. (1960). *The process of education*. Cambridge, MA: Harvard University Press.

Camus, A. (1991). *The rebel: An essay on man in revolt*. New York: Vintage Books. (First French edition 1951).

Canagarajah, S. A. (2002). *A geopolitics of academic writing*. Pittsburgh, PA: University of Pittsburgh Press.

D'Ambrosio, U. (2006). *Ethnomathematics: Link between tradition and modernity*. Rotterdam: Sense Publishers.

Derrida, J. (1993). *Aporia*. Palo Alto, CA: Stanford University Press.

Derrida, J. (1998). *Writing and difference*. Chicago: University of Chicago Press. (Original French edition 1967).

Descartes, R. (1993). *Meditations on first philosophy in focus*. Edited with an introduction by Stanley Tweyman. London: Routledge. (Original Latin edition 1641).

Descartes, R. (1998). *The world and other writings*. Cambridge: Cambridge University Press. (First published 1664).

Descartes, R. (2001). *Discourse on method, optics, geometry, and meteorology*. Indianapolis, IN: Hackett Publishing Company, (First published 1637).

Eco, U. (1995). The search for the perfect language. Oxford: Blackwell.

Foucault, M. (1977). *Discipline and punish: The birth of the prison*. Harmondsworth, England: Penguin Books..(First French edition 1975).

Foucault, M. (1989). *The archaeology of knowledge*. London: Routledge. (First French edition 1969),

Foucault, M. (1994). *The order of things: An archaeology of the human sciences*. New York: Vintage Books. (First French edition 1966).

Freudenthal, H. (1973). *Mathematics as an educational task*. Dordrecht, The Netherlands: Reidel.

Gutstein, E. (2003). Teaching and learning mathematics for social justice in an urban, Latino school. *Journal for Research in Mathematics Education, 34*, 37–73.

Hersh, R. (1997). *What is mathematics, really?* Oxford: Oxford University Press.

Kant, I. (1933). *Critique of pure reason*. London: MacMillan. (First German edition 1791).

Kilpatrick, J. (1981). The reasonable ineffectiveness of research in mathematics education. *For the Learning of Mathematics, 2*(2), 22–29.

Levi, P. (1989). *The mirror maker*. New York: Schocken Books.

Lyotard, J.-F. (1984). *The postmodern condition: A report on knowledge*. Manchester: Manchester University Press. (Original French edition 1979).

Macedo, D., Dendrinos, B., & & Gounari, P. (2003). *The hegemony of English*. Boulder, CO: Paradigm Publishers.

Makarenko, A. (1955). *The road to life: An epic of education, I-II* (Original Russian edition 1932–1935) http://www.marxists.org/reference/archive/makarenko/works/road1/index.html http://www.marxists.org/reference/archive/makarenko/works/road2/index.html

Mukhopadhyay, S., & Greer, B. (2001). Modeling with purpose: Mathematics as a critical tool. In B. Atweh, H. Forgasz, & B. Nebres (Eds.). *Socio-cultural aspects in mathematics education: An international perspective* (pp. 295–311). Mahwah, NJ: Lawrence Erlbaum.

Nietzsche, F. (1974). *The gay science*. New York: Vintage Books. (Original German edition 1882.)

Nietzsche, F. (1998). *On the genealogy of morality: A polemic*. Indianapolis, IN: Hackett Publishing Company.

Nietzsche, F. (2003). *Twilight of the idols* and *The anti-Christ*. London: Penguin Books. (Both books from 1888).

Rawls, J. (1999). *A theory of justice*. Oxford: Oxford University Press.

Phillipson, R. (2009). *Linguistic imperialism continued*. Hyderabad: Orient Black Swan.

Said, E. W. (1994). *Culture and imperialism*. London: Vintage Books.

Sen, A. (2009). *The idea of justice*. London: Penguin.

Skovsmose, O. (2009). *In doubt: About language, mathematics, knowledge and life-world*. Rotterdam: Sense Publishers.

Skovsmose, O., Alrø, H. and Valero, P. in collaboration with Silvério, A. P. and Scandiuzzi, P. P. (2008). "Before you divide you have to add": Inter-viewing Indian students' foregrounds. In B. Sriraman (Ed.), *International perspectives on social justice in mathematics education. The Montana Mathematics Enthusiast Monograph 1* (pp. 209–230). Charlotte, NC: Information Age Publishing.

Skutnabb-Kangas, T. (2000). *Linguistic genocide in education or world-wide diversity and human rights?* Hyderabad: Orient Longman.

Smith, L. T. (1999). *Decolonizing methodologies*. New York: Zed Books.

Whitehead, A. N., & Russell, B. (1910–1913). *Principia mathematica, I-III*. Cambridge: Cambridge University Press.

Ole Skovsmose
State University of São Paulo, Brazil
Aalborg University, Denmark

Brian Greer
Portland State University
Oregon, USA

Annica Andersson is a Swedish scholar who recently completed her PhD. She is enrolled in the International Doctoral School of Technology and Science at Aalborg University in Denmark and there part of the Science and Mathematics Education Research Group. For seven years, she has also been a mathematics education lecturer at Malmö University in Sweden and there part of the NMS research group. Prior to coming into lecturing and research, Annica was an upper secondary mathematics and psychology teacher in Sweden.

Bill Atweh is an Associate Professor of Mathematics Education at the Science and Mathematics Education Centre at Curtin University, Perth, Australia. His research interests include socio-cultural aspects of mathematics education and globalization, the use of action research for capacity building, and critical and socially responsible mathematics education. His co-edited books include: *Action Research in Practice* (1997), *Socio-cultural Research on Mathematics Education* (2001), *Internationalisation and Globalisation in Mathematics and Science Education* (2008), and *Mapping Equity and Quality in Mathematics Education* (2011).

Sikunder Ali Baber works as an independent educational consultant. He concluded his post-doctoral research at the Autonomous University of Barcelona, Spain in 2011. His PhD (2007) is from Aalborg University, Denmark, in multicultural education and mathematics education. He has also worked at the Aga Khan University Institute for Educational Development, Karachi, Pakistan. His research interests include: looking at learning and teaching in general, and mathematics learning in particular, in multicultural educational contexts with focus on learning trajectories of students (especially immigrant students). He is also curious to understand how the processes of globalisation and transnationality, and associated discourses, are shaping and re-shaping the forms and contents of social institutions such as schools, and how, in turn, these changes are affecting individual lives. He is also critically analysing the features of neoliberalism through the lens of governmentality and how this concept can be connected with the calculated rationalities that are dramatically affecting ordinary life under the changing conditions of modernities.

Tony Brown is a senior lecturer in adult and organisational learning at the University of Technology, Sydney, Australia. He teaches in the history and political economy of adult education and work and researches education and learning in trade unions and other membership-based organisations.

Marta Civil, formerly at The University of Arizona, is the Frank A. Daniels Distinguished Professor of Mathematics Education in the School of Education at the University of North Carolina, Chapel Hill. Her research focuses on cultural and social aspects in the teaching and learning of mathematics, equity, linking in-school and out-of-school mathematics, and parental engagement in mathematics. Her work is located primarily in working-class, Latino/a communities. She has directed several initiatives in mathematics education engaging parents, children, and teachers.

Munir Jamil Fasheh was born in Jerusalem, Palestine, 1941, driven out in 1948 with his family to Ramallah where he lived and worked most of his life. He taught mathematics for 14 years, was Head Supervisor of mathematics instruction for 5 years (West Bank schools), and Director of the Arab Education Forum at Harvard University for 10 years. The single most influential event that transformed his relationship to mathematics, learning, and knowledge was the "discovery" of his illiterate mother's mathematics, which liberated him mainly from two beliefs propagated through education: the belief in a single undifferentiated path for progress, learning, and knowing; and the belief that life can be understood by the intellect and expressed by language completely.

Uwe Gellert is a Professor of Mathematics Education in the Faculty of Education and Psychology at the Freie Universität, Berlin. His research interests include social inequalities in mathematics education, cross-cultural studies, microanalysis of classroom interaction, sociological perspectives on mathematics, and mathematics teacher education. He is active in the Mathematics Education and Society group (MES) and in the Commission Interationale pour l'Étude et l'Amélioration de l'Enseignement des Mathématiques (CIEAEM), and co-edited with Eva Jablonka *Mathematisation and Demathematisation: Social, Philosophical, and Educational Ramifications* (2007).

Maisie Gholson is a doctoral student at the University of Illinois at Chicago. She has worked as a graduate research assistant in the Learning Science and Research Institute at UIC for an NSF funded curriculum development project and as a graduate research assistant for the College of Education to support and evaluate afterschool mathematics programs in the Chicago Public Schools. Maisie recently received a fellowship from the NSF Graduate Research Fellows Program in STEM education. Her proposed research relates to how classroom talk in 9th grade Algebra classes develops African-American children's sense of self racially and academically, and affects student's participation patterns and mathematics achievement.

Brian Greer worked for most of his career in the School of Psychology, Queen's University, Belfast before moving to a position in mathematics education at San Diego State University in 2000, and thence to Portland, Oregon where he now works independently. His work has evolved from psychological studies of mathematical cognition, through work on aspects of mathematics teaching/learning, towards a critical stance and an interest in the cultural and political contexts in which mathematical education takes place.

Eric (Rico) Gutstein is a professor in curriculum and instruction at the University of Illinois-Chicago. He focuses on Freirean approaches to critical and culturally relevant education and on education policy. He has taught middle and high school mathematics in Chicago public schools, wrote *Reading and Writing the World with Mathematics: Toward a Pedagogy for Social Justice*, and co-edited *Rethinking*

Mathematics: Teaching Social Justice by the Numbers. Rico is a founding member of Teachers for Social Justice (Chicago) and is active in the education for liberation movement.

Eva Jablonka is Professor of Mathematics Education at Luleå University of Technology, Sweden. She earned her PhD at the University of Technology in Berlin in 1996, and worked for many years at the Freie Universität in Berlin. As a member of the Learner's Perspective Study (LPS), she spent one year in Australia helping to set up the International Centre for Classroom Research at Melbourne University. She has lectured in a variety of undergraduate and graduate programs on mathematics education. Currently, she is engaged in an international study on the emergence of disparity of achievement in mathematics education.

Gelsa Knijnik works at the Graduate Program on Education at *Unisinos*, a Brazilian Jesuit university. She holds an M.Sc. Degree in Mathematics and a Ph.D. Degree on Education in Brazil. She is a researcher of the Brazilian National Research Council and coordinates the *Inter-institutional Research Group Mathematics Education and Society*. Her major research interests are related to education of peasant social movements, politics of knowledge and ethnomathematics. She is the editor of the *Unisinos Journal on Education.*

Danny Bernard Martin is a Professor in the College of Education and Department of Mathematics, Statistics, and Computer Science at the University of Illinois, Chicago. Prior to these appointments, he was Instructor and Professor of Mathematics at Contra Costa College, California. His primary interest is in the mathematical experiences and development of Black learners. He is the author of *Mathematics Success and Failure among African American Youth* (2000) and editor of *Mathematics Teaching, Learning, and Liberation in the Lives of Black Children* (2009).

Swapna Mukhopadhyay is an Associate Professor at Portland State University. As a mathematics educator, her scholarly interests focus on issues of critical mathematics education and cultural diversity. The main thrust of her work is in realizing that mathematics is a socially constructed mental tool that is accessible to all. Using the framework of ethnomathematics, she works towards unifying research and curriculum design, an act that is synonymous with activism. She co-edited *Cultural Responsive Mathematics Education* (2009). She is also a part-time potter.

Alexandre Pais, after teaching mathematics for nine years in several public Portuguese schools, decided to stop and wonder. A PhD project appeared as the true excuse to do all the reading, thinking, and writing that being a full-time teacher strategically inhibits. Submersed in readings, he stumbled into what someone has called the Axis of Evil: Hegel-Marx-Lacan. At the moment, his work consists in developing a dialectical materialist critique of education, using mathematics education as a case study.

Núria Planas is an Associate Professor at the Department of Mathematics and Science Education, University Autonoma of Barcelona. She has a degree in Pure Mathematics and a PhD in Mathematics Education. Prior to her permanent position at University, she served as a mathematics teacher in a highly multilingual secondary school. Her research interests focus on discourse and learning processes in mathematics classrooms. She is currently working on two community projects on issues of social participation and mathematical argumentation, with data from secondary urban multilingual classrooms. These projects are funded by the Spanish Ministry of Science and Innovation and the Catalan Agency for Research, respectively.

Ole Ravn is an Associate Professor at the Department of Learning and Philosophy, Aalborg University. With a background in mathematics and philosophy, he defended his PhD thesis in the field of mathematics and science education. His research is primarily related to the philosophy of mathematics and science education. In addition, he is doing research in the philosophy of mathematics and on the design of problem-based and project-oriented university educations.

Mamokgethi Setati holds a PhD in Mathematics Education from the University of the Witwatersrand. She is a Full Professor and Vice-Principal of Research and Innovation at the University of South Africa. She is also Honorary Professor of Mathematics Education at the University of the Witwatersrand. She served as co-chair of the ICMI Study 21 entitled *Mathematics and Language Diversity*. In South Africa she is also well known for her community work and visionary leadership in mathematics education development.

Ole Skovsmose has a special interest in critical mathematics education. He has investigated the notions of landscape of investigation, mathematics in action, students' foreground, and ghettoising. He has been Professor at the Department of Learning and Philosophy, Aalborg University, but is now retired and is living most of his time in Brazil. He has written several books including *Towards a Philosophy of Critical Mathematics Education, Dialogue and Learning in Mathematics Education* (with Helle Alrø) (2004), *Travelling Through Education* (2005), *In Doubt* (2009), and *An Invitation to Critical Mathematics Education* (2011).

Fernanda Wanderer works at the *Universidade Federal do Rio Grande do Sul*, in Brazil. She holds a M.Sc. Degree and a Ph.D. Degree in the field of Mathematics Education from the Brazilian Jesuit University *Unisinos*. Her work is focused on Mathematics Teaching Education of primary and secondary pre-service and in-service teachers, and Ethnomathematics. Her recent research projects have been developed in different time-space German descendent rural communities of Southern Brazil.

Keiko Yasukawa is a Lecturer in Adult Education at the University of Technology, Sydney. She teaches in the areas of adult literacy and numeracy, work and learning, and vocational education and training. She is an activist in her union branch at the university, and in the professional organisation for adult literacy and numeracy practitioners. She has been researching in theories of critical mathematics, literacy and numeracy teaching practices and teacher development, and union education.